T0214973

Availability
of World Energy
Resources

Availability of World Energy Resources

D C Ion

ENERGY INFORMATIONAL SERVICES
McGraw-Hill, Inc.

Energy Informational Services
McGraw-Hill, Inc.
1221 Avenue of the Americas
New York, N.Y. 10020

©D C Ion 1975, 1980

Softcover reprint of the hardcover 1st edition 1980

Library of Congress Catalog Card Number 81-80984

ISBN-13: 978-94-009-8729-6   e-ISBN-13: 978-94-009-8727-2

DOI: 10.1007/978-94-009-8727-2

Contents

More detailed chapter breakdowns can be obtained from bold entries in the index

LIST OF TABLES

LIST OF FIGURES

Introduction

Energy will be a most important topic in the 1980s. The speed with which a dozen or more trends will develop will be critical. Most of these trends are interdependent and interacting, and include:
- the degree of constraint on oil and gas supplies exercised by the producers, whether inside or outside the OPEC, as they each attempt to match production to their own energy needs and the funding of their own economic growth from exports,
- the depth of the appreciation by industrialized countries that energy supplies will be tight and fossil fuels will be very expensive at least until the end of the century,
- the actions taken by those countries to ameliorate this situation,
    in exploration for new oil and gas sources,
    in exploitation effort for new coal supplies,
    in acceptance of the need for expansion of nuclear energy supplies,
- the balancing of energy supply and demand in centrally-planned economies,
- the rate of development within developing countries, including China,
- the development and adoption of unconventional energy sources,
- the adaptation of the world financial system to new situations.

These examples highlight some of the continuing problems in the energy field. These problems will be discussed in all sorts of meetings of all sorts of people in all sorts of places and through all forms of the communication media. Other trends will materialize and take the centre of the stage, often only for a short time. There are, however, basic facts and principles to which all trends are subject. Therefore, rather than pick only some or all of the trends mentioned and treat them in turn, I have chosen to retain the format of the first edition of *Availability of World Energy Resources* for this second edition. The horizontal treatment of all the energy sources, from resource base through to the supply/ demand balance, stresses the basic principle of the interdependence of all energies and all countries. Therefore into that format have been melded the vertical treatment of coal, nuclear energy and the natural gas of North-West Europe of the first supplement to the original edition, the suggestion in the second supplement which arose from the reports which flooded 1977, together with the facts and developments which 1978 and 1979 brought along. Hence I have again attempted to provide a thought-provoking reference book on energy with which to be equipped for the 1980s.

The gap of ignorance between the operators in the energy industries and others, whether in other industries, in political or academic life or elsewhere has narrowed only slightly between 1975 and 1980. Therefore the main aim of this second edition, as of the first edition, is to

'assist the understanding of the numbers, concepts and factors
on which judgement of energy resources problems must be based,
whether by an individual, group or government.'

15 December 1979   D. C. Ion

1
Resource
definitions and measurement

The resources which man can adapt to his own use are those which result direct from the sun (solar energy), from the moon (lunar energy) and from the earth (nuclear and geothermal energy), and indirectly from the sun, either historically in the fossil fuels in carbon or hydrocarbon form, or currently in wind-power, water-power and by biological processes through the flora and fauna. The direct sources create the renewable sources, for one might say that nuclear energy is only a technique or two away from becoming virtually renewable. The indirect sources of the wind, etc, are renewable but limited to being of local rather then global value. The fossil fuels are the non-renewable resources.

The carbon, hydrocarbon and nuclear resources are of the earth and their occurrence is conditioned by the history of the earth. The earth is a dynamic body, not a static one. The oceans are fundamentally different from the continents. The crust under the continents is thick, about 35 km, and consists primarily of acidic rocks, such as granites, whereas the crust under the oceans is thin, about 6 – 8 km, and formed of basic rocks, like basalts. The continents are old and of complex structure, the ocean floors are young, being mostly less than seventy-five million years old and simple in rock type and structure.[1] New oceanic crust is formed from upwelling material along the mid-ocean ridges, with the older, outer oceanic plates being forced under the light continental plates and being redigested into the material of the mantle around the core of the earth. The continental plates have suffered many contortions. Mountains have been formed, suffered erosion from sun, wind and water and the resulting sediments deposited into topographically low areas. Igneous rocks have been injected into the sediments and volcanic rocks ejected through and onto the sediments. Forests have flourished, been submerged and buried as the earth's geography changed. Fauna have flourished and died. There is not a cubic metre of the continental crust which has not had a unique history, different from that of every other cubic metre.

The waters of the oceans cover the oceanic crustal segments and, to varying extents, the edges of the continental crustal plates. Hence although the oceans cover 71% of the surface of the globe[2] (363×106 km^2) the submerged continental shelves take up some 55×10^6 km^2 of that and must be thought of as submerged land, not as ocean.[3] The variations in extent of the submerged continental edges are great. Chile has virtually no shelf; the USSR has over 3.43×10^6 km^2 of adjacent shallow water of less than 100 fathoms; France with a coastline of 2,560 km has 140,600 km^2 under 100 fathoms and 32,600 km^2 between 100 – 1,000 fathoms, whereas Spain has a coastline of 2,766 km has 68,600 km^2 under 100 fathoms but 157,700 km^2 between 100 – 1,000 fathoms. These submerged lands have as varied geology as their adjacent land masses.

Our knowledge of the earth, geology in the broad sense, has grown rapidly in the past two hundred years through field and laboratory work, observation and deduction. New tools are always being developed such as the dating of rocks by measuring their radio-isotopes, deep oceanic drilling and mapping of palaeo-magnetic trends from which developed the plate tectonic ideas of the structure of the oceans. Our knowledge, however, is still incomplete and fragmentary. There are many areas of the earth virtually unexplored. We do know that they will be different from anything we do know. They may be similar to areas we know and perhaps we may be able to use analogy, but such is the infinite variety of nature that homogeneity is a virtual impossibility. This is a basic thought with which one must approach the discussion of the earth and its resources. A second is that current knowledge is incomplete and one should think and act in the humility of acknowledged ignorance and not with the arrogance of unconscious ignorance.

Definitions

Much of the misunderstanding about natural resources is due to lack of appreciation of our ignorance and to the misuse of words. Definitions of resources and reserves have been developed in the different energy industries in different terms for different reasons because of different uses. These definitions have been mangled by other workers in different fields who have confused and are confusing understanding. However, the problem is not simple; because of the variety of circumstances, the lack of homogeneity and the gaps in our knowledge, no precise standardization is possible. The best that can be done is to define the terms being used and then be as consistent as possible.

In this book the terms used will be:

Resource base: The total amount of the energy source occurring in the world in commonly recognizable form.

In the renewable energy resources a time factor has to be added to allow quantification.

In the non-renewable resources of the fossil fuels the amount of the coal-in-place or oil-in-place is finite because the current formation rates are infinitesimal. In the nuclear or fissile resources there is a finite amount of radioactive material in the earth and when used in current reactors the fuel may be considered as non-renewable; when used in breeder reactors the energy content is so expanded that the energy source becomes, in effect, a renewable source.

Resources: The total amount of the resource base which is estimated to be probably recoverable for the benefit of man. This estimate will be based on both knowledge and reasonable conjecture regarding location and probable recovery techniques but is a very imprecise term.

Reserves: The total amount of the resource which can be defined as recoverable in stated terms of economic and operational feasibility. Whenever possible the degree of feasibility will be given by qualifying the term 'reserves' as:

> **Possible reserves**: the amount about which geological knowledge is insufficient to give any but the most vague recovery costing or indicate optimum recovery method, yet is still within the range of possibility. This is again imprecise and dependent on individual opinion.

> **Probable reserves**: the amount about which geological and engineering knowledge is insufficient for an explicit statement that it could be recovered under current economic and operating conditions, but can be judged as becoming economically recoverable with only a slight increase in knowledge of either the deposit or operating techniques or both.

2

Proved reserves: the amount that is reasonably certain could be produced in the future under current economic and operational conditions from deposits established on known geological and engineering data.

These categories are listed in increasing feasibility, both economic and technical. The concepts are summarised in figure 1.

For most resources past production is the best known part of the resource base, but with solar energy and terrestrial heat flow the resource base is best-known and the past production small and least-known. This point is developed under the methods of measurement section (p. 12ff).

In 1974 the US Bureau of Mines and the US Geological Survey agreed to standardize reserve definitions and adopted the same basic concept[4] but subdivided total resources into identified and undiscovered, with further subdivisions as shown in figure 2.

Figure 2 may well become increasingly adopted but it implies greater knowledge, world-wide, than is valid. This is illustrated by the rigidity of the rectangular framework and the definition, in the glossary accompanying the original diagram, of sub-marginal as being 'the part of "sub-economic resources" that would require a substantially higher price (more than 1.5 times the price at the time of determination) or a major cost-reducing advance in technology'. The price of crude oil quadrupled in 1973/74, yet it is impossible as yet to assess in concrete terms what quantities have moved thereby from submarginal resource to para-marginal resource or from resources to reserves. Therefore, yet whilst noting the proposed standardized nomenclature and fully agreeing that the mutual goal should be standardization, the diagram of figure 1 with its suggestion of expansion and fluidity is preferred.

An attempt was made to progress towards this goal in the spring of 1977 by a small group which was assembled at the Centre for Natural Resources, Energy and Transport (CNRET) of the Department of Economic and Social Affairs of the United Nations.[351] One of the recommendations of that Group was that:

RESOURCES CONCEPTS   Fig. 1

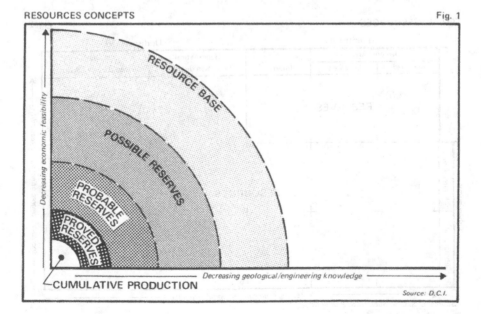

3

'The United Nations should be concerned that the definition of scientifically important and practically collectable crude oil and natural gas data is the next logical improvement of information about oil and gas resources.'

(The more detailed recommendations of this group are noted towards the end of this chapter.)

In early 1979 an Expert Group on Definitions and Terminology for Mineral Resources was convened by the Secretary-General of the UN and was organized by CNRET. Their main recommendations[416] were that:

- the word 'resources' should be used exclusively, eschewing 'reserves',
- a letter-number system would remove ambiguities of descriptive terms,
- three basic categories of all in-situ resources should be used according to a decreasing level of geological assurance but which might be of economic interest within the next two to three decades.

The three basic categories were identified as R-1, R-2 and R-3.

Each of these categories could be subdivided and designated:

- 'E' - Those in-situ resources that are considered exploitable in a particular country under the prevailing socio-economic conditions with available technology (in brief 'economic').
- 'S' - The balance of in-situ resources that may become of economic interest as a result of forseeable economic or technological changes (in brief the sub-economic of some classifications).
- 'M' - a special category may be possible to designate those 'S' resources which may become exploitable in the more immediate future as a result of normal or anticipated changes in economic or technological circumstances (which some would term 'marginal').

The Group acknowledged that some resource estimates for some minerals, such as oil, natural gas and uranium, are more commonly reported as estimates of the recoverable material, which more closely approximates the quantity that may

TOTAL RESOURCES   Fig. 2

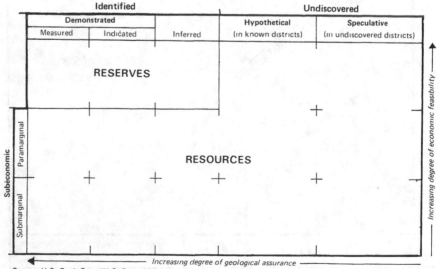

Source: U.S. Geol. Surv./U.S. Bur. of Mines, 1974

appear as mineral supply; therefore they recommended a parallel set of categories and definitions for recoverable quantities, which would be shown as r-1, r-2, r-3, with the notations 'E', 'S' and 'M' also used where possible and appropriate.

Yet another classification system has therefore been launched with the authority of the UN. Yet time alone will tell whether there is world-wide acceptance. There are many and obvious problems which will still make full comparison between minerals difficult, after the definitions have been adapted to each mineral. Figure 3 is the Group's illustration of their concept. The dangers of overly rigid a system into which inadequate data might be squeezed have been avoided. Perhaps the number-letter system may lessen the tendency for a reader to assume that a term carries his personal definition, and force the reader to check with the official definition, but that is a moot point because most readers are lazy. However, the reader of this book might note that R-1, R-2, and R-3 are the resources in-situ from which may be recovered r-1, r-2, and r-3, which are the proved, probable and possible reserves as used herein. The Group stressed the need for periodic review of the nomenclature and there can be little quarrel with that idea. My main criticism is that the scheme does not give any real indication of the dynamism of the whole concept of resource classification and therefore the 'ripple diagram' of figure 1 is still considered to be the best starter.

RESOURCES CLASSIFICATION CATEGORIES   Fig. 3

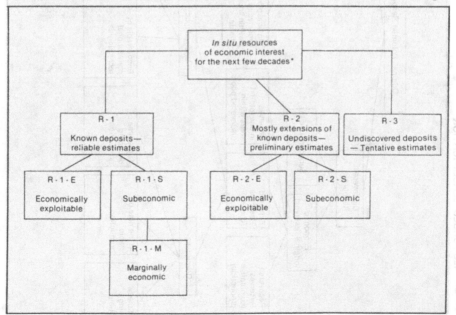

*While the capital 'R' denotes resources *in situ,* the lower case 'r' express the corresponding *recoverable* reserves for each category and sub-category, such as r - 1 - E.

Source: United Nations Economic and Social Council, E/C.7/104, The International Classification of Mineral Resources, Report of the Group of Experts on Definitions and Terminology for Mineral Resources, Annex 1, p.1.

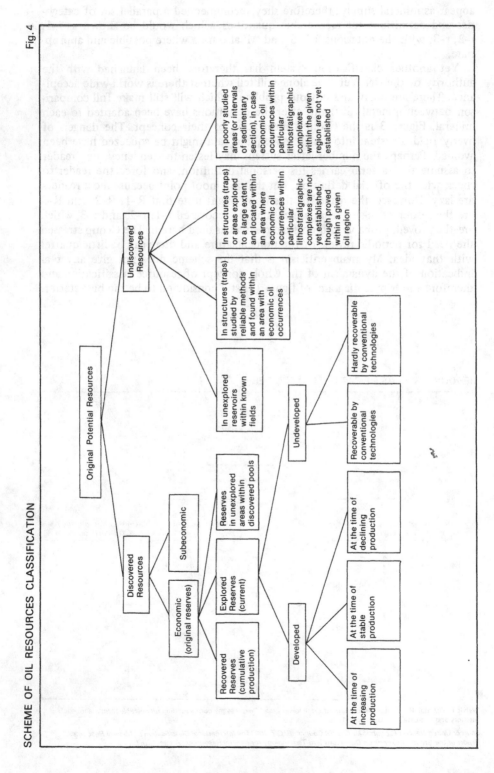

Fig. 4

SCHEME OF OIL RESOURCES CLASSIFICATION

Original Potential Resources

Discovered Resources

Undiscovered Resources

Economic (original reserves)

Subeconomic

Recovered Reserves (cumulative production)

Explored Reserves (current)

Reserves in unexplored areas within discovered pools

In unexplored reservoirs within known fields

In structures (traps) studied by reliable methods and found within an area with economic oil occurrences

In structures (traps) or areas explored to a large extent and located within an area where economic oil occurrences within particular lithostratigraphic complexes are not yet established, though proved within a given oil region

In poorly studied areas (or intervals of sedimentary section), in case economic oil occurrences within particular lithostratigraphic complexes within the given region are not yet established

Developed

Undeveloped

At the time of increasing production

At the time of stable production

At the time of declining production

Recoverable by conventional technologies

Hardly recoverable by conventional technologies

Although the USSR does not publish its oil reserves figures, its scientists have recommended standardization of reserve classifications for many years. Their long-established alphabetical system is well-known and comparisons have been made in many papers.[351] A. M. Kalimov and M. V. Feigin[476] put forward a new system at the Tenth World Petroleum Congress in September 1979, as a reconciliation between the USA system, figure 2, and Soviet current thinking, figure 4.

The 'explored reserves (current)' are the equivalent of proved or measured reserves, but also include 'undeveloped, recoverable by conventional technologies', which would be part of the proved reserves of the Middle East where simple structures and stratigraphy make extrapolations easy, but not so in more complex areas according to strict rules of the American Petroleum Institute, and would be classed nearer to the 'inferred' or probable category. The weakness of this part of the system, in my opinion, is, however, the further inclusion of the 'undeveloped, hardly recoverable by conventional technology' reserves, for these are verging from the probable to the possible in my reckoning. On the other hand, the subdivision, where possible, of the 'explored reserves, developed' according to whether they are estimates made whilst production is building to its designed peak, or during its peak plateau or when production is declining, is an important suggestion because it emphasizes the dynamism of the system and the essential concept that proved reserves are only a working inventory of what is currently available.

The Canadian Petroleum Association in 1979 produced a metric version of their Statistical Handbook. Crude oil and natural gas were re-measured in cubic metres and sulphur in tonnes. They also changed their reporting nomenclature. Their 'probable reserves' had included 'proved' reserves, but both had been reported. For the end-1978 data the term 'established' replaced 'probable' and separate reporting of the 'proved category', closely allied to the API 'proved' category ceased. The new definition was:

'**Established Reserves:** Those reserves recoverable under current technology and present and anticipated economic conditions, specifically proved by drilling, testing or production, plus that judgement portion of contiguous recoverable reserves that are interpreted to exist, from geological, geophysical or similar information, with reasonable certainty.'

'Initial established reserves' are those prior to deduction of any production, and 'remaining established reserves' are those less cumulative production. 'Initial volume-in-place' replaces 'original in-place reserves', and more sensibly now approximates the term oil-in-place as most of the industry understands the term. There seem some advantages in the new system but it does mean an extra burden.

In general there seems to be increasing closeness, at least amongst oilmen, with semantics the main barrier to full understanding and thence accepted standardization. The task of the 1979 UN Expert Group to standardize nomenclatures for all mineral resources was, perhaps, an impossible task at that time. The working group established by the International Energy Agency in 1976 under the guidance of the National Coal Board in London, started the attempt to bring greater understanding of resource and reserve concepts within the world coal industry. It was suggested at the 1979 World Petroleum Congress that a multilingual, multi-disciplinary international team might study the oil problem and produce a consensus report to the 1983 Congress in London.

Resource Bases

The resource bases of the world's energy potential are immense, but it is worthwhile considering them in turn as the starting point towards the more practical sides of the problem of availability.

Solar. The energy reaching the earth from the sun is a very popular introduction to discussions on energy and it is of some interest to list examples of different ways of expression. 'One hundred years ago man's rate of consumption of fossil fuels was equivalent to about $1/1,000,000$th of the sun's energy falling on the earth; today's is about $1/10,000$th...'[5] 'The solar flux on earth is 2×10^{17} W..'[6] 'The total solar imput to the earth is about 5,200 quads/year....'* Other examples[8-13] also illustrate these points; the solar base is very large, it has little practical value and there are differences between the facts and measurement points in the various modes of expression.

Lunar. The tidal effect of the moon on the solid land of the earth is too slight to allow practical utilization. The effect on the waters of the earth in the oceans, as tides, is great. One estimate is that 3×10^{12} W of tidal energy is dissipated on the earth.[14] But this energy is of use only when focussed to abnormal variations of level in confined spaces. These places are surprisingly few in number. Present technology is capable of dealing only with tides of over 2 m in height. One calculation[15] is that only $2 = 3\%$ of the world's tidal energy, perhaps $2 - 3$ million TJ/year is useable. 'With a 20% recovery as electricity this would mean a world capacity of only 13,000 to 20,000 megawatts or less than 1% of the maximum production achievable from conventional hydraulic resources'. In brief, until a new technology is developed the total tidal energy is a figure of little significance.

The effect of tides, currents and winds, a combined lunar and solar effect, on waves to create the transfer of energy across a line as wave power can be calculated and is said to indicate 'prodigious amounts of power'.[16] Again, until a technology is developed to utilize that power, and work is being done in this field, the total immensity, the resource base, is only of academic interest.

Terrestrial heat. The heat from the hot core radiates outwards and its flow is influenced by the geology of the rocks in its path to the surface. The average value of terrestrial heat flow by conduction on the continents is about 6.3×10^{-6} J/cm^2/second, and where magna from the earth's mantle layer is near the surface the flow may be ten to twenty times the average.[17] Part of the heat carried by the magma is dispersed in volcanoes, part heats the rocks by conduction, and part heats the fluids within the rocks, which, in porous rocks, then move the heat by convection. The hot waters may reach the surface as hot water springs or steam geysers. If the upward convection currents are stopped by an impermeable rock layer, then a closed circuit may be formed. This may be naturally fractured or artificially pierced by a well. The heat then released may give a heat flow some 200 to 1,700 times as great as the average terrestrial conductive heat flow. Such amounts of heat can be used for transformation into electricity or as direct heat. Abnormal conditions, providing possibilities of readily utilizable energy, occur in broad belts across the earth, but such is the variety of the geology that actual probes must be made before the value at any one place can be assessed. Hence the total is of far less significance

---

*Quad (Q) = 10^{18} Btu = 25×10^9t oil = 200×10^9 barrels oil = 1.26×10^{24} J)

than the specific amount at any point and this is true even when the normal, rather than abnormal, heat flow and temperature differences are used in the heat pump.

Terrestrial water. Whether one states that 'about 1200 Q/year of the total solar input to the earth enters the hydrological cycle',[7] or that the total annual energy in flowing streams is approximately 290×10^{18} J and is about equal to the present annual use of all commercial energy in the world,[18] one has figures of little more than conversational value. The resource base is great but adequate concentration to form resources is comparatively limited.

Terrestrial air. The winds are ubiquitous, except in the doldrums area of the North Atlantic. From the earliest days of man moving out to sea, winds have been harnessed to sail ships. Holland's windmills illustrate their use on land. Gliders in peace and war are a development from the legendary Icarus. The upper atmosphere global jet streams assist high-flying aircraft. In total, as with the tides, the winds have an immense resource base, but the problems of variations in strength, unpredictability except in rare places and, more particularly, cost of conversion restrict their use also.

Fissile fuels. The resource base of nuclear energy is immense because of the wide dissemination of radioactive material. Although one may be sceptical of the validity of the basic knowledge of the granites of the earth's crust, yet the size of the figures involved indicates an order of magnitude well beyond any necessary accuracy. Thus an average granite is said[19] to contain 4 g/t or 48,000 $tU/mile^3$. Another figure is that the average uranium content of the earth's crust is about 2.8 ppm by weight, so that the top kilometre of the crust contains material which, if used in a fast reactor, would provide energy of 4.2×10^{25} J and the seas 1.263×10^{23} J.[7] For the distant future it has been calculated that the oceans also contain thermonuclear energy by fusion techniques to the values of 10^{31} J from deuterium, 5×10^{27} J from lithium plus 5×19^{23} J as a conservative estimate of lithium on land.[20] Another mode of expression is that 'if we were to extract the energy content from only 1% of all the deuterium available in the oceans, it would supply the total energy needs of the world in the year 2000 for almost 100 million years.'[21] As with solar energy, such figures indicate a resource base which is virtually infinite and, so noting, warrants no further discussion.

Fossil fuels
Carbons. The carbon resource base is usually confined, and probably rightly so, to the coals. Recently, however, particularly in circles where 'hydrogen economics'[22] and very long-range energy problems are being discussed, mention is made of the carbon locked in the carbonate rocks, the limestones, chalks and dolomites, which form a considerable part of the continents. Others[23] would make calculations about the total mass of sedimentary rocks in the earth, the percentage of shale and the percentage of those shales which are younger than the Precambrian age when land life became abundant some 500 million years ago. Such figures are not repeated here because the scientific basis for estimation is so sketchy that they are of negligible value.

The resource data for the coals as published are an aggregation of estimates moving from the known to the unknown, in contrast to the resource base data mentioned before. These estimates, as will be shown later, vary greatly in method of measurement, though these variations are often ignored. Yet the totals are so great that, as far as is known, no one has seen a need to calculate a world resource base for coal, though a figure often used for proved reserves plus undiscovered resources is of 10 teratonnes.

Hydrocarbons. In contrast to the coals, there have been a number of attempts to estimate the resource base for the hydrocarbons or, more specifically, for ultimate crude-oil-in-place, with, in some cases, figures for natural gas liquids. One estimation in 1965[24] gave 10×10^{15} US barrels of crude oil, 30.5×10^{15} cu.ft., natural gas and 8.30×10^{11} US bl. natural gas liquids. Others estimate ultimate recoverable reserves, such as the proved plus undiscovered resources figure of 1.73×10^{12} US bl. which at 7.4 US bl. to the tonne would be 233 Gt and, on similar definitions, 200×10^{12} m^3 of natural gas.[25] Such figures are resource figures and will be discussed later.

The resource base of heavy oils, and oils in oil sands and oil shales is undoubtedly large. Duncan and Swanson, in 1965,[26] estimated as 4.22×10^{24} J the total world combustion energy content of shales containing 10–65% of organic matter. They admitted major assumptions and, in truth, such figures have as little practical value as an undefined figure for solar flux. Let it be simply appreciated that there are indeed very large resource bases, greater than those of fluid hydrocarbons, behind the heavy oils, the viscous oils in oil sands and the solid kerogen in oil shales, and discuss in the chapter on resources the few large areas where deposits are known.

Resources and reserves – units of measurement

Within each resource there are variations in nature or form; in the fossil carbon resources from peat to anthracite, in the hydrocarbons from heavy oil to light gases, and in the fissile fuels in the richness or concentration of the ore. The definitions of 'quality' were developed by the industries according to end use.

Coal is classified by rank according to percentage of fixed carbon and heat content, calculated on a basis of being free of other mineral matter. Coals of different ranks are adaptable to different uses. The anthracites have the most fixed carbon. The low-volatile sub-bituminous coals, which are the best for producing coke, have a heat content of over 7.2 MJ/kg (over 15,000 Btu/lb), being double that of lignite. Heat content is measured sometimes 'as mined', sometimes 'as burned', that is after washing and sorting. Coals are also classified by grade, largely according to the content of ash, sulphur and other non-carbon constituents. This is a classification which has become increasingly significant with growing public concern over clean air conservation. The common measurement is by weight.

Crude oil is usually measured by volume, usually the US barrel of 42 US gallons, but in Iran, for the first thirty years or so of production, measurement was in UK gallons and in Japan is in kilolitres. The quality is judged in a number of ways. The specific weight is expressed either on a scale standardized by the American Petroleum Institute, degrees API, or as specific gravity figures. Usually the lighter the oil, the better the 'quality' because of a higher proportion of easily distilled gasoline, currently the highest-priced fraction. The nature of the oil, whether naphthenic or paraffinic, is important. The amount of contained impurities is also important, and sulphur in one form or another is the most common impurity and expressed usually as a percentage by weight. Petroleum gases are measured by volume but the heavier the gases the more valuable, in contrast to crude oils, in general, but the non-combustible content of inert gases can be substantial.

The richness of uranium and thorium ores may be measured in grammes per tonne but resources are measured by weight of elemental metal or specific oxides.

Nuclear energy sources are therefore easily expressed in tonnes of equal content. All carbon resources can be expressed in tonnes but the quality of those tonnes can vary and their types should be specified where possible or, if aggregated, should be designated 'of all types'. It is very much more difficult to convert the volumetric units of the hydrocarbons into mass, because of the very wide and important variation in specific gravities. These vary up to 10% in the averages of country crudes and even more between individual crudes. Most oil-men are accustomed to working in barrels, in billions (10^9) for reserves and thousand or million barrels/day for production and there is a massive store of information in the literature expressed in US barrels. Furthermore, the handling of hydrocarbons, in liquid or gaseous phase, is by volume, whether in pore spaces in the reservoir or flowing and being metered in a pipe, or static in a tank. However, where the characteristics of crude oils are sufficiently known, then volumes have been converted to mass in some instances. The basic point must not thereby be overlooked, that all tonnes of coal or all tonnes of oil are equal only in mass and not in any other characteristic.

The transfer to units of energy is fraught with even greater dangers and yet many theoreticians consider that this is an absolute necessity for comparison and understanding of total energy resources. This transfer tends to obscure differences between resources and ascribe precision and comparability to statistics which do not possess them. Philosophically, there is here a conflict between the physicist or engineer who strives for precision and is only happy when dealing with measured quantities, the statistician to whom a figure is a figure, and, on the other hand, the geologist who appreciates the infinite variety of nature and is unhappy when unnatural precision is attempted and unnatural distortion results.

There are two main areas liable to distortion. There are variations in order of magnitude of end-use according to the method of conversion from primary to secondary end-use. This is usually recognized in the nuclear field and the conversion method, e.g. thermal or breeder reactor, used to give an energy content is often quoted. Yet there are wide variations in end-energy value when coal is used in a household fire or converted into manufactured gas , whether crude oil is burned under a boiler or converted into a range of fuels used in different engines. Moreover, the natural energy resources have other non-energy uses and, particularly for the hydrocarbons, these uses may have even greater value than use as raw energy.

There can, indeed, be practical difficulties and misunderstandings arising from lack of care in conversions from mass to energy content. Two crude oil reserve figures of near equivalence of mass but different energy content according to conversion method can illustrate the difficulty. If both are converted using a standard factor of 43.24GJ/t for an average crude oil of SG 0.949, this would confirm the rating of the reserves by their mass. If, however, conversion factors closer to the average crude values of the two countries are used then their rating is reversed. In both calculations the variation in 'value' caused by differences in sulphur impurities is ignored, even though it might tend to restore the former rating, since many Venezuelan crudes have a higher sulphur content.

It is in the least known, the most speculative areas of resource bases and resources that the least practical, the most theoretical of the persons concerned with energy resource problems are most insistent on energy content values. This can be most misleading, for generalised factors with no sound base are concocted for one purpose by one author and by the repetitive use by other authors acquire a verisimilitude of 'truth' and 'accuracy'.[27] However, such is the pressure, that, to avoid misuse of figures, conversions into joules have been made for some world figures, but the factors are given.

In the better-known areas towards the working stocks of proved reserves, each resource is assessed separately. Substitution, between energy sources, though common, cannot be universal and each specific case must be considered separately. Indeed some consider, with considerable justification, that the better measure of availability of resources is the real cost of resource products, rather than volumes of specific ore reserves or numbers of acres.[28] First, however, let us look at how the cubic metres or hectares or tonnes are measured.

Table 1
Comparison of crude oil reserves by mass and energy content

	Proved* reserves Gt	Energy** content @ 43.24 GJ/t	S.G.	Conversion factor GJ/t	Energy content on col 5 factors
Algeria	5.977	258.45	0.806	46.28	276.6
Venezuela	6.093	263.47	0.904	44.23	269.5

Sources: * 1974 Survey of World Energy Resources, Table IV-4, p. 108
        ** idem, Table IX—2, pp. 253 & 255

Methods of measurement

The methods of measurement of the physical size of the energy resources are important and they vary widely. Some might say that the most precise quantities are those of the resource bases of solar energy and the earth's gross heat flow for they are the result of physical/mathematical calculations corroborated by some observations. Yet their proved reserves are imprecise because their points of utilization are scattered and often unreported because small. The proved reserves of crude oil in the USA are probably the most precise and yet their resource base is most imprecise.

Geothermal energy. The calculation of the amount of energy in a geothermal reservoir rests on assumptions of the base temperature, porosity, etc, existing in the reservoir. These are only known during the maturity of exploitation, can only be very approximate for newly-discovered fields and only very speculative for non-drilled areas.

The US Geothermal Steam Act[29] defines a known geothermal resources area as 'an area in which the geology, nearby discoveries, competitive interests or other indicia would, in the opinion of the Secretary (Interior), engender a belief in men who are experienced in the subject matter that the prospects for extraction of geothermal steam or associated geothermal resources are good enough to warrant expenditures of money for that purpose.' There are a number of implications, philosophical and practical, arising from that definition. However, the US Geological Survey have apparently given the following parameters for a geothermal reservoir to have appreciable potential for exploitation: a relatively high temperature, greater than 65°C to 205°C depending on the use and production technology; a depth shallow enough to permit economic drilling, currently about 3.5 km or less; sufficient rock permeability to allow the heat transfer agent, water and/or steam, to flow continuously at a high rate; and have a sufficient water recharge to maintain production over many years.

The definition of known geothermal resource areas and the technical guidelines for the USA illustrate clearly that, as will be noted in the resources chapter, although the broad areas of potential geothermal energy are known and methods of measurement for drilled and exploited reservoirs are known, there are no known methods of measuring the unknown.

Fissile fuels. The measurement of the fissile fuel resources is midway in method between that of the renewable resources and the fossil fuel resources. On the one hand theoretical calculations can be made from global physical 'measurements' as to the amount of uranium and thorium in the earth's crust and in the oceans. These, however, assume unnatural homogeneity of both the land and seas but the numbers are so great that they will never be tested except by new physics and are outside practical energy consideration.

Uranium and thorium are both widely distributed in minerals of varying complexity, especially in acidic igneous rocks. The granites of the earth may contain 2–4 g/t, in finely disseminated form and in veins and pegmatites, in more concentrated form. It is, however, when the granites are weathered and eroded and concentrations of uranium oxides are possible in sandstones and conglomerates and concentrations in placer deposits of the heavy thorium mineral monazite, a rare earth thorium phosphate, that ores are formed. It is only such concentrations as ores that are economically mineable, and can be beneficiated and sold. Some ores can also be produced as a by-product in the mining of other minerals, as along with gold in South Africa.

Examples of concentrations can be taken from the USA where the uranium contents, g/t, are reported as 3,500 in Colorado Plateau ores, 30 to 300 in phosphate rocks, 10 to 100 in black shales and 4 in granites.[30] One might note here that a gram of uranium in a non-breeder reaction has an energy content of the order of 860 MJ compared with a gram of average bituminous coal with 29 KJ, or about 30,000 times greater.

S. H. U. Bowie[31] differentiated between 'ore reserves' as material in ore bodies that is likely to become commercially viable at some time in the near future, and 'resources' as all material likely to be available at a realistic price but which is unlikely to be exploited until reserves have been largely depleted. He stated that 'in order to calculate ore reserves it is widely accepted that material available at a price of up to $20/kg U_3O_8 in concentrate form classifies as ore.' He noted that

'errors in calculating the amount of ore in the ground can range from a few to as much as 50% of the actual tonnage, depending on the nature of the ore body. Relatively homogenous deposits of uranium in flat-lying sediments are by far the easiest to assess, veins and pegmatites the most difficult. Fortunately, over 70% of reserves of uranium are disseminated in quartz pebble conglomerates and sandstones so that the estimate of tonnages can be accepted with a reasonable degree of confidence.'

Even so, the measurement of the ore bodies is no more precise than that of any other mineral ore bodies, even though the detectable radioactivity is an additional exploration and measurement tool. Bowie's more detailed description of Canada's uranium reserves, ranging in price up to $20/kg U_3O_8 as reported in 1973 is illustrative. The reserves were given as 185 kt U, of which 150 kt U were in ore bodies of elongate or tabular shape in quartz-pebble conglomerates which vary greatly in size, with the economic horizons varying from 2 to 12 m. Some of the ore bodies are deep-seated. The remaining 35 kt U are in vein-type deposits and perhaps 2 kt in granite pegmatites. Canada has reportedly 190 kt of estimated additional resources. Such a description illustrates that the dollar sign does not give more than a pseudo-precision to the measured reserves. However, for each mine the assessment by drilling to place the next year's development does give the operator a necessary closer appraisal of the next few hundred metres to be mined.

Thorium reserves are much less measured, except in the current mines.

Fossil fuels. The hardest facts of the fossil fuel resources are those of current production, the next are past production and proved reserves and the softest facts, the least-known, are the resource bases. There is no known method to estimate directly the total coal-in-place or oil-in-place; one must extrapolate from the known to the unknown. There are, however, two basic differences between these resources which cause differences in method of measurement. The causes and conditions of formation of the coals are few and well-known and coals are immobile solids. In contrast, the causes and conditions of formation of the hydrocarbons are many and ill-known and the most easily and commonly exploitable types, crude oil and natural gas, are mobile fluids, from genesis through migration and accumulation to production.

Carbon fuels. The formation of coal depends on luxuriant vegetation in gently sinking, widespread swamps, with subsequent burial by sediments and resulting subjection to compaction and heat. Usually these deposits have been subjected to no violent horizontal earth movements though vertical movements by faulting are common in many coalfields. There has usually been a cyclical rhythm to the movements of the basins as the coals were accumulating and these are easily recognizable. Rank of coal is a function of the weight of original overburden, age and local deformation, in effect an expression of successive stages in the formation of the coal, as volatile matter and moisture are expressed from the material. In some areas the carbon-bearing rock sequences are thick and widespread, even globally, e.g. the coal measures in Britain are similar in age to the coals of eastern North America. In other places only thin stringers of coal were formed over short distances. In some areas the coals are buried deep, in others are now near the present surface. Hence there is great variety, indeed more than many appreciate, in coal deposits.

In producing coalfields there is a mass of information to aid measurement, including mine maps of individual seams and logs of exploratory boreholes. The local degree of continuity and homogeneity can set reasonable parameters for extrapolation. The US Geological Survey Bulletin 1136 (1961) set limits for its extrapolations, e.g. for 'measured reserves' the necessary observation points of mine or drill hole are 0.8 km apart, for 'indicated reserves' about 1.6 km but up

14

to 2.4 km apart in areas of 'good continuity', whilst 'inferred reserves' are more than 3.22 km from an observation point. This definition of limits was a conservative step, for earlier estimates had assumed greater continuity in individual beds. It was also an important step away from the traditional coal industry habit of describing only original reserves in the ground before mining, with no allowances for past production and losses or future losses. However, there was still no real economic parameter, indeed it was noted that '. . .the word reserves means estimated quantities of coal in the ground within stated limits of minimum bed thickness and maximum overburden thickness. As thus defined, the word has no economic connotation, although the selected limits of the various reserve categories have economic application.' Some 135 possible categories were evolved by breakdown by rank (5), thickness of beds (3), thickness of overburden (3), relative reliability of estimates (3). Two most important points concerning coal reserves can be illustrated by consideration of the thinnest category, the reasons given for inclusion and the size of the calculated total.

Seams of between 36-71 cm thickness for the ranks above bituminous coals, were adjuged suitable for hand mining but of 'little present economic interest'. However they were estimated as such and included because:

(a) prudence dictates that occurrences of marginal reserves of coal should be recorded for possible future use, just as the marginal reserves of other useful minerals are recorded, (b) some coal in this category is mined, (c) the information is obtained with little additional effort . . . and would be lost if ignored. The minimum of 35 cm also permits comparison with older estimates which generally employed the same figure.

This category of thin seams comprised 44% of the reserves of the fifteen states in which sufficient information was available for segregation and in the report considered that sample 'is fairly large however and the observed distribution should show the general order of magnitude of distribution for the United States as a whole.'

In the sense in which the words are used in this book these thin seams would be, in bulk, relegated at best to possible reserves because, apparently, only hand-mining is possible and no other alternative currently economical recovery method is known. Yet it is implied that they form some 44% of the measured reserves of the USA. The setting of the limits from observation points had reduced earlier estimates by 12% but if this thin category of seams had been discounted as 'reserves', perhaps the reduction would have been by 50%.

Even greater caution must be employed when dealing with the less well-known reserve categories. Yet, outside the existing, operating coal mines of the world, most of the so-called reserves are at best in the 'inferred' category of the 1961 US definition, being more than 3.2 km from an observation point.

One example is the recently indicated Oxfordshire coalfield in Britain.[32] Even in one of the best geologically mapped areas of the world, where the Geological Survey has been involved with coal for 140 years, new discoveries can be indicated by meagre information but must be allocated to the appropriate reserves category and not just accepted as 'x million tons of new coal' being added to the immediate working inventory. Barren upper coal measures were encountered in the area in a borehole nearly 100 years ago, and again in a second borehole in 1913. Four other boreholes, in 1913, 1960 and 1965 proved the absence of coal measures to the east and west. Two bores completed in 1972, however, provided good correlation with the rocks of one completed in 1961. The distances between these boreholes are 20 km and 16 km, forming a flat triangle. The seams encountered are considered workable. An area of some 1165 km² is now estimated to contain reserves of the order of 10 Gt of moderate to low quality coals, but of simple geological structure, perhaps well-suited to automatic machine recovery. The method of measurement and the information available would seem to put these deposits into a low probable reserves category, perhaps even only possible reserves.

In summary, perhaps the traditional coal industry reserves concepts might be illustrated by fig. 5 which should be contrasted with fig. 1, to which, as the discussion following will show, the hydrocarbon industry concepts are much closer.

COAL RESOURCES CONCEPT   Fig. 5

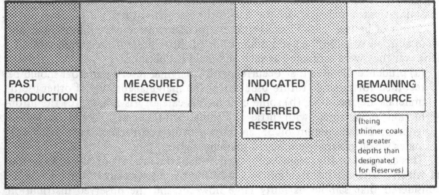

Source: D.C.I.

Hydrocarbon fuels. For the hydrocarbon fuel resources it is again the fully delineated producing field which can provide the best estimate of reserves. However, the methods of measurement are affected by the two major differences with the carbon fuels which have already been noted; crude oil and natural gas are mobile fluids with great variety in their mode of occurrence. They may have moved distances of up to many kilometres horizontally and hundreds of metres vertically from their place of origin. They are commonly associated with water, are under at least hydrostatic pressure and occur in the pores and fractures of rocks, not as discrete homogeneous bodies. Any type of rock, if porous and permeable or fractured, may act as a reservoir if overlain by an impervious cap rock, forming a seal. There are many types of tectonic trap caused by deformation of the originally flat sediments by folding or faulting or the upthrusting of salt masses. There are many kinds of stratigraphic traps due to lateral changes in the characteristics of the sediments. Because of the broad geological history of a petroliferous basin there can be similarities between the fields and prospects of the same age in the same basin, but there can be marked differences between fields in different parts of the basin. The fields of Iran on land in the foothills of the Zagros mountains are very different from those of the Arabian peninsula though they are both huge and geologically simple with comparatively homogeneous reservoir rocks. In other areas, as in the central Romanian plains or the Niger delta, the reservoirs are multiple lenses with rapid lateral and vertical variations. In the Middle East, once a structure, already delineated geologically and/or geophysically, is proved petroliferous by a well, only very few and widely-spaced wells may be needed to indicate millions of tonnes of crude oil or thousands of millions of cubic metres of gas. In other areas of smaller, less widespread and less simple fields, many closely spaced wells may be needed. Hence the methods of estimation must depend on the local circumstances, and various factors must be known. These include the original reservoir pressure, the porosity and permeability of the rocks, the type of drive (whether from water from below or only by expansion of dissolved or gascap gas), the viscosity of the fluid in the case of oil fields, and in gas fields whether the gas is dry or has heavier hydrocarbons which are liquid at normal temperature and pressure. In a normal test in an oil well there are about seventeen items which must be measured before a sound judgement of the flow characteristics can be made, e.g.:

depth of interval tested, openhole perforated, time of test in production history, type, duration, tubing/drill stem internal size, size of surface choke, downhole restrictions, bottom-hole shut in pressure when fully built-up or gauge depth sub-sea level, bottom-hole flowing pressure, wellhead pressure, gas/oil ratio with H_2S content (if any), water/oil ratio with salinity or specific gravity with any H_2S indication, method of metering, type of formation, specific gravity of oil, sulphur content and pour point, flow rate or quantity of fluids recovered.

In those countries where the mineral rights belong to the landowner and land-holdings are small, there is greater caution by the oil producer in estimating reserves, particularly where the geology is complex. Thus, the American Petroleum Institute's definition of proved reserves is:

'These are volumes of crude oil which geological and engineering information indicate, beyond reasonable doubt, to be recoverable in the future from an oil reservoir under existing economic and operational conditions. They represent strictly technical judgements. . . . Both drilled and undrilled acreage are considered in the estimates of proved reserves. However the undrilled proved reserves are limited to those drilling units, immediately adjacent to the developed areas, which are virtually certain of productive development, except where the geological information on the producing horizons insures continuity across the undrilled acreage.'

In the USA the drilling units may be less than 0.64 km between centres. In the Middle East, in contrast, both mineral rights ownership and the geology/engineering are simple. Extrapolations of many kilometres can be made early in the life of a field. Hence annual revisions and extensions are much less important there than in North America.

In North America there is also much more caution than elsewhere in including in measurements the results from possible stimulation of the reservoirs by any of many types of so-called secondary recovery. The API exclude 'oil that may become available by fluid injection or other methods from fields on which such operations have not been applied.'[33] In the USSR it has been common practice, for thirty years in many areas, to supplement the original energy mechanism by the injection of water almost from initial production. The distinction between primary and secondary recovery is therefore blurred. In all countries, with growing concern about the conservation of energy, natural resources and the environment, there is now also an increased tendency to re-inject any natural gas, being produced with the oil, back into the reservoir, which is another stimulation mechanism. In the USSR water flooding produced 240 Mt of

17

the 452 Mt produced in 1974 and the ultimate recovery from waterflooded reservoirs is estimated to average more than 50%.[34]

These considerations in the measurement of those hydrocarbons in known reservoirs are unknown in respect to the other energy resources and therefore alien in thought to those in other energy industries. It is difficult for many to appreciate why annual revisions of previous figures from last year's producing wells and from extensions by out-stepping development wells and/or initiation of secondary recovery seem to form such an important part of the North American reserve measurement while Middle East 'reserves' move in big jumps. To the oilman the reasons are obvious.

There are, however, some considerations common to all energy industries which apply to methods of measurement. No matter whether a field is new, or sufficiently old that its reservoir characteristics are known and a declining production curve established, it is the scientist on the field who knows what assumptions have been made in estimating the oil-in-place. It is true that in coal mines or uranium mines, the coal-in-place or ore-in-place or richness of the ore and the amount recoverable are best known to the man-on-the-spot. As these field estimates go up the chain of command, whether company or government, there is a tendency to round the figures, up or down, as they are aggregated with those of other fields or mines. There is undoubtedly the opportunity for pride or prejudice to amend the numbers and certainly the original assumptions or approximations in basic formulae are lost on the way up the chain or ignored. This is a most important point to remember in examining any resource figures and indicates the need to collect any such figures as near to the source as possible.

The public rarely has the opportunity or the knowledge to appreciate the significance of energy resource measurement but has to accept the interpretations of the communicators of the 'mass media'. Discoveries of new deposits of coal, oil, gas or other minerals are usually rated as newsworthy. In the case of oil, only two or three of the seventeen information items, noted as being necessary to assess an oil discovery, are usually released to the Press. Hence they tend to make bricks without straw and sometimes completely mislead just because they do not understand the methods of measurement.[35]

The problems have been stressed which arise from the variety in occurrence, of definition and usage and lastly the possible distortion in aggregating resources of even the best-known deposits under production. Yet these last are the firmest resource figures and one can only move towards estimating the less well-known and so into the unknown by analogy with the known resources.

Estimation of petroleum prospects is one of the most difficult, most misunderstood and most controversial in the field of energy resource estimation and therefore warrants special attention. There are many who make estimates; oil companies, consultants, government agencies, joint industry/government groups, multi-government groups (like the Organization for Economic Co-operation and Development, the European Economic Community, and the new International Energy Agency), investment analysts, bankers, trade journals and journalists. Exploration and/or producing companies have their own commercially confidential data and the published data. They wish to know where to go to find oil or gas to maintain or expand their own particular circumstances. The circumstances of one company may be very different from those of any other company, their needs, motivation, available manpower and finance, level and spread of expertise within the organization, policy as to hiring or breeding expertise, etc. It is a common but foolish idea that oil companies are uniformly motivated.

Consultants have their own data and may be employed on a number of different angles, one of which is often to prepare an impartial report, subject usually to some scrutiny by a regulatory body, in order to attract or advise investors. Government agencies, most often the national Geological Surveys, may have unpublished commercial information as well as published material and their objectives are to assess the national resources. There are other government agencies, particularly in the non-centrally planned countries and in the petroleum field, who are increasingly interested in the estimates of reserves and resources and future revenue by participation or taxation. Most of these interested parties have to rely on others who have, or claim to have, special knowledge. No matter what the purpose or what the knowledge, it is from the base of the known, which is far from precise, that even the 'expert' must move into the speculative area of measuring the less known.

Estimates of world petroleum resources have been made since the industry really started in the USA over 100 years ago.[33] In the past thirty-five years there have been, essentially, four methods, the geological, the mathematical, the geological/probability and the pragmatic/production method.

The geological approach or method is epitomised by the early work of Lewis G. Weeks, originally around 1940, when Chief Geologist to the Standard Oil Co. of New Jersey (now Exxon Corp), and subsequently in many publications. The basis was a basin-by-basin study of the world's sedimentary basins, assessing the geology and the prospects for petroleum by analogy with known areas. Weeks has increased the numbers with increasing knowledge and as new techniques have developed, the biggest development being the knowledge and ability to move offshore. Similar work by the US Geological Survey was published in 1965.[36] Weakness in the detailed analogies, particularly in the USGS circular, was pointed out by me in 1967.[36] The major problem is that this approach, when responsibly done, relies on so much detailed information that cannot be economically published, for it would take years to assemble in publishable form and fill many volumes.

An organization like the US Geological Survey has the many volumes of unpublished material. The basic USGS assumption appears to be that just as much of a resource will be found in the future as in the past, when geologically similar sedimentary basins are subjected to equivalent exploratory effort. The most thorough recent examination of the potential of the USA for petroleum was that carried out by a task force under Ira H. Cram for the National Petroleum Council with the collaboration of the American Association of Petroleum Geologists and published in 1971.[37] Cram, like most geologists, is an optimist and he said:

'The fund of geological and production knowledge proliferates and, equally or more important, the geologist's understanding continues to grow — understanding of what kinds of geologic situations result in crude oil and natural gas accumulations. As a result, the modern geologist is less inclined to condemn areas than his predecessors. Instead he is more likely to reduce the area of the 'impossible', to expand the area of the 'possible' and to forecast more crude oil and natural gas in the basinal areas, which are, and are likely to continue to be, the sites of major production.'[38]

In this study estimates were made of future potential reserves of discoverable hydrocarbons, noting that what will be discovered and developed depend upon the interplay of economic, political and technological factors in the future. After aggregating figures for the original oil-in-place in known, probable, possible and speculative resources and ultimate recoveries at different factors, they used an arbitrary category of 'expectable' reserves being the known + probable + possible + 50% of the potential of 199×10^9 barrels in addition to the 233×10^9 barrels of ultimate recovery from known reserves at a 60 recovery factor. The total original oil-in-place for future reserves was 436×10^9 barrels and for known reserves 388×10^9 barrels. In other words, more was still discoverable than discovered, but the judgement that only 50% of the speculative was really discoverable and 60% would be the ultimate discovery factor, reduced the future to less than the past. This is an excellent example of the type of judgement factor which must play an important role in the geological method after all the preliminary painstaking collation of evidence and analysis and synthesis.

The mathematical method is epitomised by M. King Hubbert who, since the late 1940s, has been stressing the finite limits of any non-renewable resource, in opposition to the euphoric open-ended approach. King Hubbert used past drilling effort and discovery success in the mature US industry to construct a symmetrical, logistic curve of production. Hubbert's method, for petroleum, basically assumes that the reserves figure added by each additional unit of drilling is a function of cumulative exploration drilling and little else.[39] Criticisms, including those from myself, have been continual, since its inception, of this illustrative curve with its exaggeration of the mathematical simplicity and inevitability, and under-emphasis of the results of variation in effort. Nevertheless the basic truth of limits and therefore the need for realism is one which must never be forgotten and King Hubbert's name will rightly long be remembered for his emphasis of this truth, even though it is only one of his contributions to the applications of science to the petroleum industry. Charles L. Moore favoured a Gompertz curve as more nearly approximating future actuals for it does, in effect, use the past to assess future probability.[40] A. R. Martinez favoured a controlled curve, 'adjusted' from a logistics curve, for Venezuela and this has proved a good forecast of actual production but is essentially a pragmatic approach.[41]

The third method might be termed the probabilistic exploration-engineering analysis method. It is common practice now in many circles to try to use probability mathematics and computer techniques 'to take the guess out of guesstimates'. Risk profiles are constructed, random samples taken, simulation models developed and much thought given to the probability of higher or lower values than those expected. The input obviously depends on the parameters chosen. These must be based on one or more of the obvious circumstances in which petroleum is found, the area or volume of sediments, geological analogy, exploration and drilling effort and technological development, discovery chances and analogies, production and reserve analysis and analogy. The proponents believe that such methods are better than the 'best guess' method. The only virtue is really reproducibility and continuity. The input is based on the same ignorance of basic facts, though it has to be numerated in order to be mathematically handled with machines. The input to an experienced, trained human brain need not be in numbers. The brain makes subconscious as well as conscious judgements. These are not reproducible nor capable of being used by another operator. This is a grave disadvantage, for an essential feature of all resource evaluation is that it must be a continual process. An interesting comment on this method was made by John D. Moody (1975).[42]

'Efforts have been made to locate and isolate certain independent variables that could be quantified and applied to resource estimation. This attempt was not proved feasible as there are so many complexly interrelated variables which act independently. However, certain basic parameters can be applied in analogue extrapolation to arrive at a potential figure when engineering data is scarce.'

Some of these figures are used in his paper and quoted later in this book but the 'certain basic parameters' are not disclosed.

The Canadian Geological Survey embarked on an annual updating of reserve estimates based on a basin-by-basin examination. Cumulative probability curves are drawn according to best judgement in relation to known petroleum occurrences and reserves, probable petroleum 'plays', volume of sediments in basins and a yield factor based on analogy with other areas. The curves are asymmetric and a mean is computed to allow comparison and summation. This may be termed the current normal method but the important point will be the continual updating and refining of the method. Present differences with other estimates are shown below and these can be taken as typical of many other similar studies. The most noteworthy feature is the variations between the two Geological Survey estimates B and C in the different basins. This is a hopful sign of humility and courage in changing estimates as new facts appear or new interpretations become more reasonable.

Table 2
Estimates of Canada's ultimate recoverable potential of crude oil,
(cumulative production + proved reserves + recoverable potential)
(gigatonnes*, rounded).

Basin	A	B	C	D
1. Arctic Is. & Coastal plain (N)	5.9	6.6	2.7	3.9
2. Beaufort Mackenzie	} 6.4	2.0	0.8	1.1
3. Western Canada		3.9	3.0	3.4
4. Offshore East Coast	3.3	5.2	6.4	3.0
5. Hudson Platform	0.4	0.2	0.2	(1)
6. Eastern Canada Onshore	0.3	0.2	0.2	0.2
TOTAL	**16.3**	**18.1**	**13.4**	**11.1**

A. Canadian Petroleum Association, 1969
B. Geol. Surv. of Can. 1972, Estimate I
C. Geol. Surv. of Can. 1973, Estimate II
D. Can. Soc. Petr. Geol., 1973 (1) included in area 1 total

* Barrels converted to tonnes at 7.4 b/t

Source: *An Energy Policy for Canada, Phase 1*, Vol. 1, Table 3, p.91

The fourth method, the pragmatic production approach, is to acknowledge that the final objective of all reserves estimates is to forecast future production. Past and current production is used as a check on productive capacity and estimates of possible future production as both a check and a guide to potential reserves. This method attempts to take in all factors which can affect production. The most important aspects to be accepted or considered are:

1. The resource base is finite but unknown.
2. The resources estimate must be continually reviewed, in the light of possible future recovery methods and economics to as great detail as the available facts warrant about each basin.

3. Proved reserves are the best known totals but each pool and each province is finite, and the size and number of the individual reservoirs, whether small and many or giant and few, has an important bearing.
4. Past and current production are a guide to productive capacity but the economic/political constraints or incentives which may have affected those rates must be understood.
5. Immediate future production will depend on the inertial momentum from the past on the working inventory of the proved reserves.
6. Further ahead, production will depend on the effort applied to exploration and development as well as on the estimated size of ultimate discoverable reserves.
7. The factors determining that future effort must be assessed, in as great detail as possible, for they vary widely from place to place, e.g. the strength of a drive for self-sufficiency, the distance from market and hence the threshold figure for economically exploitable reserves.
8. The forecast production curve must 'make sense' in a projected global picture. Supply equals demand. However, for example, the proved reserves of only one country might sustain a production equal to a total world demand yet one knows that this will not happen because the production equipment is not available and vested interest would not permit such.
9. The whole scene is dynamic and the whole scene must therefore be under continual review and reappraisal.

This pragmatic production approach was used much more widely than perhaps many of the more statistically minded would acknowledge. It was, and is, a valid approach. A. D. Zapp (1962), for instance, discussed the use of productive capacities as a method of measurement.[43] This is only one aspect of this approach. With the improved tools for calculation and modelling, developed with the computer, many of the factors can be better quantified and more alternatives calculated, but the quality of the input is still the most important factor, whether considering crude oil or natural gas or, indeed, any resource.

Throughout this discussion on methods of measurement of the various resources, there has been continuous emphasis on the need to test the validity of each method and each resultant figure. This testing can be done by comparison with other figures relevant to the circumstances and not necessarily run through a computer. Perhaps one of the greatest problems in the world energy scene today was highlighted by the oil 'crisis' of 1973/74. The experience of most people involved, particularly in governments, was inadequate for them to make 'back of the envelope' calculations to check whether the figures being produced by themselves or others 'made sense', and many self-nominated 'experts' were born in those days!

A Panel Discussion at the 1979 World Petroleum Congress considered the estimation of both proved reserves (DeSorcy), and ultimate resources (C. Bois et al.). The former concluded that 'the best accuracy which can be expected in proved recoverable reserve estimates for individual reservoirs is about ± 10% and in many situations the accuracy will be much less'. Four classes of estimation methods for ultimate resources were described, broadly similar to those already noted, with the conclusion that 'a review of these different methods shows important causes for error and that great care must be taken in considering the figures for ultimate reserves put forward by different experts, even when they are convergent'.

Many of these points were summarized in the recommendations of the United Nations Expert Group[351] in 1977, even though their main consideration was oil and gas resources; namely:

1. A more uniform series of well-defined terms to designate categories of petroleum resources should be adopted, as a means of increasing understanding and communication among geologists, engineers, and various segments of the public.
2. Assumptions, data sources, regional and stratigraphic information, and methods of making resource estimates, should be published with the estimates so that scientific credibility will not be in doubt.
3. Methods of assessment should be flexible; no single method is applicable to all provinces, basins, stratigraphic sequences, or stages of exploration effort.
4. More than one method of assessment should be used for all areas so that reliability can be enhanced and reasonableness of estimates can be tested.
5. Where possible, estimates should be stated as ranges of probability.
6. All resource data for crude oil and natural gas be reported in metric quantities and their energy equivalent be provided in joules, using contemporary conversion rates.
7. Measurement should be made under standardized conditions of temperature and pressure.
8. Natural gases may contain various inert constituents (N_2, H_2S, CO_2 etc.). In most cases, where these quantities are insignificant and do not affect the use of the gas, no correction is necessary. If, however, the quantity is sufficient to render the gas less usable, then an appropriate reduction in gas volumes should be made.
9. It is preferable that natural gas liquids be estimated and reported separately from crude oil and natural gas. Recognizing that this will not always be possible, crude oil and natural gas liquids may have to be combined in the data. However, crude oil data including natural gas liquids should carry notification of this fact.
10. Frequently, there are additional details about resource estimates that may be available from respondents at the time they submit their resource estimates. Reports of high, low, and most likely estimates, accompanied by probabilities, are more useful than point estimates. It is helpful to the analyst if a distinction can be made between associated and non-associated gas, and if crude oil is reported according to gravity classification. Other descriptive information that may be provided includes the geographical distribution of resources, field size information, and a listing of resources according to reservoir types or lithology.'

Michel Grenon of International Institute of Applied Systems Analysis, in a presentation on 2 April 1979 to the American Association of Petroleum Geologists in Houston Texas, stressed that though there have been many forecasts of potential resources, they have essentially been developments of the four main methods discussed earlier, with a marked tendency for accretion rather than fundamental new work behind each forecast.

The Delphi Poll conducted by Dr. Desprairies in 1976/1977, discussed in the later chapter on petroleum supply, was a unique attempt to test the opinions of a wide range of persons, conditioned though they obviously must be by current knowledge and ideas.

In February 1979, P. W. J. Wood[418] presented a more optimistic picture of

23

potential petroleum resources than some people had been doing for some time. He pointed out that, whilst more of the petroleum originally in place is in small reservoirs which are proportionately more uneconomic the smaller they are, yet continuing technological advances are reducing the size which is the limit of being economic. He argued that, taking proved reserves plus cumulative production at 1×10^{12} barrels, after analysing nineteen former forecasts, there was a 90% probability that the economic petroleum base contained 1.5×10^{12} barrels and a 50% probability that it contains at least 2.2×10^{12} barrels – say 300 Gt. Comparing discovery rates on five-year averages, Wood suggested that, outside the Communist countries, the trend from 1950-1975 was encouraging. Taking another tack, he noted that the USA, as the mature industry area, had only 36% of discovered petroleum in giant fields. The world currently has 75% of discovered petroleum in giant fields. As the world approached the USA standard of drilling maturity and non-giant accumulations approached a 64% share, a figure of 2.1×10^{12} barrels might be attained. Even the USA is not fully explored and the world outside is very inadequately explored. The concept of crustal plates has evolved rapidly in the past few years and led to a classification of areas prospective for petroleum which, one might note, were not considered in the early work on basins by Pratt and Weeks in the late 1930s. Using such concepts and data published by Huff[419] and Nehring[420] Wood noted that:
- some 55 Gtoe are in divergent margin basins (23% of world total)
- some 22 Gtoe are in convergent margin basins (9% of world total)
- some 164 Gtoe in the more accessible plate interior basins (68% of world total).

The Mexican discoveries since 1971 in the southwest, onshore and offshore, are in a divergent marginal basin. Many similar ones in the world are waiting to be explored. Of the convergent margin type, basins like the Sumatra basin may be the most prospective and are in a very early stage of exploration world-wide, though lateral slip basins, such as those of California, have already provided almost 20 Gtoe but may be less common. Exploration in plate interiors because of ease of access, is the most mature, yet big discoveries are still being made, e.g. the West Pembina Field in 1978 in Canada, and the Wyoming Thrust Belt fields in 1978/79 in the USA. Furthermore, technology and scientific ideas are in ferment and great progress is being made. The optimism shown by P. W. J. Wood is in line with my own optimism as to the potential, but has to be tempered by the fact that the lack of political will from 1975 to 1979 in many areas, and particularly in the USA, has made the more difficult the task of quickly demonstrating that potential.

2
Resources

Solar

Whilst one need not discuss the resource base of solar energy because of its immensity, at the other end of the scale there is little installed plant capacity, or 'proved reserves'. Hence the potential of solar energy must be discussed in the broad category of resources, even though there is little expectation that solar energy will be a major contributor on a world scale until well into the twenty-first century.

On a local scale, however, solar energy can be an important contributor even without discussing the biological aspect of photosynthesis. The utilization of solar energy does not require a scientific breakthrough such as that required for the utilization of atomic power. It requires the type of technological break-through which miniaturized valve radio sets or computers. The low intensity of energy received requires that the collectors be cheap and robust and its inter-mittency requires storage facilities. Much work is in hand in many countries both on the thermal farm, or major collecting, storage and transmission centres, and on the small individual cooking utensil with, in between, individual house systems, energy generation for satellites, or specialized solar furnaces, as in the French installation in the Pyrenees.

Geothermal

Geothermal resources lie in the great belts of volcanic activity, in the major regional fault zones and in the areas of abnormal temperatures in some rapidly sinking deep sedimentary basins.

The volcanic belts with the surface seepages of hot springs and geysers are the more obvious areas and it is in these that one has both resources and, in some areas, even proved reserves. The great circle around the Pacific, the Rift Valley of East Africa, the Red Sea, Italy and Iceland are the prime areas. For instance, in the Western USA, there is a large, long-known, hyperthermal area of the bulk of Oregon, the whole of Nevada and parts of Idaho, Utah and Arizona with only the Geyser area of California in the reserves category.[50] In New Zealand the geothermal belt in North Island is about one hundred and fifty miles long and thirty miles wide, narrowing to ten miles at each end, with the Wairakei area of about fourteen square miles only being in the proved reserves category. Endo-genous hot fluid sources are known in Algeria, France, Greece, Hungary, Turkey, in the Phillipines, Taiwan and around the Pacific to Chile. In all these regions hyperthermal areas have been mapped or there are plans for mapping. The United Nations has conducted technical assistance programmes since 1966 in six countries and since the UN symposium in Pisa, Italy, in 1970, in twenty other countries. Systematic surveys have been carried out, starting with a re-connaisance geological survey, possibly an airborne infra-red scanning survey, a geochemical survey of all hot and cold springs, temperature gradient surveys and electrical resistivity surveys to detect hot water in permeable rocks, micro-seismic surveys to detect minor earthquake movements and finally drilling.

The drive to find geothermal sources, as illustrated by this activity, is obvious when the advantages of their use in the generation of electricity are appreciated. Geothermal installations are uncomplicated in layout and simple to operate; the capital cost is less than that of a conventional steam power plant; incremental costs are low and thus can provide cheap power during off-peak hours; the source is indigenous and so no import or foreign exchange problems are involved; power generating plants can be small and installed in step with demand, so avoiding high initial investment; the absence of economies of scale are offset by low capital cost and simplicity; the source is continuous and therefore ideal for baseload generation. Such considerations are of particular interest to developing countries, but the problem, as mentioned before, is that it is impossible to get more than an indication of the resource value before delineating the specific proved reserve.

This problem is even more acute in the case of the potential energy in some sedimentary basins. As yet there has been no exploitation to serve as a yardstick. Theoretically, however, the potential is sufficiently high to warrant dicussion in this resources section. This is said even though the geochemical and hydrodynamic processes are not fully understood which in some basins create fluids of abnormally high temperature and pressure. The Mississippi delta basin is the best-known area.[51] There the Neogene deposits, because of high deposition rates and burial off the Louisiana coast, reach a maximum of 13,700 m of fine-grained, generally undercompacted sediments and subjected to contemporaneous faulting on a grand scale. These are the elements which have given rise to the anomalies shown up in wells drilled for petroleum, which Paul H. Jones[52] summarized as follows:

> The energy in the Neogene deposits, stored in the form of very high pressures and intense heat, is potentially available for large-scale water supply and thermal power development. The superheated fresh or brackish water in the geopressured zones, produced from deep wells that reach deeper than 10,000 to 14,000 feet in some parts of the basin, could be used both to pump and distill saltwater from shallow aquifers. Throughout most of the Gulf margin, superheated salt water from the geopressured aquifers could also be used for self-desalination, thermal power, or steam flooding of pressure-depleted petroleum reservoirs.

In a US report[53] these geopressured zones were said to 'occur in continuous belts for hundreds of kilometres.' The US Geological Survey, in the same report, are said to have designated about 1.8 million acres in the western US as being within 'known geothermal resources areas, (KGRAs), with a further ninety-six million acres as having prospective value for geothermal resources (PGRAs)'.

The Niger delta, like the Mississippi delta, has petroleum resources but it may also have geothermal resources from geopressured zones. Perhaps other similar areas, perhaps the Amazon delta or other fossil deltas, may have a potential geothermal resource of this type.

One practical viewpoint of the geothermal resources of the USA and their near-term utilization is illustrated in a 1974 US report.[54]

> The Task Force believes that a logical, orderly programme should be carried out in those areas where geothermal power is practical and logical. Some have suggested a 1985 goal of perhaps 7GWe (Nat. Petr. Council – US Energy Outlook, Dec. 1972). This would be a very large expansion over the installed geothermal capacity today and it is difficult to see it being accomplished. However, even if done, it would be extremely small in relation to the nation's power needs, being the equivalent of about 0.2 million barrels crude oil per day.

Turning elsewhere, another US report[55] stated that 'the total geothermal resources of the Soviet Union have a capacity of 22 million cubic metres of hot water and 430,000 tons of steam daily. If fully utilized, this would save about 40 million tons of fossil fuel annually. The annual output set for 1975 is to produce about 15 million cubic metres of hot water and 470,000 tons of steam.' Obviously in this case the word 'resources' means 'installed capacity', which are two very different things by the definitions of this book; nonetheless, the Soviet capacity, if correctly reported, shows the potential of its proved reserves which are, presumably, only part of its resources.

Hydraulic
The theoretically available hydro-energy is the quantity of water multiplied by the distance of fall or head, hence both water and height differential are needed. The Amazon has five times the water of the second largest river, the Congo, but has less hydraulic capability because of less favourable topography. Every river system has a gradation from high head and low volume near the source to low head and high volume in its mature, lower reaches. This feature shapes the type of installation needed to extract maximum use; large reservoir capacities in the headwaters, less regulated stations downstream and run-of-river stations in lower stretches. Not only must there be sufficient water but it must be available as a regular daily load to give a base load. This usually necessitates artificial storage, and, therefore, dams needing favourable sites and areas available for flooding. Cascade stations downstream of the first dam may be one way of increasing output. Sites for pumped storage may be needed to offset fluctuations in demand. All need careful geological and engineering examination and construction. Economies of scales are considerable in the hydro-electricity industry and large schemes may have parallel benefits in improvements in irrigation, navigation, flood control and water supplies for communities and industry. Many massive schemes in the USSR are multi-purpose, initiating new industrial growth and exploiting natural mineral resources. The cascades of stations along the Dneiper and the Volgakama are examples. The plans for the Vaksh, Naryu-Daria and Syn-Daria rivers in Middle Asia and, particularly, the projects for the coal-rich Angara and Yenisei river basins in Siberia will change these lands considerably.[56] These are examples of large-scale projects on major rivers. The basins of the fifty largest rivers of the world cover 40% of the land surface of the earth. The big stations are naturally on these rivers with, therefore, the bulk of world potential. Some sixty projects, with capacities of more than 1000 MW, are in hand in twenty-one regions.

World hydraulic resources are not easy to assess, because the information available is often vague and imprecise. Two categories of capacity are in common use, G_{95} being the maximum generating capacity useable 95% of the year and G_{av} being the capacity useable under average annual flow conditions. The differences between these depend on flow pattern and storage potential. These differences can be important, as table 3 illustrates, yet often the type of capacity is not designated except in specialized publications. The 1974 WEC Survey attempted to gain data for resource estimation. From the data obtained a table showed the twenty-five nations with the largest hydraulic resources, which totalled 80% of the world total. The first six nations accounted for 50% of the world total and the statistics are given in table 3.

The differences between the developed and developing countries and their potential are important and were summarized in 1977 by Ellis L. Armstrong[424] as in table 4.

Table 3
World probable annual generating capacity of
hydraulic resources

	G average		G 95	
	TJ	% world	TJ	Rank
China	4,752,000	13.5	1,728,000	2
USSR[1]	3,942,000	11.1	1,440,000	3
USA	2,523,600	7.2	777,960	6
Zaire	2,376,000	6.7	2,234,880	1
Canada	1,926,720	5.5	867,240	5
Brazil	1,869,480	5.3	1,372,000	4
	35,288,640	100	15,964,083	

	G average		G 95	
	TJ	% world	TJ	Rank
Developed* nations	12,415,140	35	5,287,791	33
Developing nations	22,873,563	65	10,676,292	67

[1] 18% in European USSR
* Southern Africa, Japan, Europe, USSR, USA, Canada, Australia, New Zealand
Conversion factor: 1 GWH = 3.60 TJ

Source: 1974 WEC Survey, Table VI-7, p.173, 1974

Table 4
Estimated probable hydroelectric development as probable annual
average energy from installed facilities for year indicated
(thousand terajoules)

	1976	1985	2000	2020	Total developable
OECD Countries	3,776	4,493	5,369	7,800	8,214
Centrally planned economies	719	1,200	2,880	8,700	9,990
Developing countries	1,172	1,973	4,490	11,800	16,717
WORLD TOTAL	**5,667**	**7,666**	**12,739**	**28,300**	**34,921**

Source: *World Energy Resources, 1985–2020,* Executive Summaries of reports
to the Conservation Commission of the World Energy Conference, 1978.

K. R. Vernon[57], using somewhat older figures, scaled to his own assessment
of practical values, illustrated the same point with the figures in table 5.

Table 5
World hydro-resources
(terajoules)

	Potential 10^{18}J	Developed 1970
Developed countries	4.68	65.0 %
Centrally planned economics	6.47	10.0 %
Developing countries	18.00	3.0 %
Total	29.15	14.5 %

Source: Table 2 — 'Hydro (including tidal) energy', K. R. Vernon,
Phil, Trans.R.Soc.Lond., A 276, p.486, 1974

Vernon considered that a figure of 82.8×10^{18} J, used at the 1968 World Energy Conference, was an unrealistic estimate of the world potential. He preferred a figure of 29.15×10^{18} J, which is of the same order as the 1974 WEC Survey figure of 35.29×10^{18} J at G_{av} which are of installed and installable capacities.

In the developing countries in which there are the greatest potentials, there has in the past been a strong tendency towards massive schemes, but, as the People's Republic of China is proving, small, locally engineered, constructed and operated hydroelectric schemes can play a most important part in development.

Finally there can be confusion in the comparison between hydraulic energy and thermal energy equivalence of other energy forms. Hydraulic resources are normally used only to produce heat or electricity. One joule of the potential or kinetic energy of hydraulic resource can be converted almost completely (85-90% efficiency) to one joule of electric energy. One joule of thermal energy in fossil fuels can be converted to heat also, at about 85% efficiency directly. However, in producing electricity by steam, one joule of the thermal energy of the fossil or nuclear fuel only produces 0.3 to 0.4 joules of electricity because of thermo-dynamic loss. The overall average equivalence will be between these two values as most fossil fuels are used in both ways.

Winds
Wind power is an obvious resource, with a resource base of less practical value even than that of water power. Yet hydraulic potential can be measured in specific areas, whereas there is no yardstick for the estimation of effective wind power except at specific points. Hence for this resource the basic concepts of reserves applicable to the other resources cannot be used. However, interest has recently revived speculation about harnessing the winds, see chapter 6.

FOSSIL FUEL RESOURCES

The coals
The greater part of all solid fuel resources is in areas north of 30° N. latitude. Three countries, USSR, USA and China, have almost 90% of the 10.8 Tt commonly accepted total of the world's coal resources.[59]

Table 6
World resources of solid fossil fuels
(gigatonnes)

	Total resources[1]	%	Remaining mineable[2]	Recoverable reserves[3]
USSR	5,713	53	1,200	136
USA	2,926	27	1,506	186
China	1,011	9	1,012	38
Canada	109	1	87	6
Europe (excl. USSR)	608	6	603	127
Other	388	4	232	58
TOTAL	**10,755**	**100**	**4,640**	**551**
Total 1968 WEC Survey	8,736			

Sources: 1 — *1974 WEC Survey*
2 — 'Coal Reserves of the USA, A progress report,' Paul Averitt,
US Geol.Surv.Bull. 1136, 1 Jan 1960
3 — *1974 WEC Survey*

As stressed under methods of measurement, the total resources of this table are close to the resource base, but as they have some limitations of depth or thickness, varying between countries, they are being considered here as resources. The 1974 WEC Survey indicated only some 1.2 Tt as exploitable and 0.55 Tt as recoverable under present conditions, that is proved reserves, of the 11 Tt total.

In the 1968 WEC Survey the total resource figure of 8.7 Tt included 5.5 Tt for the USSR, 1.5 Tt for the USA and the increased estimates for these two countries make up 1.6 Tt of the 2 Tt difference with the 1974 WEC Survey figures. The 1.506 Tt remaining mineable is similar to the revised figure of 1.556 Tt used by Averitt in 1967[60] and is the total all ranks of measured, indicated and inferred reserves of more than 0.35 m seam thickness and less than 914 m overburden. Of these, some 394 Gt are the measured and indicated reserves only, in seams more than 0.7 m and less than 305 m overburden.[61] The 1974 WEC Survey[62] gives a total resources figure of 2.920 Gt of which 360 Gt is the total in place of known reserves with less than 305 m overburden and seam thicknesses of more than 0.7 m for anthracite and bituminous coals and 1.5 m for sub-bituminous coals. One might also note that 61% of the sub-bituminous coals, 47% of the anthracites and 27% of the lignites within the total are recoverable by strip or opencast mining.

The total and remaining mineable resources of the People's Republic of China are at 1.011 Tt and 1.012 Tt. This is an estimate made in 1913 and would appear to have a fixed value in the coal literature. The Chinese, in fact, undertook extensive exploration in the 1950s and were said to have increased their 'verified' reserves from 30 Gt in 1950 to between 80 and 100 Gt in 1958, with discoveries in central-south, south-west and north-west China, to a total resource figure of 2000 Gt.[63] The Chinese do not themselves publish reserves figures, but the regime's drive towards energy self-sufficiency and self-reliance through small local exploitation units, undoubtedly means the utilization of coals which in other countries would be considered unrecoverable, economically or technically. Hence perhaps a figure of 2 Tt should be adopted for the coal resources of China, which involves a lot of coal.

Finally, an illustration of the loose way in which resources are discussed can be taken from a 1974 energy conference paper.[64] In one place a figure of 180×10^{21} J was used for the coal component of major world energy resources and in another 177.5×10^{21} J was designated as 'recoverable resources'. Converted at 23.28 GJ/t, being a common factor used for a mixture of coals then some 7.7 Tt would be the implied tonnage. This is a very different figure from the 551 Gt recoverable reserves in table 6, and mid-way between the total resources and remaining mineable resource.

These discussions of one of the best-known and one of the least-known of the major coal resource countries shows the wide variation in our knowledge and the energy conversion example illustrated the danger of loose thinking. The validity and comparability of even the best-known sectors of the best-known coal industries in Europe will be discussed under proved reserves. Meantime a warning bell has been struck that coal resource figures are big but no one knows really how big.

The hydrocarbons

In earlier discussion of the definitions of resources and reserves, the resource bases and the units and methods of measurement, emphasis has been placed on the peculiarities of the hydrocarbons, and on the problems of collecting and the diversity of the data. The components of the hydrocarbon resources are more diverse than the coals, being natural gas, natural gas liquids, crude oils, heavy oils, viscous oils in oil sands and solid kerogen in oil shales. Conventional crude oil resources are the best known but not really well-known.

Crude oils. It is common practice to illustrate the variety and range of estimates of crude oil reserves by lists of estimates without any description of the parameters used in the calculations. Torrey in 1962 collected information on estimates of the original oil-in-place in known reservoirs which totalled 219 Gt $(1.6 \times 10^{12}$ US bl). This figure was important because he was able to demonstrate that in addition to the world proved reserves from primary production of 40 Gt, which was the approximate generally accepted figure at that time, there was an additional 58 Gt possibly available from known reservoirs by the application of secondary recovery methods.[65] The figure of 219 Gt is not comparable with the 849 Gt figure of Hendricks, 1965,[24] which is an estimate of ultimate discoveries from a total oil-in-place of 1,350 Gt in known plus unknown conjectured reservoirs. The Hendricks 849 Gt figure is also not comparable with the ultimately recoverable resources estimates of Weeks, 1962, of 273 Gt, or Ryman, 1967, of 285 Gt,[66] who have both taken the step of forecasting what will be recoverable, a step which Hendricks would not take, stating, 'How much might be economically recoverable is dependent upon many factors related to the geologic occurrence, geographic accessibility, and economic developments and would be variable between the United States and other countries as well as between countries.' Nor can one say that Weeks's 273 Gt is recoverable from Hendricks's 1,350 Gt total-in-place, because Weeks might well have a very different figure for total oil-in-place.

Many writers over the past ten years, probably mainly through repetition of the same figures producing verisimilitude, appear to have gained an affection for a round figure of 2,000 US billion barrels (273 Gt or perhaps, say, 275 Gt as a 'rounder' figure). This is the total used in the most recent study of crude oil resources from which table 7 is taken.

Table 7
World estimated ultimate crude oil recovery, 1 Jan 1974
(gigatonnes)

		Ultimate recovery from discovered fields	Undiscovered potential		Total expected recovery
			Expected value	Range (rounded)	
Russia, China et al		20.5	47.8	14—123	68.3
N America	USA	21.0	10.4	3—24	31.4
	Canada	1.9	9.7	2.5—21	11.6
Total N America		22.9	20.1	6—45	43
Middle East		68.1	17.9	7—68	86
Others	North Sea	2.7	4.1	1.4—12.3	6.8
	W Europe	0.5	2.0	0.5—4.8	2.5
	N Africa	6.3	5.2	1.4—2.1	11.5
	Gulf of Guinea	3.7	3.4	1.4—6.8	7.1
	Other Africa	0.5	3.1	0.3—6.4	3.6
	NW Latin America	9.2	7.2	1.6—15.0	16.4
	Other Latin America	2.1	5.2	1.4—12.1	7.3
	Indonesia et al	4.0	6.1	2.0—10.9	10.1
	Other Far East	1.0	6.6	1.4—10.2	7.6
	Antarctica	—	2.7	0.1—13.6	2.7
Total others		30.0	45.6	12.3—109	75.6
Total World		141.5	131.4	38—300	273

Source: 'An estimate of the world's recoverable crude oil resources', J D Moody,
IX WEC Preprint PD6(2), 1975

J. H. Moody emphasises the importance of the contribution of giant fields. This is now well-appreciated and is illustrated by the fact that fifteen fields, with an estimated 1.4 Gt ultimate recovery, account for 35% of the total world ultimate recovery from known fields. Moody cautions that in maturely explored basins undiscovered potential will be smaller where the ultimate recovery is mainly in smaller fields but that, in the less explored continental margin basins in different geological circumstances there may be little correlation between offshore and onshore in the discovered ultimate and the ratio of giant field reserves to total reserves.

In table 7 the wide range in the undiscovered potential in the USSR, China and other centrally controlled communist countries is due lack of knowledge. V. V. Semenovich in 1971[67] noted that about half of the USSR territory, some 12 M km^2 on land and 3.5—4 M km^2 offshore, is prospective for oil and gas with occurrences throughout the stratigraphic column. In the years 1961 to 1971 explored oil reserves increased by 1.7 times and gas five times. A. A. Karayev, 1974,[68] stressed that over 2,500 geological and geophysical crews are permanently employed in prospecting and exploring and about 2,200 deep exploratory wells are completed each year, 'more than 40% of them productive'. This high success ratio may probably be attributed to care in selection of sites, single-minded reason for drilling, namely to find oil by following up the best scientific leads, as well as the richness and size of prospective areas. The USSR operator is not subject to extraneous pressures such as an obligation to drill to retain mineral rights or exploration rights over certain acreage, though all systems contain pressures of some sort.

Moody, for his paper to the 1975 World Petroleum Congress, had a mass of information available on which a continuing effort had been employed for years and a probabalistic/engineering method of appraisal had been used. Hence this study was not merely a re-heating of old figures. When examining the expected values one must note the ranges given; these illustrate both the scanty evidence for the undiscovered potential on which one must rely and also the need for constant revision.

A joint paper by Moody and Halbouty to the 1979 World Petroleum Congress revised the figures dated 1 January 1974, by giving estimates to 1 January 1976, but in a somewhat different form — as is so common and so exasperating to the analyst. Table 8 compares the 1974 figures with those for 1976 and then adds the cumulative production to the proved reserves at 1 January 1979 together with the 1978 annual production.

The authors have rounded up the figures of 1974 for the expected values of the undiscovered potential, and those of the ranges except for those of the 'other eastern hemisphere', which they have reduced. The cumulative production to 1 January 1974[482A] was some 41 Gt, so leaving 100.5 as the proved and prospective reserves from discovered fields, which they had increased to 115 Gt by 1976, despite 7 Gt of production. Hence the later view is a more optimistic one of the world's crude oil resources. The authors suggest that nearly two-thirds of the undiscovered commercial oil will come from basins which are productive in 1979. It should also be noted that they have used an overall recovery factor of 40%, which is the same as that suggested in the study by Desprairies for the World Energy Conference.[372]

World offshore crude oil. In the earlier work Moody suggested that some 67% of his expected 131.4 Gt (range 38–300 Gt), ultimate recovery from as yet undiscovered fields would come from offshore. Since then there has been marked progress in the techniques for drilling in ever deeper water, and also changes in some ideas which might offer greater prospects at great depth.[483] The areas of the central ocean region were considered unfavourable, but the prospects of the continental margins were thought favourable with the small ocean basins very prospective.

An important 1974 review[69] estimated that some 23.5 Gt of recoverable oil had been found offshore. Though information varied in quality yet these could be classed as measured reserves, forming therefore 26% of total world proved reserves with 90% having been found in sixty giant fields of 500 million barrels (68 Gt) or more.

The first real offshore oil discovery was in 1947 from a mobile platform off the coast of Louisiana, USA. In the ten years thereafter, between two and six giant oilfields were found each year except 1959.[70] This is a remarkable achievement. Some seventy major finds were made to 1972, of which 78% are in geological settings related closely to contiguous onshore conditions. It is since 1964 that the 22% of finds have been made in settings which are not closely related to onshore conditions. Whilst this latter category may be expected to expand as exploratory techniques improved and need increases, yet such do add to the difficulties of estimating beyond discovered deposits. Only two countries

Table 8
World crude oil resources
(gigatonnes)

	Estimates at 1.1.1974[1]					Estimates at 1.1.1976[2]				Estimates to 1.1.1979[3]		
	Ultimate recovery from discovered fields	Undiscovered potential		Total expected recovery (rounded)	Cumulative production to 1.1.1976	Proved and prospective reserves	Undiscovered		Total to 1.1.1976	Cumulative production to 1.1.1979	Proved reserves 1.1.1979	Production 1978
		Expected value	Range (rounded)				Expected value	Range (rounded)				
USA	21.0	10.4	3–24	31.4	16	7	11	5–28	34	17.5	4.4	0.4
Other Western Hemisphere	13.2	22.1	5.5–48.1	35.3	8	12	25	6–50	45	9	6.9	0.38
Russia, China et al.	20.5	47.8	14–123	68.3	8	14	50	14–129	72	10	12.8	0.69
Middle East	68.1	17.9	7–68	86	12	68	19	7–70	99	15.3	50.3	1.05
Other Eastern Hemisphere	18.7	33.2	8.5–77.1	51.9	4	14	36	6–68	54	5.2	13.7	0.53
Total	141.5	131.4	38–300	273	48	115	141	40–365	304	57	88	3.08

Sources: 1. 'An estimate of the world's recoverable crude oil resource', J. D. Moody, IX WPC Preprint PD6(2), 1975
2. 'World ultimate reserves of crude oil', J. D. Moody & M. T. Halbouty, X WPC Preprint PD12(4), 1979
3. *BP Statistical Review*, 1978, published June, 1979.

reported any additional offshore resources to the 1974 WEC Survey, and these only for specific small areas, whilst all the others said they were unknown or confidential.

Much has been learned about the continental margins and deep oceans since the first survey of the Joint Oceanographic Institutions Deep Earth Sampling Project (JOIDES Project) in 1965. A notable symposium about the east coast of North America was held in 1974.[71] The results of new ocean surveys are being published, as is information, from time to time, from the exploration for petroleum offshore by some eighty countries in 1974. Our knowledge is increasing but it is still scanty.

The 1974 review, mentioned earlier,[69] ended with six salient conclusions about offshore exploration since 1947 and those applicable to the resource aspect may be summarized as follows:

– Five areas overshadow all others as to amounts of petroleum discovered – Persian Gulf, with 56% of the total offshore 23.5 Gt oil; Lake Maracaibo; the Northern Gulf of Mexico; the North Sea; and offshore Southern California. The Safaniya field, offshore Saudi Arabia and the Neutral Zone, are estimated to have some 3.68 Gt recoverable reserves which may be compared with 2.45 Gt which was the estimate given for the whole of the North Sea.
– From 1965–1972 (and probably through 1974), annual discovery has averaged close to 10 bb, a sign of the early phase of the search and also of the high proportion of giants.
– The area of the submerged continental margins of the USA are amongst the world's most extensive. The need for more domestic production should ensure active exploration of the 95% which has not been sufficiently studied to make meaningful estimates possible.
– The largest reserves have been found in the small ocean basins, in areas of extensive evaporite deposition (salts and anhydrites), in thick sequences of deltaic sediments and in drowned intermontane basins. Other such areas may well be found in the unknown continental margins.

The North Sea is one of the more important new offshore areas and because of its closeness to the energy market of Europe has added importance. North-West Europe, excluding the USSR, has only about 1.4 Gt of crude oil resources on land. By 1965 when the first natural gas was found in the southern basin of the North Sea, in the West Sole field, the prospects for oil to the north were beginning to be discussed within the interested oil companies but as being highly speculative prospects. The first oil shows were indicated in July 1967, in the second well drilled in Norwegian waters. The first oil deemed commercial was found in 1970 in the Ekofisk field, also in Norwegian waters. By 1974, three groups of fields, essentially in rocks of different ages, had proved commercial oil and in June 1975,[74] the quantities were reported as established recoverable reserves, south of the 62°N parallel, in the UK sector as 1.9 Gt and in the Norwegian sector as 0.8 Gt, with 5.72 Gt as the 'ultimate reserves'. As will be discussed later, there is at least one commentator who considers such estimates as misleadingly conservative and would prefer a figure in the range 11–19 Gt. In the 1974 WEC Survey no additional figures were given for either the UK, credited in 1972 with 1.5 Gt oil-in-place and 0.5 Gt recoverable, nor Norway, credited in 1973 with 0.238 Gt recoverable from a confidential amount of oil-in-place in known reserves.

This discussion about the North Sea emphasises points made or implied earlier, that the offshore game is a new game and even technical information is still too scanty to make any accurate assessment of the obviously rich potential. However estimates must be under continual review and they will be valuable as long as they are recognized as the speculations they are and not hard or even soft facts.

Natural gas

All estimates of natural gas resources are even more nebulous than those of crude oil because of four problems connected with origin. The first is that whilst much natural gas, whether found with crude oil, associated gas, or as dry gas with no crude oil nearby, non-associated gas, does have a common origin with crude oil, yet the factors determining whether gas and/or oil are generated are still controversial. The second is that some natural gas has an origin from materials which do not contribute to the formation of the heavier hydrocarbons of crude oil. The third is that the utilization of associated gas is dependent on the crude oil and its value as a resource is not clear-cut. The fourth is that many natural gases contain non-hydrocarbon elements.

The controversy as to whether gas or oil will be found in a petroliferous province is an old but continuing one.[72] The problem has been heightened by the growing objection to pollution of the environment by sulphur emissions, which puts sweet gas at a premium over sour gas (sulphur containing) or sour crude oils. Gas normally becomes dominant in occurrence over oil as depths increase and temperatures rise but there are exceptions due to original local and depositional circumstances. Increased weight of overburden reduces pore space and leads to both diminished recovery of liquid hydrocarbons and compression of greater volumes of gas in the pore spaces available. K. K. Landes observed[73] that hydrocarbon density decreases with increasing depth. At reservoir-rock temperatures of more than $350°F$ ($177°C$), it is unlikely that liquid hydrocarbons can exist. Liquids would not then exist below 5,640 m with a thermal gradient of $2°F/30.48$ m. This is an important point when considering prospects on the continental slope and rise with different thermal gradients from those of the continents. Despite these problems, many of the estimates of the world's natural gas resources are based on some figure such as 1,245 m^3/t, (6,000 cu ft/barrel) of crude oil estimated. This is a very arbitrary figure, for there are wild variations between ratios of gas to oil in associated circumstances.

Natural gas resources which have their origin in terrestrial vegetal materials, as in Western Siberia and the southern basin of the North Sea, are not, or should not be involved in such calculations and yet they are of more than local importance. The gas of the southern North Sea is considered to have originated in second stage changes in the coal measures as the weight of overlying sediments and the heat of downward burial expelled the gas into overlying beds where the porous components formed reservoirs and impermeable beds above formed the traps. In the southern North Sea these may be of the order of 1,415 km^3 (50×10^{12} cu ft) of natural gas.

In the northern North Sea the natural gas has a common origin with the crude oil found there. Some fields are non-associated and so can be considered an exploitable natural resource but in others the natural gas is in solution in the crude oil and/or forms a gas cap. In the latter case the gas is an integral part of the original energy in the reservoir and essential to the production mechanism. The dissolved gas has to be produced with the oil. In the case of the Ekofisk field in Norwegian waters the gas is separated from the oil and was despatched in 1976 by submarine pipeline to Emden, Germany. The oil goes by pipeline to the UK. From the Forties field in the UK sector pipelines took the oil and gas to the UK in 1976. Such gas is obviously a resource, indeed a reserve, but in some associated fields the gas may be re-injected into the reservoir for pressure maintenance and only after the oil has been recovered would the gas be available. In estimates of the gas resources of the North Sea the non-associated gas, established and probable, of the southern and northern basins are added together, giving on one estimate[74] 3,052 km^3 to which no associated gas has been added. Obviously there is often confusion or misunderstanding between associated and non-associated gas, but most commonly they are both aggregated. The 1974 WEC Survey sought information as to both types, has recorded the individual estimates, but did not receive sufficient information to give world totals by category.

The fourth problem which confuses the usefulness of natural gas resource estimates is that in some instances non-hydrocarbon components can form high percentages of the volume. These components are mainly nitrogen, (up to 90% in some wells in the Danish sector of the southern basin of the North Sea), carbon dioxide (up to 50% in the northern fields of Mexico) and helium. These non-hydrocarbon components should be excluded but most commonly they are not. Further erroneous statistics are derived when an average heat content is ascribed to such resources when converting to a common energy content for comparison with other energy sources.

An authoritative review of the world's natural gas resources, expressed as total ultimate recoverable reserves, and proven recoverable reserves is by T. D. Adams and M. A. Kirkby to the Ninth World Petroleum Congress, May 1975.[75] They note the lack of information but have arrived at a personal estimate of the world's remaining proven reserves of some 2,300 tcf (65,090 km^3). From this they then argue that an ultimate recoverable reserves figure would be about double, indeed close to a figure which Weeks deduced in 1958, but has since increased, rather than figures three, four or even five times as great (9,500 to 12,00 × 10^{12} cu ft), which they list as estimates over the past twenty years. There is an air of pessimism which permeates their forward look which might be questioned. They call on plate tectonics to dismiss prospects of the deep oceans but make no mention of prospects in the continental slopes and rises where there are thick sediments and high thermal gradients. Many would not consider South America, Southern Asia or most of the Far East as declining exploration provinces nor Indonesia as yet having reached a mature age of exploration. They do stress, probably rightly, that 'the only way in which the distribution and magnitude of world reserves can be altered significantly is through the discovery of many new giant gas fields'. Whilst diminishing the wild claims that have been made for the gas discoveries in the Khuff beds of Permian age at Kangan, Iran, and elsewhere in the Persian Gulf, they do predict possibly 8–10 giant fields for coastal Arabia (570 km^3), 8–10 fields in Fars province, Iran (1,700–2,000 km^3), with perhaps a total in the Middle East of the order of 28,000 km^3, which is 44% of their estimate of current proven world recoverable reserves. For Western Europe they take a figure of 5,134 km^3 as the proven recoverable reserves which includes some 1,964 km^3 for the North Sea, whilst the estimate quoted earlier was for 3,052 km^3. It is important to note that three North Sea gas fields were quoted as giant gas fields in the world list of 1970.[76] With the later discoveries in the northern basin and if one includes the associated gas fields, perhaps six more had been added by 1975. The April 1975, official estimates of the UK Department of Energy[77] for total recoverable crude oil reserves were three to four times their estimate of proved reserves of about 1 Gt. A paper[484] in 1979 was much more optimistic as to world reserves, suggesting some 6950 tcf (200,000 km^3) for proved plus probable plus potential reserves of gas and 24 Tt of natural gas liquids, and stressed the importance of coal-based gas. (A further discussion can be found in the chapter on proved reserves.)

Heavy oils, oil sands and oil shales

Hydrocarbons occur in a number of forms which cannot be produced in the conventional way from boreholes. Heavy viscous oils grade into oil sands or asphalts in limestones and the end of the line is the solid kerogen in oil shales.

Heavy oils occur in many countries, and have been 'mined' with shafts and galleries in Pechelbronn in France and Sarata in Romania. Asphalts occur in limestones in France, Germany, Switzerland (where they are mined for road surfacing), in Italy, Sicily, Oklahoma and Texas. But these have only local value and there has been no authoritative attempt at a world wide resource appraisal. The heavy oils of northern Alberta, Canada, where, since approximately 1958, research and experimental work has been in progress, are probably the nearest to being considered reserves, though as yet no significant commercial production has been achieved. One source[78] estimates the oil-in-place as 25 Gt, but another source,[79] quoted in an official 1973 Canadian report,[80] gave 11.74 Gt. The first source stressed that the techniques applicable for exploitation in one area are not necessarily the optimum for another. This emphasises the problem of resource assessment even in a comparatively well-known area. There are many less well-known areas of heavy oils in the world. A total world resource figure is thus impossible to calculate.

The major oil sands of the world have been described many times, mainly repetitively without much new material. However one 1967 source,[81] selecting those with more than 15 million barrels of bitumen-in-place, gives a total of 915 $\times$ 10^9 US barrels for nineteen known deposits. The two largest deposits are in Alberta, Canada (710 $\times$ 10^9 barrels ascribed), the Officina-Temblador belt, Venezuela (200 $\times$ 10^9 barrels) and the next largest the notably smaller deposit of Bemolanga in the Malagasy Republic (1.75 $\times$ 10^9 barrels). The Athabasca oil sands of Alberta, Canada, with some estimated 300 $\times$ 10^9 US bls of recoverable oil, have been well and often described and can most easily be accepted as resources, even though there may be differences of opinion as to the amount which should be classified as proved reserves, as discussed later under that heading. The Venezuelan deposits were discovered in 1935 and extend in a belt some 600 km long and 85 km wide. It is only recently being extensively explored.[82] One report[83] gives oil-in-place as 700 $\times$ 10^9 bls of 8–12°API, 4% Sulphur, (111 Gt at 6.30 bl/t), which is 3.5 times the 1967 estimate but only 10% is said to be recoverable.

The oil shales of the world have been exploited for many years, particularly in Scotland, Sweden, Estonia and China. As mentioned in chapter 1, the resource base is very great. The United Nations has attempted, since 1965, to collect information world-wide and to stimulate research into this resource. An important symposium was held in Estonia in 1968.[84] As with other little-known resources, repetitions of early estimates on scanty evidence give a verisimilitude to the latest figures. However, the richest category (0.1–0.4 litres/kg) in the 'identified resources' as given in the 1974 WEC Survey, Table V–3, quoting US Geological Survey Paper 820, 1973, gives a total of 101 Gt. The main contributions are from North America (60.6 Gt), Africa (15.5 Gt), Asia (14 Gt), Europe

(10.9 Gt). In the less rich category (0.04–0.1 litres/kg) the estimated 'identified resources' are said to be 350 Gt with North America having two thirds and South America one third. These figures can only be very approximate but they are the latest best guess.

The best-known area is some 44,000 km³ in the Piceance basin of Colorado, the Uinta basin of Utah and the Green River basin of Wyoming. The oil-in-place has been estimated at 600×10^9 barrels, in beds at least ten feet thick and yielding at least twenty-five gallons/ton of shale (0.1 litres/kg). According to a 1973 US study[85] there were some 53×10^9 barrels recoverable at current syncrude economic levels, including 47×10^9 barrels in the Piceance basin. It is most important when considering oil shale as an energy resource to appreciate that of the two methods of exploitation only mining and surface retorting to gasify the kerogen and then condense the gas to liquid, is currently technically and commercially feasible. In-situ gasification and liquefaction is not proved. The mining route, even, has two major problems, high water requirement and high waste disposal. A production of one million barrels/day (159,000 m³/day) in the Piceance basin would require the mining of 1.13 Mt of oil shale each day. This volume of about 523,000 m³/day becomes about 700,000 m³/day as spent shale after retorting. This presents an impressive disposal problem. For the same size plant, the water requirement would be some 16,000 acre/feet (197 hectares/metre), which is particularly a major problem in the semi-arid area of the Piceance basin. Such are the problems in considering oil shales as a resource.

Fissile Fuels

The resource bases of both uranium and thorium are so great that only proved reserves, or as the uranium industry say, 'reasonably assured resources', are of real significance. In the 1974 World Energy Conference Survey (WEC Survey), it was suggested that in the high cost range up to $200/kg uranium metal, the amounts of uranium available are in the tens to hundreds of megatonnes. Of the same order are the thorium resources which are far in excess of the current

Table 9
World uranium resources
(kilotonnes uranium)

	Reasonably Assured < $ 50/lb oxide	Additional < $ 50/lb oxide	Speculative over $50/lb oxide
Africa	570	200	1,300–4,000
America, North	830	1,711	2,100–3,600
America S & Central	60	14	700–1,900
Asia & Far East	37	24	200–1,000
Australia/Oceania	296	49	2,000–3,000
West Europe	388 *	91	300–1,300
Total (rounded)	2,190 **	2,100 **	6,600–14,800

* including 300 kt for Sweden's 'box-grade' black shale resources,
** These totals include 600 kt and 500 kt respectively which are considered uneconomic as costs would be over $80/kg U.

Source: 'World Uranium Resources – An International Evaluation', OECD NEA & IAEA, as summarized in reference 444.

demand of about 1 kt per annum for minor non-nuclear uses. Only when breeder reactors are in significant operation will thorium resources become meaningful.

However, the joint report of the OECD Nuclear Energy Agency and the International Atomic Energy Agency, of the position of uranium resources in the non-Communist world as at 1.1.1977, added to their estimates of 'reasonably assured resources' (proved reserves) and 'additional resources' (probable reserves), a 'speculative' category, but excluded low grade ores whose recovery costs would be over $130/kg uranium metal. Table 9 summarizes their findings, in kilotonnes of uranium metal, (kt U), as at 1.1.1976.

The report also gave an estimate of the eventual uranium resources of the Communist world of 3.3–37.3 Mt U, with prospects very high in the USSR and China, high in Czechoslovakia and DR Germany, moderate to high in Romania and moderate in the remainder of Eastern Europe. These estimates and those in the speculative category must indeed never be considered as anything but a general indicator of what informed opinion in 1978/9 thought was likely on current knowledge.

For the record it could be of interest to note below the estimate quoted in 1974 by S.H.U. Bowie of the resources of thorium in non-Communist countries.[31]

Table 10
World reserves and resources of thorium
(Non-Communist countries)
Price range up to $20/kg.ThO$_2$,
(kilotonnes Thorium)

	Established reserves	Estimated additional resources
Brazil	1.2	31.8
Canada	80	80
Egypt	14.7	280
India	–	300
South Africa	20	–
USA	52	265
Others	8	–
Total	**176**	**957**

Source: S. H. U. Bowie, *Phil. Trans. R. Soc.,* A1276, 1974, p. 503.

3
Resources to reserves

The resource bases and resources of the various energy forms of natural resources have been discussed. It is now opportune to consider the transformation of resources to reserves. This is effected by exploration, discovery and initial development. Development leads to production but the determinants of production are somewhat different and more precise than those of exploration.

The exploration effort for energy resources is affected by many circumstances which are complex and vary in relative importance with time. The intensity of effort does not vary directly with demand, nor is progress smooth but is stepwise.

The determinants or the groups of factors affecting effort include:
1. The equipment of the explorer, mental and physical.
2. Changing demand, both in absolute and relative terms.
3. Government action.
4. Finance.

The petroleum industry can furnish many examples of the impact of these factors but the principles are the same for the other energy industries.

1. The explorer

It is the explorer who finds new resources and, helped by the drillers and engineers, converts them to reserves through increasing knowledge. The tools of the explorer are both mental and physical. His mental concepts and knowledge develop with time, as do the capabilities of his physical tools. It is the co-incidence of breakthroughs in ideas and physical capabilities which lead to the sudden enlargement of resource parameters and stepwise increases in numbers. In 1936, although it was appreciated that the crude oil seeping to the surface over many square miles south-east of Lagos, Nigeria, was coming from seaward, exploration had to go landward because there was no capability to go onto and under the sea. However, developing ideas about the offshore geology of the USA coast of the Gulf of Mexico and the success of a mobile drilling rig there in 1947 started the burst of effort for offshore petroleum which has produced many offshore fields including those of the Niger delta. Mention has already been made of the co-incidence of the capability of very deep ocean drilling and the revival and refinement of the ideas of the movement of the continents and ocean floors which have made the continental slopes and rises worth considering. An earlier example was the development of geophysical techniques to locate salt domes in southern Texas, together with new understanding of the principles and mechanisms of salt movements or diapirism, which led to the burst of activity there in the 1920s. The discovery of light oil in 1932 in Bahrain, in the Persian Gulf, is an example of a sudden change in concept for that discovery reversed the geological thinking of the time that only heavy oil, if any, would be found on the west side of the Gulf. Another example is that only after the banishing of both the psychological mystery of the Sahara and the physical deterrents of desert exploration, by the reconnaisance aircraft and the camelfoot, low pressure, automobile tyre, were the revolutionary concepts of a Swiss geologist as to the petroleum prospects of North Africa listened to and an opportunity seen to increase French interests.

Elliot and Mathieu[473] noted that seismic technology had advanced markedly but steadily and undramatically between the early 1960s and 1979, in data acquisition, processing and interpretation, but that continued micro-miniaturization of electronic and computer components, the application of digital radio telemetry and image reconstruction techniques, used successfully in medicine, might well presage a conceptual breakthrough in which 'transmission' rather than 'reflection' might be the spring for an abrupt forward leap in applied geophysics.

The uranium ores in the rich Colorado-Wyoming area of the USA were thought, until 1953, to be syngenetic and limited to certain beds of Jurassic age. Dating of the ores by new techniques showed that some were younger than the host rocks. This widened the search to rocks of all ages in the region and led to major new additions to resources.[31]

Prospecting for uranium can also illustrate the close cooperation needed between government organizations, service companies and the mining iindustry in order to equip the explorer with constantly developing geological concepts and improving tools. Knowledge of the mode of emplacement of uranium ores is still sketchy and new ideas may have greater future significance than improved equipment. However, equipment will be improved, as in methods of preliminary surveying by aerial photography and satellite imagery. Choice of increasingly specific ground equipment is widening and scope for appropriate instrumentation is available if, perhaps, not always used.

Interest in uranium exploration in less developed countries (LDCs) increased after 1973 as shown by the increase in requests for assistance from the International Atomic Energy Agency (IAEA). Training schemes set up on request by IAEA for LDCs increased in cost from $65,000 in 1973 to over $200,000 in 1978. The commitment of developing countries has been classified by IAEA[454] in five categories of effort in uranium exploration and production, of which the following table lists the two top classes.

Commitment of developing countries to uranium exploration and production.

Class I	Class II
Algeria	Bolivia
Argentina	Central African Republic
Brazil	Chile
Gabon	Colombia
India	Pakistan
Iran	Yugoslavia
Niger	Zambia

Class III has 16 countries, class IV some 17, class V, minor effort, has some 18 countries listed. Most LDCs are in tropical countries with thick soil and often forest cover which has required special adaptations of exploration techniques as practiced in more temperate lands. Furthermore exploration is inhibited in some of these countries by the lack of any policy or regulations. Table 11 gives estimates of the resources of the six more important countries and figures 6 and 7 give the locations of the major deposits and occurrences as known in mid-1979.

Mention has been made of the leap in exploration for geothermal sources when concept and capability joined to move from 'seepage' hunting to defining possible concealed geothermal reservoirs.[17] The development of the Earth Resources Technical Satellite allowed synoptic imagery of large areas under

URANIUM DEPOSITS AND MAJOR OCCURRENCES IN AFRICA AND THE MIDDLE EAST

Fig. 6

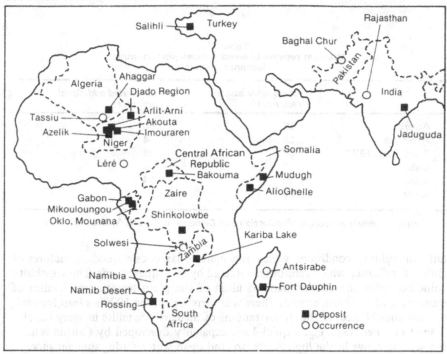

URANIUM DEPOSITS AND MAJOR OCCURRENCES IN LATIN AMERICA

Fig. 7

Table 11
Uranium reserves in selected developing countries
(kilotonnes Uranium)

	Reasonably assured resources	Estimated additional resources
Algeria	28	50
Argentina	23	4
Brazil (Dec. 1978)	62	58
Gabon	20	5
India	30	24
Niger	160	53

Source: *Uranium Resources, Production and Demand,* NEA/IAEA, Dec. 1977.

uniform lighting conditions which can indicate large-scale trends, structures of global significance which cannot be detected by any other means. This capability coincided with the development of ideas of the plate tectonic mechanism of movement of the earth's crust. There was therefore in the 1970's a leap forward, by this coincidence, in the understanding of the earth. Satellite imagery has also helped in other ways, e.g. a quick-look capability developed by Canada is helping seismic crews in the high Arctic to find open water during summer surveys. Satellite communication can offer a data relay capability for transmitters in remote areas where personally manned monitoring surveys for hydroelectric power schemes would be impossible. There have also been other developments, such as the use of hovercraft in permafrost areas which have made access possible into areas which would have been as inaccessible before soft tyres as the desert. The data storage and retrieval systems through use of computers and calculations based on simulation by mathematical models are other new advances.

In brief, the capability of the explorer in developing, or adapting to his speciality, new ideas and new techniques is an important factor affecting the intensity of effort of exploration, particularly when magnified by the common human tendency to 'follow-the-leader', to jump on the bandwagon. Perhaps the best example of this was in Romania in the 1920–1935 period when exploration would move away from one area or field leaving only one driller who could not afford to move but could drill a few more feet. His success would then bring back all the others like a flock of sheep. Successful trend or concept plays, in Canada, since the discovery of the rich oil-bearing Devonian reef of the Leduc field in 1947, provide another example of this 'follow-the-leader' factor which cannot be ignored when considering exploration effort.

2. Demand
The effect of demand on exploration effort is also complex and not as simple as so many economists and statisticians assume.

In times of surplus supply there is usually a falling off in effort. Japan during the 1950s and 1960s was able to import cheap oil to satisfy its remarkable economic and industrial growth. Japan therefore made few attempts to gain control of any foreign-based oil or to conduct its own exploration except for the venture in the Saudi Arabia/Kuwait Neutral Zone. They ignored the increase in net imports to total energy supplies from 7% to 85%.[87] In the USA during the late 1950s and early 1960s it was the surplus domestic capacity which inhibited land exploration. American companies went abroad, seeking cheap oil and higher profits. This had an important and often unappreciated effect on the domestic US companies. The exploration manager became a land manager, not a wild-catter, caution superceded initiative. That influence is not eradicated at the stroke of a pen. Lincoln Gordon[87] discussed the mental attitude which the comfortable world energy situation of the 1950s and 1960s engendered world-wide. Most energy consumers proceeded as if the age of low cost oil would go on forever. Specialists in natural resources appreciated that this would not in fact happen. Indeed we know that some forecast that the very size of the petroleum reserves of Saudi Arabia must give that country an increasing role. Some urged that significant work on the Athabasca oil sands should be undertaken. Yet these arguments often fell on deaf ears and though some attempts at diversification were made they were often half-hearted. In general it was thought that there could be a gradual transition, as the end of the century grew nearer, to other supplementing forms of energy.

The success in 1970-1972 of the Organization of Petroleum Exporting Countries in gaining sharp price rises 'shifted the concept of an energy supply transition from the realm of academic futurological speculation to the beginnings of practical action' by governments and industry. In the period October 1973 to January 1974 by the initiative of the Organization of Arab Petroleum Exporting Countries production levels were reduced, selective embargoes on exports by destination and swingeing price increases were imposed. It was possible therefore in 1974 to write in the first edition of this book:

'One result is a drive in the USA, Europe, Japan and large developing countries, like India and Brazil, for a lessened dependence on the Middle East. This drive is a many-pronged attack affecting all forms of energy. There is increased exploration for domestic petroleum, where possible, and/or for new sources outside the Middle East. There is increased exploration for and production of coal, even though the coal mining industry is prone to labour problems, particularly in the USA and UK, as improving living conditions contrast with dirty, dangerous mining conditions. Interest in geothermal and more hydroelectric possibilities has been revived along with interest in oil shales and oil sands. Most marked, however, is the growing appreciation of the need for more nuclear power. The realisation that new sources of uranium will be needed has reversed the apathy which reduced exploration between the late 1950s and late 1960s. There had been a burst of activity in the early 1950s, when almost all the currently known reserves were found. Some thought then that nuclear power was the panacea for all energy problems. Later material problems and problems of techniques slowed progress and lessened demand. Bowie[31] suggests that "accumulative requirements as forecast as being about 1.4 MtU in 1990, and with an eight-year forward reserve this will mean the discovery of more than twice presently known reserves by 1982".

This is a formidable exploration task which will require increased basic research both into the geology of known deposits and the regional and local controls of ore deposition, as well as development of means of detecting ore bodies at depths of tens to hundreds of metres. Once again one sees the need for breakthrough in both exploration concepts and equipment to meet indicated demand.'

However, demand did not rise because of world economic recession and, for nuclear power in particular, strong environmentalist opposition inhibited expansion.

Apathy and complacency returned and exploration was not pushed with the effort which the longer term prospects warranted. In late 1978 Walter Levy[471] and in early 1979 Walt Rostow[472] both used the phrase 'the years the locust hath eaten' for papers bemoaning the waste of the years in which the necessary effort was not made.

The problems of forecasting demand are many, as will be discussed in the chapter about demand, but one aspect that must be mentioned here is the influence of pressures from environmentalists on the pattern of demand as it affects exploration and indeed the relative values of types of energy resources, within as well as between resources, and hence the exploration targets.

Three different aspects can be distinguished: the desire to preserve the natural landscape, the fear of pollution through accidents and the wish to avoid pollution in conversion and end use.

Strip-mining of coals can scar the countryside particularly when possible on the vast scale of the western United States. The coals there constitute the bulk of USA remaining coal resources, and some 90% of those with a sulphur content of 1% or less. They are in almost flat seams of sub-bituminous and lignite coals up to 30 m in thickness and with as little as 10 to 15 m cover. They are ideally suited to strip-mining with low capital cost and high productivity per man. Restitution of the surface adds considerably to the cost. These coals range in sulphur content from 0.2% to over 0.7% and theoretically can meet the Clean Air Act (1970). The statistical need for these western coals is plain. A 1973 US Bureau of Mines report estimated coal shortfalls in supply/demand balance, because of pollution abatement standards, in 1975 at 133–173 million tonnes (Mt), in 1977, 128–183 Mt and in 1980, some 124–202 Mt, for the utility sector alone; the lower limits assume supplies of additional low sulphur coal or faster availability of acceptable abatement technology. Here, therefore, is an example of the environmentalists having won a rigorous Clean Air Act but preventing the least expensive conformity because some of the environmentalists object to the disturbance of the land.[88] The other low sulphur coals in the USA are in the southern Appalachian area, but production cannot fill the gap from its deep mines even at relatively high cost. The pressure to use natural gas is baulked by the increasing awareness that, despite increased exploration, supplies are not to be sufficient for what is a non-premium use in electric utilities as against direct domestic use. Another alternative is to import low sulphur coal, as is being done from Poland, South Africa and Australia. This obviously increases the need for exploration for further resources in those countries. In the Rhineland brown coal area of Germany, producing some 96 Mt of the total German production of 100 Mt annually, the cost of environmental measures has been reduced to 9% of production costs by concentrating on reducing the number of mines and increasing their capacity. The division of these costs is reported as surface clearing 42%, reclamation 20% and mining damages 38% (sags, minor changes 6%, ground water withdrawal 32%, e.g. compensation water to water stations, regulation of ditches, etc.).[89]

Landscape problems are also involved in the exploitation of oil shales and progress is being inhibited in the semi-arid lands of Colorado and its neighbours by environmental concern, As noted earlier, the retorting of the shales after mining requires large quantities of water, as does the consolidation of the spent shale dust, which is of greater volume than the mined rock. Oil shales in a wet climate could therefore be a worthwhile target if the logistics were satisfactory, even though as yet no optimum process has been developed. On the other hand, these surface problems encourage the development of in-situ recovery processes in deeper beds. Some environmentalists may criticise geothermal exploitation even in areas where natural steam seepages are common and accepted. Certainly the most natural energy resource of hydroelectric power from man-made reservoirs is criticised where the water areas are not considered 'natural' or the strand lines of the shore are thought to be ugly. The International Commission on Large Dams (ICOLD) considered in depth at its Eleventh Congress, Madrid, June 1973, the three categories of the consequences on the environment of the construction of large dams, namely the local, downstream and regional effects. The real concern for environmental opinion is shown by the following quotation from a report by the chairman of ICOLD:[90]

'ICOLD notes also that the efforts undergone in the past to improve the impact of dams on environment and to cope in their design with ancillary effects, should be greatly intensified in order that the building of dams leads finally, for the river basin concerned, to a passage from the existant equilibrium, with its advantages and its disasters, to a new equilibrium, offering man an improved quality of life.'

Only in areas with extremely sensitive natural environments does the exploration phase for energy resources have any significant effects on the environment. The permafrost areas of the Arctic region in the USSR and North America are, however, very sensitive areas. In the USSR the problem is less well-reported but is also less in the later stages of production and transportation because only gas has yet been found and exploited and not hot oil. In the Canadian Arctic there have been no major problems and exploration has been inhibited by lack of major success rather than by environmental considerations. In northern Alaska, ignorance and urgency to get the work done, did lead in the late 1950s to some damage to the thin protective layer of the tundra surface. Considerable concern was also expressed about the effects of exploratory and development drilling but, more particularly, about the effect of the hot oil pipeline. Although the delay of almost ten years in getting the oil to market from the North Slope of Alaska is to be measured in thousands of millions of US dollars, yet the knowledge of the environment, the permafrost, the vegetation, the salmon, the caribou and other flora and fauna, has been increased greatly. Further exploration and later phases will now be possible with minimum environmental disturbance. Hence permafrost areas, whether in the USSR or North America, are now explorable.

In the storage aspect little can be done about hiding solid coal stocks, but in many areas it is not only less obtrusive but also more economical when very large volumes are concerned to store petroleum, as liquid or gas, underground in natural or man-made stores. There has developed, therefore, considerable interest for environmental reasons, as well as the obvious military ones, in the finding of suitable locations for underground storage of crude oil and natural gas. Yet even here there can be objections, as the British Gas Corporation found when wishing to store gas in suitable strata in southern England, mainly because of concern for safety.

The second group of environmental pressures stems from fear of pollution by accidents. These normally restrict exploration, and more particularly exploitation, in some areas more than others. The closing of work off Santa Barbara, California, after an accident and the inhibition to explore for petroleum off the east coast of the USA are the most obvious examples of direct environmentalist influences which have forced exploration elsewhere. One item in

Project Independence, launched by President Nixon in March 1973, was to lease ten million acres of the outer continental shelf for petroleum exploration. Even by April 1975 there was not the activity required to make this item contribute to the Administration's objective of energy self-sufficiency. Explorers were inhibited by the environmental lobby fearing accidents, even though opinion polls in the east showed that 80% of the residents favoured exploitation, but also, in the west, by opposition from the US Navy. The exploitation of offshore petroleum does have its hazards but they are appreciated and safety measures under codes of safe practice are enforced. There is one hazard which, however, may be difficult to combat and that is deliberate sabotage or 'hijacking'. The navies of the world are increasingly concerned about this because governmental action would be necessary, as has been shown already in incidents to aircraft. Somewhat similar, but even more suicidal, artificial risks apply to nuclear power stations. It is, however, more normal risks of accidents which have, particularly in the USA, inhibited the expansion of nuclear power and so slowed-down exploration for new supplies. Some concern has been shown for the possibility of radioactive material being moved from or to nuclear stations falling into unauthorized and inexperienced hands when in transit. It is, however, in the petroleum industry that the transportation side is under constant criticism. It is in the self-interest of the petroleum industry to avoid accidents. As said so often, the oilman does not wish to see any oil from the oilfield to gasoline tank in an automobile except the few cubic centimetres that he needs in order to measure its characteristics. Larger quantities outside the storage are a waste both materially and economically, Hence the petroleum industry in particular expends much effort in research and development to avoid waste by accident and was doing so long before the sudden access of public interest in the 1960s.

The third group, consideration for the environment through the avoidance of pollution in conversion or end-use, is concerned mainly with emissions to the atmosphere. Possible thermal pollution of rivers from power stations and oil refineries is not a major problem and can be avoided without great difficulty; indeed in some cases the heat can be beneficial. Excessive emission to the atmosphere of particulates and sulphur and nitrogen compounds can be harmful and have led to the effective downgrading of high sulphur coals, oils and gases and the enhancing of the value of low sulphur coals, oils and gases. Exploration for such resources has therefore been encouraged because the removal of sulphur in particular is expensive no matter at what stage in the conversion, but least of a problem with natural gas, much of which is sweet. Hence there has been pressure for natural gas to be used everywhere. This has led to an expansion of demand which has led to exploration in lands distant from markets. Long-distance transportation of natural gases, particularly in cold tankers as liquids, can be both practical and economic. The enhanced value of natural gas has also led to a lessening of flaring of gases both on oilfields and at refineries, for the conservation not only of the environment but of the natural resource and in both aspects government are involved.

3. Governments
Governments have influence on exploration and initial development, the steps to convert resources to reserves, whether they have a neutral or uncertain policy or a positive one, which usually implies a drive to self-sufficiency, or conservation of energy resources or of the environment.

So interested have governments become since 1940 in the energy industries that, whereas in earlier times, entrepreneurs welcomed a negative/uncertain attitude which allowed them to work with minimum interference, nowadays a negative attitude reduces activity. Two examples can be taken. In the British sector of the North Sea the original gas discovery won a price from the monopoly buyer, the nationalized British gas industry, sufficient to encourage exploration. Continued success in the early 1960s, however, brought lower price offers which, undoubtedly, inhibited exploration for gas in the southern basin and, in the late 1960s, encouraged exploration in the northern basin. Success in oil, which has no statutory monopoly buyer as yet, has further encouraged exploration northwards and there was minimal exploration activity in the British sector of the southern North Sea. In Canada in June 1974, it was reported that twenty-three geophysical crews and their equipment had left Canada since November 1973, seven more were due to depart soon, in addition to six sets of equipment without crews, so bringing the operating crews to forty, which the Alberta Association of Professional Engineers, Geologists and Geophysicists considered as critically low to mount any sort of major exploration drive. The reason for the loss in effort was given as the 'depressing effect of government taxation and royalty problems on Canadian oil industry'.[91] 'The greatest uncertainty facing the Canadian oil industry has been created by the conflict between the provinces and the federal government over the sharing of oil and gas revenue. The provinces have sought to maximize their share of revenue by increased royalties . . .and to protect its share of revenue the federal government has proposed . . .to eliminate the tax deductibility of provincial royalties paid by oil and gas producing companies.'[92] This example also illustrates that in federal systems, the constituent states or provinces have their effects as well as the senior partner.

In Canada western crude oil could not be taken east of the Ottawa Valley to compete with imported oil into Quebec, but this gap would make all the more valuable any oil found in the east, particularly in the St Lawrence Bay where dogged exploration has been carried on for many years without major success. The USSR also suffers from the great distances of its size. Hence it makes up a gas deficiency in the south by imports of natural gas from Iran whilst exporting to the west from its West Siberian gas fields. In West Germany, France and Japan the lack of indigenous oil in significant quantities has led to increasing participation in foreign exploration. The UK is still in this position though the successful discoveries in the North Sea should alter this situation in the mid-1980s — but perhaps be again reversed in the 1990s, and even in the 1980s quality/product needs will undoubtedly lead to both import and export of different crude oils. In all these countries there will be a drive to find more indigenous supplies of all energy resources. The UK perhaps has the best chance of getting a significant contribution from coal, because its industry is perhaps the strongest in Europe. New sources, also, can still be found, as shown by the 1973 publication of new reserves in the north of Yorkshire. West Germany reports higher coal reserves than the UK but there is some doubt, as will be discussed later, as to comparability of these reserve estimates.

In the USA the original objective of Project Independence was self-sufficiency in energy in 1980. This would seem impossible to attain and perhaps a capability to self-sufficiency by 1985 might be a closer practical goal. There was in 1975 still a strong drive for exploration to find more oil, to find methods of utilizing alternatives and, by conservation in various ways, to reduce demand. Variations on the theme of USA self-sufficiency are being developed and studied in every

US government agency, research foundation, university and major energy industry corporation. Their findings are voluminous and varied. All in some form stressed the importance of finding more indigenous oil, gas, uranium ores and geothermal resources. Enthusiasm however waned until 1979.

The self-sufficiency energy programme which appeared to guide China in 1975 had a number of interesting facets.[63] The basic economic policy has been described as aiming at self-reliance which includes 'self-sufficiency' and has meant a refusal to import massive, expensive foreign equipment and seek economies of scale. Rather, as is most obvious in hydroelectric schemes, sites close to consuming centres capable of being engineered and maintained locally from local resources have resulted in thousands of small schemes. Some engineers and economists would dismiss these as uneconomic and would favour a few big schemes. These small schemes are, however, just what China needs, as also in the coal industry, which meets 80% of China's requirements. China has big reserves in North China and Manchuria, far from the consuming centres in the south. Exploration in the 1930s led to discovery of thirty coal beds in Central South, South-West and North-West China. Since the Cultural Revolution (1966–1970) the emphasis has been on exploration and exploitation throughout the whole land, and they claim that rich deposits have been found in 'a thousand and one places'. Small local pits are currently producing perhaps only 30 out of 300+ Mt per annum. Seven or so large centres in North and East China are producing a total of 100 million tons. However the small locally-operated mines have an importance to the economy greater than their statistical weight. Exploration for these small deposits will continue and indeed this policy in both hydroelectric and coal resources might be a useful one for many developing nations.

Venezuela evolved a policy of conservation of its petroleum resources in the late 1950s which has produced virtually a production of over 1×10^9 barrels/year (137 Mt) since 1957, peaking at 1.35×10^9 barrels in 1970. New territory for exploitation has been released very sparingly and the state company has increasingly taken over the operational control. Exploration has been restricted. In Kuwait oil production rates have been more or less controlled since 1967 but little more exploration remains in that small country. Libya has virtually controlled its oil production from 1969 and has deliberately inhibited exploration which it would seem is far from complete. Mexico nationalized its petroleum industry in 1938 and has been striving ever since to find new resources. It was handicapped, until comparatively recently, in that its self-reliant attitude for many years inhibited the import of funds and expertise; this slowed its growth. Recent successes however should encourage even more extensive exploration in a very rich petroliferous region. Two new oil producing countries, the United Kingdom and Norway, have been added to those which pursue production control.

Finally, governments have also a role as the regulatory body for the protection of their nation's environment, as was discussed earlier, and another role in finance, which is discussed below.

4. Finance

A fourth factor affecting the practical transformation of resources to reserves is finance, the availability of the necessary funds. Accountants do not find or make new reserves but the essential exploration and initial development cannot be done without adequate funds.

Prior to 1940 the primary energy industries of coal, crude oil, natural gas manufactured gas and hydroelectricity were virtually self-financing from retained profits. Since 1945 that position has been gradually eroded as governments have taken an increasing part in the energy industries. First the number of centrally controlled economies increased, e.g. Romanian oil. In Western Europe, Britain nationalized its coal, gas and electricity industries; Italy virtually nationalized its oil industry; France adopted a tight control and set up state industries; the State Mines of the Netherlands took a share in the natural gas industry; Germany tried to encourage exploration for German-controlled oil outside Germany. As importantly, the ease of taxation of petroleum and its products, whilst the raw material was cheap, set limits to profits. The nuclear power industry, because of its military origin, is nationally-controlled in most countries. The volume of energy demand, particularly in the petroleum industry, increased the high cost of individual pieces of equipment, whether an offshore drilling ship or platform, or submarine or land pipeline. This means that greater sums of money are now essential for exploration and production than hitherto. Since 1951 exploration has been restricted and subject to regulation as governments in the oil producing countries assumed a stronger role. In some countries exploration and production has been taken over. In other countries which wished to attract foreign capital and expertise the system has been changed from concessions to contracts. Even in the latest new area of the North Sea even Norway and Britain are increasingly deepening their interests. Lastly, the political action of the Arab oil producing countries in October 1973 not only confirmed governments as the controlling influence in the petroleum industry, but by quadrupling the prices raised all costs. This not only ended the era of cheap energy but also ended previously orthodox financing.

Throughout 1974 and 1975 there was considerable discussion as to the effects on the world monetary system of the massive transfer of wealth from the consuming, industrial, highly populated countries to the oil producing, developing, low population countries, many of which cannot absorb all the money coming to them. There were talks of recycling the 'petrodollars', of floor prices to safeguard new ventures from falling prices, of special funds to aid, wholly or substantially, the developing countries like India dependent on imported crude oil, and of other funds to aid the hard-hit developed countries linked within the OECD organization. These and many other schemes have been a feature since November 1973 and, significantly, these and the new International Energy Agency are all government matters. Indeed even the chairman of the Chase Manhattan Bank in August 1974[93] made an appeal for more governmental management. He said that the volume of business was rising so fast as to be growing more than private institutions could bear. He noted that in the first half of 1974 commercial banks in Europe and North America had made about $20,000 million dollar balance of payment loans. He suggested joint ventures in the energy field between producing and consuming countries. The oil producing countries, on their side, are estimated to have supplied $7,000 million in 1974 to under-developed countries, a sum six times their aid in 1973. Many other figures have been bandied about as to aid and investment placings and plans. These examples from the financial turmoil created by the Arab actions in October 1973 are most relevant to this subject of transformation of resources to reserves, because of the need for new investment for which somebody must provide.

Three points are important. The Middle East countries already have sufficient proved reserves without having to invest in a conversion from resources. Some have spare capacity but will need new facilities in a few years. Others are not so fortunate and new funds will be required. The drive towards self-sufficiency in the USA could absorb a lot of money, as will the moves of other governments, like those of the UK and Norway, deeper into the energy industries based on conventional sources. Thirdly, there are the investments necessary in research and development into non-conventional energy resources. The security of the funds for the two latter points depends to a large extent on the level of energy set by the oil producers, hence the talk about 'floor prices'. The Shah of Iran was interested in investing in oil sands exploitation as an insurance that Iran's oil would last into the time when its true and higher value from its chemical characteristics will be appreciated.

Even within the conventional petroleum industry massive funding is essential even for the next ten years. A study in March 1975 by the Chase Manhattan Bank indicated the scale of those funds.[94] The Bank estimated that $400,000 million would be needed for funding and development, plus $370,000 million for refineries, pipelines, tankers, etc, plus $400,000 million for working capital, repayment debts, etc, making a total of $1,200,000 million in 1970 dollars. Even at an inflation rate of 10% p.a., to 1985 the sum becomes $3,100,000 million. The Bank estimated that the $1,200,000 million (1970 dollars) might be obtained, in the following way: $230 thousand million from the capital markets, $260 thousand million from capital recovery and $460 thousand million from profits. Inflation at a 10% p.a., rate from 1970–1985, which implies a much lower rate towards the end of the period, would involve twice these amounts from each source. This exercise illustrates not only the direct financial cost of the transformation of 81 Gt (600×10^9 bl) of oil from resources to reserves but also the continuing cost to transform that oil proved reserve into use. It also stresses the need for the industry to generate its own funds, for which the term profits is used but in the new circumstances will not be profits in the old sense.

The other energy industries obviously have also the need for large sums of money and, as with the petroleum industry, there is going to be a problem in raising those funds. It is extremely important for the world monetary system to regain stability in no matter what new pattern. This subject of finance will be discussed later in the supply and demand chapter in the specific problem of the deployment of OPEC funds.

PROVED RESERVES AND PRICE

Proved reserves are defined with a current economic parameter. There is, therefore, an obvious question as to whether raising the price would automatically move more resources over into the proved reserves category.

Uranium proved reserves are classified according to 'price' which is really cost plus a profit, and it is the increasing cost of extraction at current monetary levels which would allow the extraction of more ore. If the customer is willing to pay $25/kg uranium oxide then more would be available. But, if the extraction costs of what is rated now as costing $10/kg uranium oxide went up to $25/kg, then only that amount now rated at $10/kg would be available at the new price of $25/kg.

In the coal industry a rise in price obtainable per ton because, as recently, the price of its competitor has risen does not immediately expand coal reserves for three reasons. Firstly, the capacity of the industry is limited by manpower and machine power; because of a general rise in all costs through the rise in price of one energy source, which is so pervasive, then coal industry costs will rise both in equipment and in labour costs. This will probably be most noticeable in the underground mining sector and less in the open cast. Secondly, as will be seen later, coal reserves are not really measured with a price tag but only an implied rather vague cost by the depth and thickness parameters which are unlikely to be affected by even a tenfold price increase. Thirdly, deep mines abandoned because they are uneconomic cannot be reopened easily. However, where, as in UK, very rigorous cost factors rule proved reserves and coal seams not being worked in present mines are not included, then a doubling of price might well increase the reserves substantially. There are the problems of safety, water flooding, etc, which even cessation of a few weeks during labour troubles soon accentuate in many workings. In shallow or surface opencast mines it might be possible to cut deeper into thicker overburden but again only rarely would these affect the stated coal proven reserves.

It is, however, in the petroleum field that the question is most acute because of the 1973/74 fourfold increase in price which some consider a fivefold increase, depending on their base. The cost of recovery of oil from its reservoir depends on the characteristics of the reservoir and the method of extraction. There is increasing cost with the degree to which the primary energy which expels the oil has to be augmented with additional energy. This may be by drilling infill holes, by stimulating flow by injecting gas or water or solvents, by fracturing or flooding or heating, etc, or by pumping. At first glance therefore it would seem possible to calculate that for any field $x\%$ of the oil-in-place can be recovered at a cost of $\$a/t$ and $x + y\%$ could be recovered at a cost of $\$a + b/t$. This simplistic view is one which leads to the expression of opinions such as the following quotation from *Fortune* Magazine, December 1974.[95]

'US oil reserves have been greatly underestimated. One reason is that the American Petroleum Institute itself tends to underestimate them. The API has declared that our proven reserves now stand at thirty-five billion barrels (4.72 Gt), down one billion from last year. But this figure is absurd. The API defines proven reserves as "the estimated quantities of crude oil recoverable under existing economic and operating conditions". Yet in calculating its thirty-five billion figure, the API applied economic conditions prevailing two years ago – including 1972 prices, about $3.40 a barrel.

According to a number of engineers and independent oil operators consulted by *Fortune*, a price of $11 a barrel – which is approximately the current OPEC price – would make it profitable to attempt recovery of about half of the 350 billion barrels (4.5 Gt) that remain in the ground in already discovered fields. In other words, producible US reserves probably total something like 165 billion barrels (22.25 Gt), five times the API's 35 billion barrels figure for proved reserves.'

The editorial then admitted that problems of manufacturing and deploying plant and facilities would determine the speed of production of these additional reserves but considered that inertia in government was the most important obstacle to a great expansion in US oil production and advocated unitization of fields and de-control of prices.

It may be significant that the opinions summarized were those of engineers, not geologists, and independent oil operators of unknown experience and possibly prejudiced views. Nonetheless this thinking is widespread and merits consideration.

A symposium on tertiary recovery methods in June 1974 in the USA may well have sparked the *Fortune* article and it is wise to go to the source material.[96] The figures there used were original oil-in-place 434 billion barrels (bb) less 100bb already produced, less 36bb recoverable by present primary and secondary recovery, less 5bb recoverable through future fluid injection, leaving a target of 292bb remaining beyond 32% recovery implied by these figures. Variations in reservoir characteristics were noted as leading to possible recovery factors of 13.5% to 46%. If the reported consensus of twenty-four companies that some 50–60bb (6.7–8 Gt) additional oil could be recovered then the average implied recovery factor would be 44–46% for the whole of the USA, whereas the *Fortune* figure implies a 61% recovery factor, average over the whole USA. Even the extra 6.7–8 Gt is a leap from another estimate which was of 3.35 Gt. The real problem is in the aggregating of all wells as if all reservoirs were the same. The producing wells by states even vary from 0.4 barrels/day (bd) average to 81bd in Louisiana with a US average of 18bd. It is most unlikely that the stripper wells, of which there are many thousands even in Texas, are capable of increasing their production. It is true that there is a lot of oil remaining but the cost of recovering that oil is not known. One suggestion has been that, if the three hundred and eighty-three known shallow oilfields in the USA with an overburden of less than 150 m were mined, their total eventual recovery factor might be 90%.[97] Another figure is that 'if new techniques of tertiary recovery are developed, about twice as much additional oil might be produced as has come from past cumulative production, perhaps yielding 14 Gt for the USA and 39 Gt for the world.'[97] This last phrase '39 Gt for the world' sets the standard of guesswork. This is just a hypothetical exercise, of value only as an indicator, but essentially saying that if you can recover more oil then more oil will be recovered. The danger however lies in, first, unfounded suspicion being aroused that the more conservative figures are deliberately misleading and, secondly, unfounded hopes and unfounded complacency that one only has to increase the price and the oil is there to be produced.

The raising of the price of oil will undoubtedly encourage exploration for new oil and the development of many methods to increase the recovery factor. In the Middle East, for instance, there would seem to be scope for some new technique to be applied to the Kuwait reservoirs. In the USA it may well be that methods applicable to fields, and not wells, will encourage unit management of reservoirs (see the unitization section in production chapter). As the recovery factor of each field is increased so will the proved reserves figure be increased and obviously the giant fields will be the significant features. Proved reserves will rise gradually, not suddenly, with any increase in prices, neither therefore will they drop suddenly, if the price drops.

The proved reserves of natural gas will not increase significantly with a rise in price because of the nature of natural gas reservoirs and their recovery factors of almost 100%. The world's proved reserves of natural gas are approximately 65,000 km^3, with some 49,000 km^3 in giant fields (over 800 km^3), of which 31,000 km^3 are in simple anticlines. These are easy to delineate and, after considering growth trends, Adams and Kirkby[75] concluded that:

'The recent increases in gas price can be expected to increase the rate of growth (of reserve estimates), though it cannot increase the ultimate size of the reservoir, which is a factor determined by nature. The percentage economically recoverable will increase slightly, though it is our view that this increment will be small.'

The rise in price of crude oil since 1973 has increased interest in methods of exploitation of non-conventional hydrocarbon resources, the heavy oils, oil sands and oil shales. Yet, despite the maxim held for many years that the cost of exploiting these resources was a ceiling to the price of conventional hydrocarbons, there had been remarkably little extra production from these resources in the last half of the 1970s.

COAL

Interest in the coal industry was greatly increased and its development can confirm a number of the points already made as to the problems of moving resources into reserves.

If the proved reserves of coal are indeed 210 Gt, as argued later in the chapter on proved resources, and the 1978 coal consumption was 2.7 Gt then the coal reserves/consumption ratio was 78 years, which compares with 28.7 years for crude oil, at the beginning of 1979. This global coal ratio is no cause for complacency for most of the world; some 165 Gt of the total 210 Gt is in three countries, 75 Gt in the USSR, 50 Gt in the USA and 40 Gt in China, and, as will be discussed later, these countries will need most if not all that coal for their own domestic needs. If coal consumption had risen to 4 Gt in 2000 then over 80 Gt will have been consumed and only about 30 years supply would remain if there had not been, as there will be, a drawing of some resources into the proved category. This though, is the type of argument which has been commonly used with crude oil, but not with coal, and is equally foolish with both, because it ignores the continual accretion to proved reserves from the probables. However, this may stress the point that the major task ahead of the coalman is to convert his resources into the practical working inventories of proved reserves. The task is formidable, but the chances of success are enhanced by a quiet revolution which seems to be taking place in the coal industry, world-wide: new exploration methods, as well as, and partly because of new exploitation methods, are being adopted which are bringing new ideas and new men into the industry.

Exploration

The origin of coals is much more simple and better known than that of oil, and yet that knowledge is not perfect. More precise foreknowledge is now required for both operational reasons and to determine the optimum utilization and to meet customer criteria. Quality is becoming a more important aspect, as greater efficiency and cleanliness is required in end-use. On the one hand there are the attempts to utilize lower grades of coal by improved burning, as in the fluidized bed combustion method, and by blending different coals to produce substitutes for high quality coking coals. On the other hand, there are attempts by electricity generating organizations, particularly in the USA, to seek assurance of sufficient supplies of preferred qualities over the life of their stations, and big quantities are involved, for some 200 Mt of coal may be needed over a 40-year life for only one 1000 MW power plant. Coal classification methods — chemical, physical, and petrographic — have been refined over the past 50 years and an immense amount of data has accumulated, but it is often inaccessible and some is inadequate to modern needs. The US Geological Survey is now trying to computerize the mass of information available to them; other organizations, from the Electric Power Research Institute in Palo Alto, California,[261] to the International Energy Agency in Paris and London, are making detailed assessments of recoverable coals, with an eye to quality as well as quantity.

Geophysical methods of exploration, which were developed for the oil industry, are now being applied in different types of surface and underground surveys and in borehole surveys. Commercial coal seams are usually less than 1500 metres deep, and, therefore, the geophysicist is having to alter his objectives; he must aim for the greatest possible definition of the strata at depths which he would often consider too shallow, and would ignore in exploring for oil or gas, particularly in recent years when the search has gone ever deeper. When working in areas where the coal may be amenable to strip or open-cast mining, the important depth is from surface to a few tens of metres. In such areas, gravity surveys have to be of great precision if they are to locate, for instance, cut-outs in thick seams. Magnetic methods are now being used, for example, to map, quickly and accurately, parts of a seam which may have suffered from spontaneous combustion inwards from the coal face, which is often found in the Western USA, though whether such methods would be applicable to the burned seams in Southern Africa is doubtful. Seams with sharp boundaries can be mapped by seismic methods using waves travelling along the seam; shallow seismic reflection surveys are being increasingly used, usually to delineate rather than to discover new fields and in some circumstances electrical surveys may help. The location of large coal deposits is often known in general but imperfectly mapped in detail. Seismic surveys can provide a quick and cheap supplement to boring holes on a grid pattern and may reduce the number of slow and expensive holes by locating them at critical points only. Geophysical logging in the bore-holes and in-seam seismic measurements between holes can increase the effectiveness of both types of survey. Many different well logs can be run, measuring natural radioactivity, rock density, hydrogen (neutron) density, resistivity, temperature, variations in the size of the hole using caliper logs, which also indicates the relative hardness of the rocks, the dip of the strata, and the velocity of passage of sound waves, which depends on density. The application of petrophysics to the interpretation of the geophysical data so obtained is capable of determining ash content, calorific value, moisture content and so the rank of coal seams, as well as the nature of the overburden beyond and between the borehole control. Many of these techniques have come over from the oil industry, but need adaptation to the peculiar needs of the coal industry, e.g. coring techniques have to be refined to recover 100% of soft and brittle coal seams, whether thin or thick, and allow the detection of even small faults. The objective is to eliminate guesswork, for this is a luxury which an increasingly capital intensive industry cannot afford.

Mining

The struggle for survival in the established coal mining areas of the industrialized countries, during the 20 years of cheap oil — from 1953 to 1973 — led to fewer mines, fewer faces being worked, and the introduction of automatic cutting, roof-supporting and coal-transporting machines in deep mines, and massive earth-moving equipment in strip mines. The extreme flexibility of the miner with his hand pick, able to move around a trouble zone, and provide detailed knowledge of the geology from the many seams being worked slowly and simultaneously, was lost many years ago. Therefore the need increased for greater precision in knowledge before opening a new mine, or moving into new ground in an old mine with the greater gamble as fewer faces were providing geological data. The new methods were boosted by the dramatic change of opinion, after the oil supply restrictions of late 1973, when the need for coal became more obvious and funds were provided. Until then the coal industries of few countries had been planning new mines or expanding at a faster rate than that with which the old methods could cope. New exploration methods, especially continuous geophysical surveys, have been increasingly used, particularly since 1974, in many countries. The youth of the application of geophysical methods does result in many unresolved problems and often lack of mutual understanding of the needs of the coal miner and the strengths and weaknesses of the geophysicist. The new aids to mining have to guide the optimum production of solid coal in a very different way to that in which geophysics helps guide the production of liquid or gaseous petroleum from wells. Mistakes must be avoided, such as that which transformed the profitable mining in Wales of a good quality and thick seam of anthracite into a struggling operation because of unpredicted faulting. The inflexibility of massive cutting machines precludes improvisation; therefore foreknowledge is important from the planning stage to the cutting stage; for instance, there is need for a probing drill which can test ahead from the cutting face for up to 450 feet (137 m), keeping within the seam and not cut into roof or floor which would return rock cuttings and mislead the engineer into thinking that the seam has been cut by a fault and the coal ended against rock. Such a problem is a technical and equipment problem, and new equipment is being designed and developed or being adapted from the hard-rock mining or oil industries. Swivel-headed boring machines are now available for drilling up, down, slantwise or horizontally from the cramped galleries of coal mines. Some equipment manufacturers, who thought that coal would offer great opportunities in the late 1940s, had to turn to other industries when those expectations faded, but are now being attracted to return, bringing with them the new tools which they have developed elsewhere. Finally, computers are now replacing 'tally boards', and the old systems are being replaced by modern computer programming, simulation modelling, data banks and retrieval systems.

The industry

The revolution is not just of new techniques and new equipment, for new men are entering the industry, as is shown by the plethora of new titles and new jargon; these new men are changing the face of the industry. Strip mining needs skilled operators for the massive machines, but otherwise only little and mostly unskilled labour. Deep mining is still hard and dirty work, and the handling of most machines is robust, as are the machines, but the high manual skill of wielding a pick axe has to be replaced by more mechanical skills. There are many strongly entrenched anachronisms in the coal mining industries of all countries which have a long mining history, and it will take a number of years to update all thinking and methods; one example is that an experienced driller in a British mine cannot be paid more than his partially or wholly unskilled assistant, because 'both are doing the same work'.[259] In Britain, 50% of the cost of mining is labour cost, though in the latest project, the Selby mine, the labour cost will be only 15%. It may be many years before the value of the new technologies is fully appreciated by the work force and the new methods seen as improvements in the older mining areas of the world.

The changes in the exploration and extraction aspects of the world's coal mining industry have had and will have very varied effects in the different producing areas. The older industrialized countries of the United States, the United Kingdom, FR Germany and the USSR, will adapt with varying speeds, and one can expect an increase in proved reserves, production and productivity. The younger areas of Australia and Southern Africa should benefit more and show the most rapid growth, perhaps along with a revived Canadian industry. There are some 100 Gt of indicated, mineable coal-in-place in Western Canada, yet in a recent description it was said that geophysicists had not been a very successful tool, although well logging surveys were used,[259] perhaps this is an example of the problem of adapting geophysicists, who are oil-oriented, to become coal-oriented, when the miner may not appreciate what might be done for him. It is thought improbable, though perhaps not impossible, for this revolution in exploration and extraction, both in machines and men, to revive the declining coal industries of France, Belgium and the Netherlands.

The British coal industry can exemplify both the scale and the type of effort which is now being exerted to move resources into reserves, where the need has been grasped. In the UK the 1976 investment was to be £300 million; this was part of the biggest capital expansion ever undertaken in the UK coal industry and will probably total between £2,000 and £3,000 million over the period 1975-1985, rather than the £1,400 million as originally estimated.[291] The *Plan for Coal*, published in June 1974, divided the £1,400 million between ordinary annual capital expenditure of £70-80 million, for the ten years, and £600 million to develop 42 Mta of new capacity. This new capacity should more than compensate for the exhaustion of old mines and will be found by gaining 9 Mta from life extensions to some pits, which would otherwise exhaust, 13 Mta from major improvement schemes, and 20 Mta from new mines. In the fiscal year 1976-1977, £15 million was to be spent on exploration for new mines and extensions to old mines, £175 million on schemes to open up further reserves, in particular of coking coals for the steel industry, £80 million on underground equipment, and £30 million on other schemes, including research facilities. This pattern of programme must be maintained if the objectives are to be reached of keeping annual production of deep-mined coal about 120 Mta in the next few years, increasing open-cast production from 10 Mta to 15 Mta, and reaching towards total production of 150 Mta in 1985.

The Selby project illustrates the methods of planning for the exploitation of a new mine and the lead time necessary.[296] Deep-holes drilling in 1964-1967 disproved earlier ideas and indicated a northeastward extension of the Yorkshire coalfield. A systematic borehole programme of some 40 to 50 holes at 3 to 4 km intervals was started in 1973. By 1975 it was clear that the best seam of the main coalfield, which had split into two thinner seams, had reunited into a 3 m seam, and the recoverable reserves of this seam alone were estimated to be about 250 Mt out of 600 Mt coal-in-place, a 42% recovery factor. A seismic method for the detection of small faults was developed in 1973, and by 1974 had indicated the fault pattern sufficiently for broad planning purposes. More precise methods will be needed to detail minor irregularities before engineering work is started. Planning permission was granted in May 1976, but with a major provision that surface subsidence must not exceed 0.99 m, because of the danger of flooding agricultural land. The plan is that five shafts will be sunk, from each of which 2 Mta will be mined. The shafts will handle only men and materials, and provide ventilation, for the coal and rock, as mined, will be transported on conveyors in a central spinal system to a single point at the surface where loading and bunker facilities will be built to load 50×1000 t liner trains per day, to feed three nearby power stations. The first face should be ready for production in mid-1981, and the whole complex be producing 10 Mta by 1986. This project is not the biggest in the world, but is an example of the revival possible in an old mining industry, and will employ the latest in equipment and technology; but it will be 12 years from definition of the field to full production.

4
Proved reserves

Proved reserves are the working stocks of the energy industries on which they have to rely for the supply of energy in the near term. The major proved reserves on a world scale are restricted to those from the fossil and fissile fuels and, in a special sense, hydraulic sources.

The complacency about the size of coal resources, inside and outside the industry is, at times, almost tangible. In Britain the coal planners were arguing in the early 1960s that the British coal industry should be maintained at a high level because 'oil was rapidly running out'. This was not a convincing argument at that time, but the milder proposition, that coal would be needed to augment oil supplies in the medium and long term, was more reasonable, and was accepted by many in the oil industry. However, those who have been in the coal industry for over twenty years now feel that their attitude at that time is vindicated; the workers feel strengthened in their claims for higher wages and better conditions as 'coal is coming back into its rightful place'; their political representatives feel strengthened in their opinions on general energy matters; governments of countries with coal resources feel that action in finding new energy sources to replace petroleum can be shelved, because 'coal is available'. The only coal available up to 1980 was that which can be won from existing mines with the existing manpower and existing machines. Very few new mines were planned in 1965-1970 to come into production in 1975-1980. Existing mines may be capable of some expansion, but modern methods of extraction necessitate far more preparatory work than was warranted in the old days of the hand-picking of coal seams. Beyond 1980, the new mines, which have been planned since 1973, should come into production, but it is essential that there should be no misunderstanding as to what are coal resources and what are proved coal reserves, for it is only the latter which are available for mining.

Table 12
Some suggested figures for
comparable proved coal reserves — an exercise
(gigatonnes)

U.S.A.	50
U.S.S.R.	75
PR China	40
Europe	25
Rest of World	20
Total World	**210**

The US Geological Survey in May 1975, used their recently adopted standard nomenclature, and allocated 50 gigatonnes, (Gt) to their reserve category of 'indicated, measured, economic, and demonstrated'. This category is the nearest to the proved reserves category of the US petroleum industry. 50 Gt is 13.8% of the 363 Gt of known coal-in-place reported to the 1974 WEC Survey. The

USSR only reported 273 Gt as known coal-in-place, yet Professor Alexander V. Matveev of Moscow University, in May 1976[259] implied that in his category of 'explored reserves' he rated the USSR as having almost twice as much coal as the USA would have in that category; let the USSR be allocated 75 Gt, as a compromise. The People's Republic of China does not publish any reserve figures, and the 'best guess' in the 1974 WEC Survey was 80 Gt of economically recoverable reserves from 300 Gt of known coal-in-place (estimated), which one must assume was considered comparable to the 182 Gt taken as the economically recoverable reserves of the USA. Hence, if the USA proved reserves are pruned to 50 Gt, then some reduction must be made in the figures for China, in order to maintain some comparability; perhaps 40 Gt would be reasonable, allowing for the intensive exploitation in China. The disparity between the reserve concepts accepted in the UK and the Federal Republic of Germany, (FR Germany), are highlighted later and perhaps a lower figure should be taken for Europe than the 13.8% recovery factor noted for the USA; let 8% be that figure and the proved reserves of Europe become 25 Gt. However, in the rest of the world, where 145 Gt was reported for the 1974 WEC Survey as the known coal-in-place, a somewhat higher recovery factor, of say 14% might compensate for under-estimation of the coal resources, particularly in Australia and Southern Africa, so that the proved reserves become 20 Gt. The total proved reserves of coal, by this reasoning, would be some 210 Gt, which is greater, and perhaps a 'better' figure, than the 140 Gt which I argued in a preliminary exercise.[256]

This calculation could rightly be considered individualistic, empirical, pragmatic, and, by some, even both unscientific and nonsensical. However, it was noted earlier that the USA coal resource figures were really not as accurate as they appeared, and that the 50% recovery factor had only gained credence by repetition. Similarly it should be noted that the commonly quoted figure of 1 Tt for China's coal resources[257] was based on an estimate made about 1880 by von Richtofen, during his travels in China, and thenceforth enshrined in the literature. The calculation of the previous paragraph may, perhaps, not be altogether out of line with similar guesses about energy resources; it should not be dismissed out of hand, although it would obviously be far, far better if more correct figures could be assembled.

In order to put this guess of 210 Gt for proved coal reserves alongside the proved reserves of other energy sources, there are two problems to be solved first; on what basis should they be compared, and what conversion factors should be used for each resource to get them to that base? The problems of using a heat content comparison have been explained earlier, yet this is the simple, rough and usual basis for comparison, and therefore, in this exercise the standard of the heat content of tonnes of world coal equivalent is taken. First the coal heat content must be decided and with the spread from lignite at 14.7 gigajoules per tonne (GJ/t) to anthracite at 33.5 GJ/t, a weighted average of world coal-in-place was calculated out to 25 GJ/t from the data in the 1974 WEC Survey.

The published crude oil proved reserves are the amounts which can be recovered from defined reservoirs and brought to the surface under current economic and operational conditions, and, therefore, the type of reserves to which the coal reserves are being brought in this exercise. Some would object to this comparison because they would argue that there are greater technological limitations on the extent to which coal can be recovered, and also greater variations in the recovery factor over the life of a coalfield, which is longer than that of most oilfields, and also that such comparisons ignore the fact that oilfields are more difficult to find than coalfields. Such arguments are, in my view, irrelevant to consideration of what can be termed the practical working inventory of each energy resource which can be made available in the immediate future with least speculation. The conversion factors from oil to coal equivalent in the literature vary from 2.15 to 1.3, according to the assumptions taken and the purpose for which the comparison is being made, e.g. *UN Statistical Papers, Series J,* No. 17, used 1 t crude oil = 1.3 t coal equivalent (tce); No. 18 used 1.47 tce. The range in crude oils in world trade is roughly from the crude of Boscan, Venezuela, of 10° API gravity and 5.5% sulphur by weight, at 39.48 GJ/t, to a light Indonesian crude oil, of 47° API gravity and 0.05% wt sulphur, at 43.65 GJ/t. Middle East crude oils comprise some 55% of world proved reserves and have a comparatively small spread in heat content around 42.33 GJ/t. This figure is therefore taken as the average and so a tonne of world crude oil is taken as having a coal equivalence of 1.69 t.

Natural gas varies greatly in heat content, mainly according to the inert gas content, and therefore, although it is impossible to be precise, a low figure is taken from the range of figures which have been used elsewhere, e.g.

- 46.95 megajoules per cubic metre (MJ/m^3), National Academy of Sciences, USA, 1974);
- 37.68 MJ/m^3 (BP Statistical Reviews);
- 34.1 MJ/m^3 (Shell Company Natural Gas and Measurement).

The figure chosen is 34 MJ/m^3, which gives a conversion figure for 1 km^3 of 1.36 megatonnes of coal equivalent, (Mtce).

The choice of a factor is very difficult for the wide range of little known recoverable oils from the very viscous and heavy oils, from the oils in oil-sands or tar-sands, and from the solid kerogen in oil shales. The characteristics of the recoverable and marketable oils depend on the process of extraction. In the USSR and China, oil shales are often mined, crushed and burned under boilers, with the heat content taken as 27.43 GJ/t. There are also very wide ranges in the recovery factors of the oils as separated from their host rocks. It has been argued that only from the Athabasca Oil Sands should the 3.6 Gt of recoverable oil be acknowledged as proved reserves for world oil sands, with about 11 Gt as probable reserves. Oil is produced from oil shales only in the USSR, Brazil and China, and, therefore, only there are true reserves proved. Even the richest USA shales of the Piceance Basin have not been proved operationally economic. There are so many unknowns in the whole of this field of oils which cannot be produced by conventional means, that only a guess is possible. A figure of 40 GJ/t has been taken.

The working stock of uranium can be taken usually as that which is assessed by the OECD, NEA and IAEA as 'reasonably assured resources' at an economic price/cost. Taking the 1 January 1976 figures given at the end of chapter 2, but subtracting the 500 kt which were described as uneconomic, leaves 1690 kt as roughly similar to the other proved reserves, and 2,600 kt as the probable reserves, after adding back the currently uneconomic 500 kt, for the non-Communist world. The speculative or possible reserves of the non-Communist world (NCW) were given as 6.6–14.8 Mt U but in this category a guess at the total world resources can be made by adding in the 3.3–7.3 attributed to the Communist world, so totalling 10–22 Mt U.

The energy conversion factor for uranium depends on the process used for conversion to electricity; figures vary from 0.86 terajoules per kilogramme (TJ/kg), for use in a non-breeder reactor; 1.7 TJ/kg when enriched for some non-breeder reactor types; 42 TJ/kg or even 51.75 TJ/kg in fast breeders, though 83 TJ/kg has been used for 'nuclear fission'. The upper limit for non-breeder use is set by the calculated energy from theoretically complete fissioning of a kilo-gramme of uranium$_{235}$ which equals 2,800 tce, whereas the lower limit may be taken as the actual fissioning at 50% efficiency of the 1 part of U_{235} found in 140 parts of natural uranium, which gives a figure of 10 tce/kg. A practical figure used in the nuclear industry is 14.3 tce/kg and this is taken, in this exercise, for conversion of the 1.69 Mt U, which is assumed as the proved reserves, in non-breeder reactors, which are the only current commercial processes. The proved reserves of uranium therefore become 24.2 Gtce. The heat value in a fast or breeder reactor may be 50-60 times as great as in a thermal reactor, so that the proved reserves would be some 1200-1500 Gtce for the fast reactor. However this reaches into the speculative realm, and it is often these speculative figures which can be the most misleading.

Professor A. V. Matveev of Moscow compiled a world map of coal resources, using the Soviet system of coal categories, which are based on the degree of knowledge aimed at eventual exploitation, and his paper discussing the data was available in late 1976.[259] It might be noted that Matveev would put the 'geo-logical resources' or the resource base, of world coal at 30 Tt. This figure fills a gap, but such figures, as has been said earlier, are of academic interest only, because even the best known are not very well known. Nevertheless, all these speculative figures must be set down for the non-renewable energy resources, if one is going to get the coal resources into perspective. However, because these speculative resource bases are far from having the same weight as the proved reserves of practical, commercial and immediate value, it is suggested that the resource figures should be plotted on a logarithmic grid, rather than on a linear scale when each increment has equal value. This is a departure from custom, and raises objection from some in the coal industry who, perhaps, gain comfort from the very large numbers of the coal which is in the ground. The log presentation reduces the exaggerated impression of the importance of these large numbers and stresses the size of proved coal reserves with those of the other energy sources, all being stocks from which energy can be won within the lead time for new projects. In this way, the different types of resources are given the values which they have, and not those which they do not have.

Figures 8 and 9 portray graphically the figures given in table 13.

The logarithmic presentation of this exercise not only stresses the importance of the various working stocks but also should dispel the myths that the energy resources of the oil sands and oil shales could be tapped at a moment's notice, and what have been dubbed probable oil shale reserves might perhaps more aptly have been termed possible or even only additional resources. The uranium column does not, of course, illustrate the potential of nuclear energy but it does show that the current proved reserves of uranium if used only in thermal reac-tors is quite small in relation to the three major current energy sources.

In this exercise the heavy oils have been ignored because of a lack of an authoritative study world-wide. The proved reserves of hydropower and geo-thermal energy should appear in any comprehensive collation of proved reserves, but it is hoped that the major items have made the point.

Another method of attempting comparability is to convert resources to units

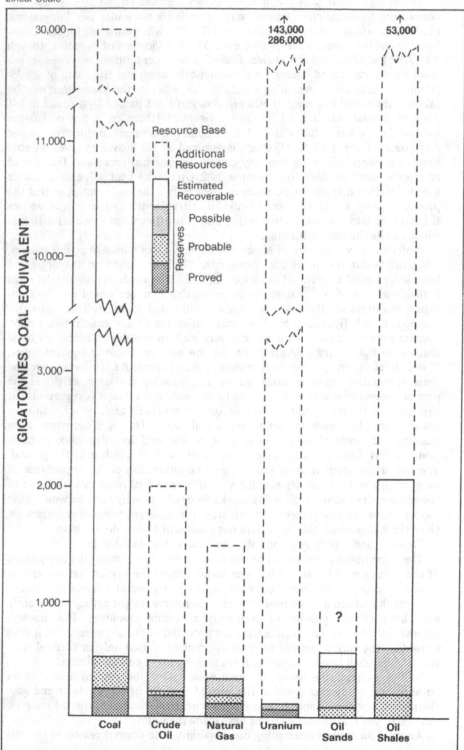

WORLD ENERGY RESOURCES
Linear Scale

Fig. 8

WORLD ENERGY RESOURCES
Logarithmic Scale

Fig. 9

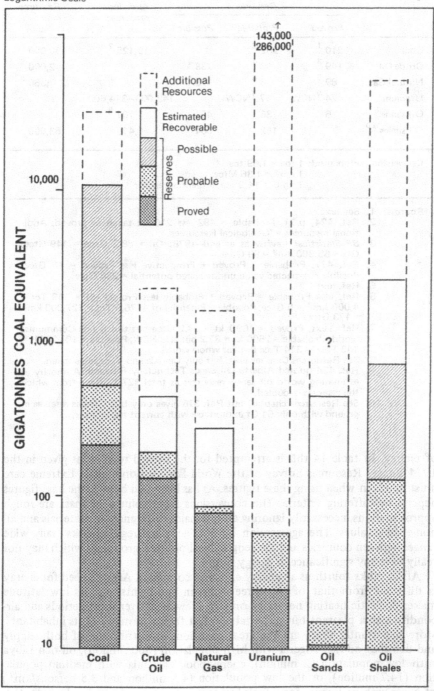

Table 13
World energy resources
(gigatonnes coal equivalent)

	Reserves			Additional resources	Resource base
	Proved	Probable	Possible		
Coal	210 [1]	282 [2]		10,125 [2]	30,000
Crude Oil	149 [3]	46 [4]	238 [4]		2,000 [5]
Natural Gas	89 [3]	5 [6]	173 [6]		1,360 [5]
Uranium	24 [7]NCW	37 [7]NCW		143,000–314,600	
Oil sands [8]	6	36	403	107 [9]	
Oil shales [10]		162	564	1,416	53,000

Conversion factors used: 1 toe = 1.69 tce
1 km^3 = 1.36 Mtce
1 kg U = 14.3 tce in thermal reactor

Sources: 1. See text
2. Ref. 424, p.71. Probable = 492 less 210 Gt taken as proved; Additional resources = 'Geological Resources'
3. *BP Statistical Review* as at end–1978: Oil = 88.3 Gtoe = 149 Gtce; Gas = 65,800 km^3 = 89 Gtce
4. Ref. 477. Probable = Proved + Prospective less Proved = 46 Gtce; Possible = expected value undiscovered potential = 238 Gtce
5. Ref. text
6. Ref. 484 Probable = Proved + Probable less Proved (3) = 148 Tcf = 4,000 km^3 = 5 Gtce; Possible = Potential of 4476.6 Tcf = 127,000 km^3 = 173 Gtce.
7. Ref. Text. Proved = 1690 kt = 24.2 Gtce but this is non-Communist world; Probable = 2600 kt = 37.2 but also NCW; Possible = 10–20 Mt = 143 Ttce – 314 Ttce but for whole world.
8. See Resources Chapter text, but only Canada and Venezuela taken.
9. Ref. 424 quoted Alberta Oil Sands Technology Research Authority as estimating world oil sands resources as total 327.12 Gtoe from which the reserves deducted.
10. See Resources chapter, but Ref. 376 gives only 676 Gtce as reserves in ground with only 51 Gt exploitable with current technology.

of energy. In table 14 this is attempted for the proved reserves as given in the 1974 Energy Resources Survey of the World Energy Conference. Extreme care must be taken when using these figures. As has been said before, the base figures depend on differing criteria. The conversions from volume or mass are rough approximations, necessarily ignoring all but major differences in materials and all non-energy values. The aggregation by continent and region masks many wide ranges between countries and are geographical/political groupings which may not really have any significance in energy terms.

Africa ranks fourth as an energy-endowed continent. Africa's need for energy is different from that of the three northern continents, e.g. its low latitude makes domestic heating necessary only in a few areas over short periods and air-conditioning a pleasant but not vital comfort factor to many of its inhabitants. More importantly there are the great variations in distribution of both supply and demand: oil-rich Nigeria with high population (500 million), oil-rich Libya with low population (2 million), energy-poor Tanzania with medium population (14.7 million), or the low population (4.5 million and 3.5 persons/km^2) of Niger with no known fossil fuels but some uranium and a hydraulic potential.

Table 14
Energy content of proved reserves
(million terajoules)

		Solid fuels	Crude oil	Natural gas	Uranium non-breeder	Total
Africa	West	5	50	47	35	136
	East	49		3		52
	Middle	18	41	14	26	99
	North	>1	465	149		615
	South	308		>1	148	456
	Total	**382**	**556**	**213**	**209**	**1,360**
Asia	China	2,344	75	25	x	x+ 2,444
	Total East	**2,371**	**75**	**26**		
	Japan	33	>1	>1	3	37
	India	272	6	3		281
	Iran	6	384	211		601
	Total Mid-South	**300**	**390**	**248**		
	South-east	18	80	25		123
	South-west	31	1,789	156	>1	1,976
	Total Asia	**2,752**	**2,334**	**456**	**3**	**5,545**
Europe	West	1,042	5	84	30	1,161
	South	222	6	9	19	256
	North	185	33	53		271
	East	1,132	16	16		1,164
	Total	**2,582**	**60**	**162**	**49**	**2,853**
USSR		3,508	352	610	x	x+ 4,470
North America	US	5,200	241	287	285	6,013
	Canada	133	47	96	161	437
	Mexico	17	18	11	>1	47
	Total	**5,350**	**318**	**402**	**446**	**6,516**
South America	(Venezuela)	(>1)	(264)	(35)		(300)
	Tropical	47	313	51	2	413
	Temperate	5	16	13	11	45
	Total	**53**	**329**	**64**	**13**	**459**
Oceania	Australia	481	9	20	105	615
	Total	**485**	**10**	**26**	**105**	**626**
World Totals		**15,112**	**3,957**	**1,933**	**824**	**21,826**

Source: Approximated from *1974 WEC Survey*, Table IX-2, pp 253-255.

It is somewhat difficult to understand to what practical use the total Africa figures can be put. The total African fossil fuels reserves, at 1,240 million TJ, have a ratio of 352 to 1972 total African consumption. The whole of Europe has reportedly total fossil reserves of 2,930 million TJ with a ratio of 54 to 1971 total Europe consumption. This means that Africa has more energy reserves in relation to its current energy demand than does Europe. Any more detailed statistical use would be wrong. It is debatable, for instance, whether Europe has twice as much energy total because, as noted below, coal is its dominant resource and its estimation may well be an exaggeration.

Western Europe, in the table of energy content of proved reserves, has a value for solid fuels of 1,042 million TJ, of which the Survey reports 968 million TJ from West Germany, whereas Northern Europe has 185 million TJ of which UK has 127 million TJ. Closer inspection of these reserves as reported to the Survey shows[100] West Germany to have 44,001 Mt of high-ranking coals-in-place with 30,000 Mt economically recoverable, whereas UK has 93,259 Mt of bituminous coals-in-place with only 3,802 Mt being economically recoverable. The general geology of the coal deposits in the two countries is known to be similar and the 1972 productions were for UK 136 Mt and for West Germany 102.5. Different parameters were being applied to determine the economically recoverable quantities.

The British coal industry had adopted a realistic attitude to its coal reserves.[101] In 1905 some 144 Gt of coal were said to be present and capable of supporting the life of the industry (though no time limit was given). During World War II the reserves which could be extracted over the next hundred years were estimated at 20.8 Gt. An assessment made over the late 1950s and early 1960s gave a figure of 16 Gt as being those reserves which were accessible to existing mines. It was anticipated at that time that no funds would be available to open new mines, because coal could not be competitive with the falling prices for oil. An assessment in 1973 gave a national total of current workable coal reserves at 3.9 Gt. This trend represented a write-off averaging more than 2 Gt/annum, being a loss of access to reserves 'directly related to national productivity and costs and subject, given good management, to increase with new technology'. The improved competitiveness between coal and oil after the 1974 oil price rises, improved technology and appreciation that the low figure of 3.9 Gt was not conforming to practice in other countries and was therefore demoting Britain's coal wealth, caused the National Coal Board to report the UK economically recoverable coal reserves as 45 Gt, to the 1978 WEC Survey.[422] This is the amount which many politicians and even some National Coal Board officials convert to be 'coal supplies for 300 years'. This phrase leads to complacency, for it masks the truth that the coal has to be extracted to have value.[256] The 1978 revision also would increase Western Europe's standing in the energy league.

Crude oil and natural gas proved reserve estimates are revised annually by the two trade journals, *Oil and Gas Journal* and *World Oil*, quoting the American Petroleum Institute for the USA reserves, the Canadian Petroleum Association for Canada, and for all other countries using their own appreciation of data which they collect and collate in as comparable a style as they can effect. Differences do occur between them and a mean between their figures is not necessarily nearer than either one because of their attempted internal consistencies. The demonstrable ignorance in many areas and of many aspects means that none of the assessments can be more than a rough guess at a rough guide of the industry's working stock. In countries as large as the USSR, Canada, China and the USA the distribution of the reserves is almost as important as the total. The Asian regions of the USSR, namely Siberia, Central Asia, Kazakhstan and the Far East 'account for over 90% of the country's fuel and power resources, including 94% of the geological coal deposits . . . whereas four-fifths of the national fuel and power output is consumed in the European part and the Urals, where economical fuels and power sources are in short supply'.[102] Western Siberia has the large Kuznets coalfield, developed to satisfy the country's need for coking coal but also producing large amounts of fuel coal. The Angara-Yenesiei region of Eastern Siberia has about '50% of the national coal resources', with most fields featuring a single flat-lying or gently dipping seam, 10-20 to 50-90 metres thick, recovered by opencast methods. 'The estimated cost of a ton of equivalent fuel will be about two roubles and the output per miner may be as high as 2,300 to 2,500 tons per month'. These productivities are difficult to compare with others which are usually quoted as tons per man per day or month of x working days; nor is it known whether the amounts apply to marketable coal or run of the mine tonnage. USA opencast mining average is 50 t per man of marketable coal per day but can reach 100 t per man per day. The Kansk-Achinsk coalfield was said to have an annual production potential of 1,000 Mt. Kazakhstan and Central Asia are said to have considerable resources of fuel coal. Although the USSR only reported coal-in-place to the 1974 WEC Survey and the 1978 revision, the Coal Report to the WEC Conservation Commission[424] gave also reserves, as in the table 15.

Table 15
Geological coal resources and technically and economically recoverable reserves
(Proved reserves for eight countries with the largest proved reserves, 1977)
(gigatonnes coal equivalent)

	Geological Resources		Proved Reserves	
	hard	brown	hard	brown
USA	1,190	1,380	113	64
China	1,425	13	99	n.a.
USSR	3,993	867	82	27
UK	164	—	45	—
South Africa	57	—	27	—
FR Germany	230	16	24	—
Poland	121	4	20	—
Australia	214	48	18	9
Total World	7,725	2,400	492	144

Source: 'An Appraisal of World Coal Resources and their Future Availability', W. Peters and H-D Schilling, 'World Energy Resources, 1985–2020', Ref. 424.

THE HYDROCARBONS

Previous discussion has been concerned with the resource base, the units of measurement with the problems of comparison by energy content, the four methods of measurement of petroleum resources, the estimates of those resources both offshore and onshore, conventional and non-conventional, the determinants in the transformation of resources to reserves, with a particular reference to the effect of prices. In this progression there has been increasing mention of the hydrocarbons. This is logical because the world which has grown so dependent on petroleum in the past thirty years will continue to be dependent for at least a decade or two. Therefore, as the working inventories of available energy are approached, the greater is the stress on the hydrocarbons and particularly on the conventional forms.

In table 16 the figures of proved reserves as published by *World Oil* and *Oil and Gas Journal* for end 1978 are listed with the latter figures displayed in groups.

Table 16
World crude oil proved reserves, end 1978
Top 21 countries
(gigatonnes @ 7.3 US barrels/tonne)

		Rank	World Oil	Oil and Gas Journal
OPEC Middle East	Saudi Arabia	1	15.5	22.7
	Kuwait	2	9.8	9.0
	Iran	4	6.2	8.1
	Iraq	5	4.7	4.4
	Abu Dhabi	6	4.1	4.1
	Neutral Zone	16	0.8	0.9
	Qatar	19	0.5	0.5
	Total Middle East		**41.6**	**49.7 (60%)**
Other OPEC	Libya	9	3.7	3.3
	Venezuela	11	2.5	2.5
	Nigeria	12	1.7	2.5
	Algeria	14	1.3	0.9
	Indonesia	15	1.1	0.9
	Total Other OPEC		**10.3**	**10.1 (12%)**
Non-OPEC Non-communist	Mexico	7	3.9	2.2 (revised to 5)
	USA	8	3.8	3.9
	UK	13	1.4	2.2
	Canada	17	0.8	0.8
	Norway	18	0.6	0.8
	Oman	20	0.4	0.3
	Egypt	21	0.4	0.4
	Total Non-communist		**11.3**	**10.7 (13%)**
Communist	USSR*	3	8.0	9.7
	China	10	2.7	2.7
	Total Communist		**10.7**	**12.4 (15%)**
Total 21 countries			**73.9**	**82.8 (100%)**
World total			**78**	**88**

* USSR reserves = 'explored reserves' = proved + probable + possible
Sources: *World Oil* 15 August 1979
        Oil & Gas Journal 18 December 1978

There are a number of important points to be learned:
- the differences between the two sets of figures illustrate the problem but each journal attempts to be self-consistent. For consistency with other data given, the *O&GJ* figures are normally used in this book.
- some 94% of the world totals are in these top 21 countries, with a range from Saudi Arabia at 22.7 Gt to Oman with 0.3 Gt; indeed 77% of the world's proved reserves are in the top 10 countries.
- Saudi Arabia has 26% of the total world (or 20% of the *World Oil* figures).
- OPEC within the 21 countries has 72% of the reserves.
- OPEC total is 69% of the world total.
- the non-Communist world outside OPEC has only 13% of the 21 country total, and the total non-Communist world has 85% of the total world reserves.
- if the Mexican reserves are 3.5 Gt, rather than 2.2 Gt then the non-Communist outside OPEC of the increased reserves of the 21 countries is 17%.
- the Communist share of both the 21 countries and the world total is 15%.
- the 1978 Annual Report of Petroleos Mexicanos (PEMEX) gave proved reserves as 25.6 bb (3.5 Gt) oil, 2.8 bb condensate and 58.9 tcf gas[487]

An example of earlier figures for North Africa can emphasise the lack of precision in even these proved reserve figures. In 1973 the following were some estimates for North Africa: *O&GJ* 5.3 Gt, *World Oil* 4.9 Gt, USGS 5.5 Gt, Warman 2.8 Gt, and the WEC 1974 Survey 10.75 Gt. Only Tunisia reported to the last and the high figure is composed from other sources, but in particular from a special 1973 assessment in the *O&GJ*[103] which estimated the proved recoverable reserves of Algeria at almost 6 Gt and Libya at 4 Gt. North Africa has caused problems before, as shown by the following figures for Algeria as reported in the *O&GJ*:

1.1.72 – 1.08 Gt;   1.1.73 – 6.35 Gt;   1.1.74 – 1.0 Gt.
1.1.76 – 7.37 Gt;   1.1.77 – 6.8 Gt;   1.1.78 – 6.6 Gt;   1.1.79 – 6.3 Gt.

Low figures were obviously favoured by Moody in 1975 as he allocated 4.5 Gt for proved plus prospective reserves.[42]

This example stresses the importance of appreciating the time of the estimates, as mentioned earlier when discussing classification and the 1979 Russian suggestion.[476] This is particularly so in the early phases of exploitation. In 1970 some 1.3 Gt were added to the proved reserves of the USA by accepting those of the Prudhoe Bay field on the North Slope of Alaska. The field was discovered in 1968. The reserves measured were such as to warrant the laying out of hundreds of millions of dollars for the pipeline and terminal facilities. These reserves helped to reverse a slight drop in US reserves at year-end 1969 from the roughly 4.15 Gt level, 1954–1968, and boost to 5.3 Gt at end-1970 and was 4.3 Gt at 1.1.77 with that 1.3 Gt still intact, for production from the North Slope only started in earnest in 1977.

One use to which reserves estimates are often put is as a basis for prorationing of production. The Organization of Petroleum Exporting Countries was formed in 1960. At the second Arab Petroleum Congress in November 1960, M. J. Sladic, an adviser to Sheikh Tariki, then Saudi Arabia's oil minister, discussed the prorationing of production, based on three factors:

1. The average annual net export achieved during any three year period, preceding prorationing.
2. The proportion of each nation's reserves to total reserves of all net exporting nations.
3. Estimate of net world demand.[538]

Sladic recognized amongst the difficulties of such a proposal the need to standardize estimates of reserves. Although there were other factors which militated against the adoption of prorationing at that time, it is understood that the impossibility of agreeing a formula for the estimation of reserves was one of the rocks on which the proposal foundered. During the production restrictions by the Arab producers in late 1973 and early 1974, straight percentage cuts were made. However the cutback in production due to reduced demand in early 1975 again raised discussion on the possibility of prorationing and probably therefore the problem of reserves estimation may well again be tackled. This problem is in fact easier scientifically, if not politically, in the Middle East than elsewhere because of the simple geology and large size of all fields, many of them giants. As noted earlier this allows more reliable estimates to be made early in the life of a field, with fewer and less significant annual revisions and additions.

Prorationing on proved reserves in North America is discussed in the chapter on production, but an allied and most important feature of proved reserves in both USA and Canada should be mentioned here. In the USA the ownership of underground oil belongs to the landowner. It can therefore be sold; it is a marketable item of value which was early accepted as collateral for loans of money by the bank or as backing for share issues. There was therefore an obvious need by both lender and borrower to estimate the amount of oil which could be recovered. This principle that oil-in-the-ground can mean money-in-the-bank has given rise to many variations in loan mechanisms and has spread to other countries. Many operators borrow money on oil-in-the-ground to finance the costs of development. The main point is that the estimates of proved reserves are obviously conservative. Incidentally, not only will banks have their own estimators, or employ consultants who may also elsewhere be arbitration experts, but a very important subsidiary industry has grown up, particularly in North America, based on this feature. In the context of this discussion on proved reserves, however, the important point is that a conservative attitude to reserve-estimation is bred into all American oilmen who wish to obtain more than one loan, or retain the confidence of the investor and the Stock Exchange regulatory bodies. Indeed a conservative attitude to the publication of reserves estimates is a logical, legitimate and proper attitude not only for industry but also governments. Their responsibilities are such that they should not adopt any other attitude.

In 1973 a US Geological Survey (USGS) paper[104] suggested that of the 50 Gt total world proved reserves, as then estimated, some 40 Gt were on land and 10 Gt offshore. A resources study [376] of 1976/77 suggested that 45% of the world's remaining resources of some 260 Gt, (the mean of the Delphi Poll) lay offshore. A 1979 study[483] noted that small ocean basins formed 8% of the surface of the earth, continental margins 15% and land only 29%, the remaining 48% being the central ocean region with few if any petroleum prospects. In brief the offshore areas with perhaps in 1979 25% of proved reserves have a strong growth capability.

The historical growth of proved reserves, despite the cautions regarding

Table 17
Development of World proved reserves of crude oil
(gigatonnes at beginning of each year)

	1960	1965	1970	1978	1979
1. W. Hemisphere	8.8	9.3	10.2	10.7	12.4
2. Middle East	25.1	29.4	47.1	50.1	50.6
of which Saudi Arabia	(6.8)	(8.2)	(17.6)	(20.5)	(22.7)
3. Africa	1.1	3.2	7.1	8.1	7.9
4. Other E. Hemisphere	1.7	1.8	2.5	6.1	5.7
5. Sub-total	36.7	43.7	66.9	75.0	76.7
6. Centrally planned economies	4.6	4.6	13.7	13.5	12.9
7. Total world	41.4	48.3	80.6	88.5	89.6
8. Annual growth rate	3.2%	10.8%	1.2%	1.3%	
9. Cum. production (Gt)	6.35	9.2	21.9	3.08	
10. Total growth (Gt)	13.25	41.5	29.8	4.18	
11. Aggregate growth (Gt)	41.4	54.65	96.15	125.2	130.1
12. Annual gross growth	5.8%	12%	5.5%	4%	

Sources: Lines 1–8 Parra [486]
Line 9, *BP Statistical Reviews* 1967 & 1978
Lines 10–12, DCI.

their validity noted earlier, are one important factor often used when trying to measure future availability of supplies. F. R. Parra[486] tabulated developments, 1960-1979, as noted above, table 17, (his volumes in barrels being converted to mass at 7.3 US barrels per metric tonne).

In the period 1960-1979 Parra noted that Saudi Arabia provided almost 15 Gt of the 48 Gt world growth in proved reserves. During that same period, however, some 40.5 Gt were produced. Therefore the gross addition to stocks was 88.5 Gt and the gross additions therefore grew at the rates of line 12 of table 17. The big increase in proved reserves, 1965-1970, was due, as Parra states, to 70% from increases in the reserves attributed to Saudi Arabia, Iran and Libya, and the gross growth rate at 4% during 1978 is less than the 5.5% from 1970-1978, but it is unwise to take a very recent year as symptomatic of a real decline because of the chance factor.

The Desprairies study[376] suggested that a reserve growth might be about 4 Gta, (new fields plus re-evaluations) to 1985, but declining to about 3.3–3 Gt by 2000. This totals roughly 65 Gt over the 21 years from 1979 which is, indeed, less than the 88.5 Gt in the 19 years from 1960 to 1979. There is a general concensus that it is unlikely that new discoveries and re-evaluations will allow the proved reserves in 2000 to support, for long, a production of much more than the 1979 figure of around 3 Gt, but the rate of withdrawal by production is also a major factor. Hence, if, during the time to 2000, production stays at or lessens below 3 Gt a and the gross increases are as suggested, then that rate of production could be maintained the longer.

NATURAL GAS LIQUIDS

NGLs are the hydrocarbon components in some gas reservoirs which become liquid at normal surface temperatures and pressures. They are also called well-head gasolines or condensates, and are part of 'wet' gases. Most countries do not differentiate NGLs but include them in their crude oil statistics or lose them in their natural gas data. The United States and a few others, including Mexico, do give separate estimates. Future statistics will probably increasingly indicate NGLs as separate items, but their estimates of reserves and production will depend on those of natural gas. Thus A. A. Meyerhoff[484] in his world estimates used a constant figure of 25 US barrels per million cubic feet of gas and arrived at a figure of 61,839 million barrels for the world's proved and prospective reserves as at 1 January 1978. This total, when converted at 9 barrels = 1 tonne, gives 6.9 Gtoe.

It must be again stressed that these figures must be taken only as indicators, based on the estimates of gas. Thus for Mexico the figure as at 1 January 1978 included in these figures is 1.075 bb of NGL based on his estimate of Mexico's gas reserves of 43 tcf. The 1978 PEMEX Annual Report[487] estimated proved reserves only as 58.9 tcf. Meyerhoff's total for world potential NGL resources was 12.44 Gtoe.

Table 18
World proved and prospective reserves of natural gas liquids 1 January 1978
(gigatonnes oil equivalent)

	Major countries	Regional totals	TOTALS
USA	0.59		
Canada	0.26		
W. Europe		0.39	
W. Hemisphere			1.19
Iran	1.39		
Saudi Arabia	0.24		
Middle East		1.94	
Algeria	0.28		
Nigeria	0.12		
Africa		0.51	
Asia–Pacific		0.34	
USSR	2.39		
E. Europe – USSR		2.43	
China	0.08		
E. Hemisphere			5.69
WORLD			6.88

Source: 'Proved and Ultimate Reserves of Natural Gas and Natural Gas Liquids in the World', A. A. Meyerhoff, PD12(5), *Proc. X WPC*, 1979 Heyden and Son, London

NATURAL GAS

Proved reserves of natural gas are much less firm than those of crude oil. Until comparatively recently only in the USA, Canada and locally in some places, has the potential value of natural gas as an exportable commodity been appreciated and therefore a resource to be measured. Much gas had to be flared, because there was no market within reach, whereas the associated crude oil with which it was necessarily produced could be transported. Very rough estimates of proved reserves and, based on approximations of gas/oil ratios, potential resources have been made in the past. As mentioned in the resources chapter, a distinction should be drawn between the gas associated with crude oil and that which is non-associated. This was attempted in the questionnaire distributed to provide the base for the 1974 WEC Survey. Unfortunately, whether through the facts not being available or misunderstandings, an insufficient number of countries provided data for any differentiation to be made.[105]

A 1974 USGS review of world offshore natural gas resources estimated the remaining recoverable measured (proved) reserves in known fields as 3,835 km^3. The review stressed that there were many ambiguities and inconsistencies in the data available and effectively acknowledged that this figure was merely a best guess. The North Sea was credited with 1,700 km^3, which indicated its importance after only a short history. Indeed the total offshore gas reserves, at only 6% of the world onshore and offshore total in 1974, must be expected to grow into the 1980s.

Table 19 gives four sets of figures. Three are for proved reserves of both associated and non-associated gas, being column 1 by Adams and Kirkby,[75] column 3 by the *Oil & Gas Journal*,[441] and column 4 by the *World Oil Journal*.[487] Column 2 is of proved and prospective reserves of non-associated gas only, as at January 1978, compiled by A. A. Meyerhoff.[484] The Conservation Commission (CoCo) of the World Energy Conference[424] accepted a figure of 57 Gtoe (2,500 exajoules), which would compare with the 68 Gtoe of the three

Table 19
World proved reserves of natural gas
(cubic kilometres)

	Adams and Kirby[1]	Meyerhoff[2]	Oil and Gas Journal[3]	World Oil[4]
Africa	4,960	5,170	5,300	6,120
Far East incl. China	3,450	4,280*	4,100*	2,200
Middle East	20,650	19,760	20,700	17,450
USSR	17,900	24,070	25,800	23,040
Europe excl. USSR	5,820	4,680	4,400	4,170
USA	7,650	5,970	5,700	5,670
Remaining N. America	2,240	3,910	2,300	3,614
South America	1,340	1,030	3,200	2,124
Oceania	1,040	*	*	2,680
Total world	**65,060**	**70,040**	**71,500**	**67,080**

Sources: 1. 'Estimates of World Gas Reserves', T. D. Adams & M. A. Kirkby, IX *WPC Preprint* PD6(1), Table 5, 1975
2. 'Proved and Ultimate Reserves of Natural Gas and Natural Gas Liquids', A. A. Meyerhoff, X *WPC Preprint* PD 12(5), Table III Proved + Prospective Reserves.
3. *Oil & Gas Journal*, 18 December, 1978.
4. *World Oil Journal*, 15 August, 1979.

proved reserves estimates of table 19. The lower CoCo figure cannot be compared exactly because of the use of a different country grouping system. However it appears to show rough agreement for North America, USSR, Europe and Oceania, but is very low for South America and Africa. However, the Commission gave a high figure for undiscovered recoverable reserves at 185 Gtoe, as compared with Meyerhoff's 127 Gtoe for non-associated gas.

The Meyerhoff figures therefore ignore the considerable amounts of associated gas and some of the omissions of known non-associated fields, like that of Bahrain, and inclusion of others as for example Kuwait emphasize that the data available on natural gas have to be treated with care.

M. A. Kirkby, in a personal communication, pointed out that the Adams and Kirkby estimates were made in 1974. Their 1979 estimates would be closer to 83,000 km^3 for proved plus probable reserves of associated and non-associated gas, with 65,000 km^3 being the figure for proved plus probable non-associated natural gas, and so comparable with the 70,000 km^3 of Meyerhoff. They had also amended the estimate of proved plus probable non-associated gas in the USSR to 27,000 km^3 which again is a closer comparison with the Meyerhoff figure of 24,000 km^3 than their 1974 total proved reserves of associated and non-associated gas at 17,900 km^3.

The Middle East vies with the USSR at the top of the list. Table 20 gives a breakdown by country, using the same authorities as the table 19. The major discrepancies are the high reserves attributed by Adams and Kirkby to Abu

Table 20
Middle East proved reserves of natural gas
(cubic kilometres)

	Adams and Kirkby[1]	Meyerhoff[2]	Oil and Gas Journal[3]	World Oil[4]
Iran	11,270	14,160	14,158	10,530
Abu Dhabi	2,830	600	566	530
Saudi Arabia	2,720	2,480	2,659	1,920
Iraq	1,500	790	787	750
Kuwait	850	960	886	1,060
Qatar	800	510	1,132	1,660
Bahrain	290		198	260
Neutral Zone	140		142	540
Dubai	100		45	50
Oman	56	56	57	70
Syria	28	90	42	49
Turkey	14		14	
Sharjah	14			
Others	3	100	2	31
Total Middle East	**20,615**	**19,760**	**20,690**	**17,452**

Sources: 1. 'Estimates of World Gas Reserves', T. D. Adams & M. A. Kirkby, IX *WPC Preprint* PD6(1), Table 5, 1975
2. 'Proved and Ultimate Reserves of Natural Gas and Natural Gas Liquids', A. A. Meyerhoff, X *WPC Preprint* PD 12(5), Table III Proved + Prospective Reserves.
3. *Oil & Gas Journal*, 18 December, 1978.
4. *World Oil Journal*, 15 August, 1979.

Dhabi, Iraq and Dubai and their low figure for Syria with its heavy crude oils associated. The Adams and Kirkby figures were calculated from aggregated individual field estimates, and not by any constant ratio of oil to gas. Hence their figures are preferred as the best indicator of the order of the proved reserves, stressing that this is all any such figures can mean.

The USSR proved reserves of the order of 20,000 km^3 figured also in a 1974 Soviet comment,[107] 'Explored gas reserves for the beginning of 1973 are evaluated as twenty trillion cubic metres in commercial categories.... Their greater part is concentrated in vast deposits. Thus 57 fields (each possessing 30 milliard m^3 and more of gas reserves) have 17.8 trillion m^3. Six out of the above gas-fields are unique from the point of view of their size: Urengoi — 3.873 trillion m^3; Zapolyarnoe — 1.945; Medvezhie — 1.548; Jamburskoe — 3.491; Orenburgskoe — 1.34; Shatbikskoe — 1.08 trillion m^3'. As mentioned earlier, giants of this size, aggregating 60% of the ultimate explored commercial reserves of the country, are an extremely important feature. One reported figure[108] for the inferred reserves of natural gas in the USSR is 113,000 km^3.

The natural gas reserves of the USA are a matter of great concern to the USA because they peaked in 1967 at 8,290 km^3 and since 1947 the reserves/production ratio has continually declined from 40 to roughly 10 at end 1978, with reserves down to about 5,670 km^3. Exxon USA suggested[488] that 1979 existing proved reserves would be supplying in 1990 only 46% of demand, new discoveries 34%, synthetic gas 4% and imports 16%, with total supply having declined from the equivalent of just under 12 mbdoe in 1974 to just under 9 mbdoe. The new discoveries must come from the probable and possible reserves. The Potential Gas Committee[489] has published since 1966 biennial estimates (except in 1974) of potential gas supply beyond the proved reserve estimates of the American Gas Association. They designate three categories: 'probable' which is the equivalent of the USGS 'inferred reserves' and is associated with existing fields; 'possible' supply which is to come from new field discoveries in known productive formations; 'speculative' supply from new pools or fields discovered in formations not previously productive but within a productive geological province or, the most speculative supply, from new field discoveries within a province not previously productive.

These represent the conventional potential supply expected to be recovered by future drilling under conditions of adequate but reasonable prices and normal

Table 21
Estimated potential supply of natural gas within the USA, 31.12.1978
(cubic kilometres)

	Probable	Possible	Speculative
Lower 48 States			
—Onshore	3,993	8,013	5,380
—Offshore	1,274	2,152	2,690
Alaska			
—Onshore	311	538	821
—Offshore	57	595	3,030
Total USA	5,635	11,298	11,921

Source: Potential Gas Committee [489]

improvements in technology. The offshore depth maximum was increased to 1,000 m from the earlier depth of 1,500 feet of water, yet the 1976 probable offshore supply was 6,088 km^3. The obvious highly speculative Arctic offshore provides 25% of the speculative supply.

The estimated probable supply is close to the 6,003 km^3 which the Committee estimated in 1972. Adams and Kirkby[75] argued in 1975 that only a further 566-1,133 km^3 would be proved in all fields discovered before 1973, (excluding the North Slope of Alaska), i.e. only 10%-20% of the Committee's estimate. They considered that the earlier growth rate of 5 times after the initial estimate on discovery no longer applied. They said that as three quarters of the gas being discovered in the USA was not associated with oil it was not subject to the increase in recovery factor commonly occurring with estimates of oil reserves; the quality of initial estimates had improved with improved technology; the annual additions to reserves in oilfields were falling. No counter to these arguments has been seen and yet one notes that in the 1977 figures included in the 1978 Survey of the World Energy Conference only 69.2 km^3 of the 6,117 km^3 of proved recoverable resources were given as non-associated.

The Committee noted that the ultimately recoverable volume of gas in the USA was some 49,894 km^3, made up of the quoted potential supply, cumulative production of 15,151 km^3 and proved reserves of 5,918 km^3, as at 31 December 1978. In Part II of the same report[489] the Potential Gas Agency cited various estimates for gas-in-place and recoverable gas from unconventional sources. In summary the ultimately recoverable gas volumes noted were in km^3

- geopressured aquifers   1,519-7,250
- gaseous coals   5,663-7,080
- tight sands   1,982-8,495
- Devonian shales   48-25,570

The geopressures aquifers were said to be probably nearest becoming a significant supply source. The wide ranges and particularly that of the Devonian shales is illustrative of the high degree of speculation in these figures.

The Canadian Petroleum Association (CPA) reported[490] in 1979 a continuing increase in remaining established reserves (i.e. proved plus probable of this book), of marketable gas, reaching 2,230 km^3 at end 1978, including Mackenzie Delta gas since 1974 and Arctic Islands gas since 1975, though it will be into the 1980s before they will be available to markets. In total some 4,700 km^3 of gas has been discovered in Canada since 1947, being the initial volume-in-place, with some 406 million m^3 of natural gas liquids (ca. 240 Mt) similarly designated.

Pakistan can illustrate two most important features in natural gas proved reserves, the influences of giant fields and the non-combustible element. Adams & Kirkby give a total proved reserves of 460 km^3, with 240 km^3 at Sui, 110 km^3 at Mari and 110 km^3 in a number of smaller fields. Sui and Mari, discovered in 1952 and 1957, are giants and listed as such in 1970,[72] but it was noted that Mari gas had a high carbon dioxide and nitrogen content. Estimates of the non-combustible contents of the gases vary considerably. Figures are published for Sui from 2.7% to 10%. Adams and Kirkby quote 34% for Mari and give figures for the smaller fields of 52%, 71%, 70%, 75%, 19%, using official Pakistan figures. Using 5% for the non-combustible element in Sui, 34% in Mari and 60% in the smaller fields, the effective energy proved reserves become 345 km^3 which, converted at 37,250 TJ/km^3, give 12.8 million TJ (see table 22). The removal of the non-combustibles, however, also requires effort or

Proved Reserves

Table 22
Pakistan's proved reserves of natural gas and their energy content

	Sui field	Mari field	Other fields	Total
1 Proved reserves[1]	240 km³	110 km³	110 km³	460 km³
2 Inert, non-combustible content (see text)	5%	34%	60%	
3 Combustible gas content	228 km³	73 km³	44 km³	345 km³
4 Energy content at 37,250 TJ/km³	8.4 million TJ	2.7 million TJ	1.6 million TJ	12.8 million TJ
5 Combustible gas consumed to remove inert content*	3%	20%	30%	
6 Combustible gas consumed to remove inert content*	6.8 km³	14.6 km³	13.2 km³	
7 Remaining combustible (3 minus 6)	221 km³	58 km³	31 km³	310 km³
8 Energy content	8.2 million TJ	2.2 million TJ	1.1 million TJ	11.5 million TJ
9 Combustible available as marketable LNG§	221 km³	45 km³	†	266 km³
10 Energy content	8.2 million TJ	1.6 million TJ		9.8 million TJ
11 Energy content of proved reserves as if 100% combustible	8.9 million TJ	4.1 million TJ	4.1 million TJ	17.1 million TJ

* Arbitrary judgement on analogy of 15% to remove the 20% inerts in Qatar.
§ On analogy that, including storage and transportation losses, only 50% of the original recoverable gases are marketable, if the inert gases constituted an original 20%.
† Arbitrarily assigned as uneconomic to process.
Source: 1. 'Estimate of world gas reserves', Adams & Kirkby, *WPC*, 1975.

79

energy. Adams and Kirkby, taking the example of gas in the Khuff formation of Qatar, Persian Gulf, state that removal of the 20% of inert, non-combustibles might use up 15%–20% of the reserves in fuels and adding storage and transportation losses, would reduce the amount marketable as liquefied petroleum gases to 50% of the original recoverable reserves.

Pakistan consumes all its natural gas production internally but its energy-effective proved reserves are, from a first figure of 460 km^3, at best 345 km^3. If the removal of the inerts uses up amounts of combustibles roughly proportional to the Qatar example, then the remaining combustible gases would be reduced to 310 km^3. If, on the other hand, Pakistan had been producing only for export, then the removal of inerts, plus storage and transportation losses, might make production from other fields uneconomic, reduce Mari to 45 km^3, and so make a total of 266 km^3 or 9.8 million TJ. In the 1974 WEC Survey, Pakistan's proved reserves were taken at 546 km^3 (1972) and converted at 3,720 MJ/m^3 to 20.35 million TJ. The calculations on the Adams and Kirkby figures suggest that Pakistan, in effective energy terms, really only has 11.5 million TJ and, with the same reserves with the same characteristics, would have only 9.8 million TJ, if exporting the whole production. These are very significant differences and only because the bigger giant, Sui, has only a low inert content are they not greater. Yet 99% of all commentators on Pakistan's reserves would ignore the inerts content.

Western Europe has had well-established natural gas proved reserves in Italy and France for many years. It was, however, the discovery of the Groningen field in the Netherlands and its early recognition as a giant which triggered the exploration of the North Sea and the delineation of the southern gas-bearing basin. These reserves have an enhanced value because they are close to the market of Europe, and merit particular mention.

The southern gas basin, though not exhaustively explored, is well over the peak of exploration for the known targets of the Rotliegendes beds of the Permian, and Bunter of the Trias. Major increases in reserves will come with developments in the northern basin where oil is the prime target. An industry figure in June 1975 was 1,064 km^3 of established non-associated gas in the south and 644 km^3 in the northern basin, where 1,652 km^3 could be the ultimate reserves,[74] with 1,400 km^3 in the south.

A very much more optimistic view is taken by Professor P. R. Odell.[110] He argued that initial estimates of discovered reserves were habitually understated, particularly by major companies. He asserts that this is proved by worldwide evidence and that initial estimates should be multiplied three or four times. In the case of the North Sea, he suggested that a 1973 estimate would be more reasonable, of 2,500 km^3 increasing to 6,400 km^3 by 1980, for the whole European continental shelf with the southern North Sea basin individually increasing from 1,250 km^3 to 2,500 km^3. This optimism is less than that which Professor Odell suggests for crude oil. Time will tell whether such optimism, basically a subjective matter, is justified. Meantime, even planners should not use these speculations, and perhaps even governments, as they learn more about the hazards, will move cautiously.

The North Sea gas reserves are continually being revised with active development. The 1978 offshore reserves, estimated and published by the British Government[422] are given in table 23. There was a fall in reserves in the southern basin from end 1975 due to production with virtually no further exploration or development. Increases in the total reserves of the northern basin more than

Table 23
Estimated offshore reserves of natural gas — Northwest Europe
(cubic kilometres)

A. United Kingdom — remaining in known discoveries at 31 December 1975 & 1978

	Proven 1975	1978	Probable 1975	1978	Possible 1975	1978	Total 1975	1978
Southern Basin								
Fields under production or under contract to BGC	478	387	28	14	42	25	549	426
Other discoveries believed commercial but not yet contracted to BGC	74	51	9	65	40	—	89	116
Other discoveries	0	—	34	31	40	40	74	71
Total Southern basin	552	438	71	110	88	65	711	613
Northern Basin *								
Fields contracted to BGC	190	173	8	—	0	—	198	173
Other significant finds *	28	40	150	166	155	314	333	520
Other gas with oil	45	55	96	50	47	10	188	115
Other possible development						59		59
Total Northern basin	263	268	254	216	202	383	719	867
Grand Total	815	706	325	326	290	448	1,430	1,480

* including reserves found in Liverpool Bay in the Irish Sea, and in Morecambe Bay
(The conversion factor assumes is 1 tcf = 28.317 km³).

Source: 'Development of the oil and gas resources of the UK', 1976 and 1979, Dept. of Energy.

B. Norway — some estimates

	Proved	Ultimate Recoverable
a. R. Bexon, BP, June 1975.	481	990
b. Exploration Consultants, London, March 1975.	368	736
c. Adams & Kirkby, May 1975[75]	776	
d. Ministry of Industry, Norway, *Financial Times*, London, 8 December 1975, p. 24.	796	
e. *Oil & Gas Journal*, 29 December 1975, as at 1 January 1976.	708	
f. *Petroleum Economist*, Vol. XLIII, no. 7, July 1976, reserves as at 1 January 1976.	801.5	
g. Norwegian Petroleum Directorate, 1 12.1.78, Pet. Econ. April 1978	710	
h. Norwegian Oil & Energy Min. *Fin. Times* 21 July 1979	400	

Availability of World Energy Resources

Table 24
Ultimately recoverable natural gas reserves on the UK Continental Shelf
(cubic kilometres)

1. Cumulative Production to end 1978	280
2. Reserves remaining in present discoveries	706—1480
3. Total ultimately recoverable reserves in present discoveries	786—1760
4. Reserves estimated in future discoveries	0—550
5. Ultimately recoverable reserves in the UK Continental Shelf (rounded)	1000—2300

Source: *Development of the oil and gas resources of the United Kingdom, 1979,* Department of Energy, London, July 1979.

compensated, and will affect the future pattern of supply. Although there was a natural shift between categories with greater knowledge, it is noteworthy that for the UK sector of the North Sea the total reserves only increased by 50 km^3 between the estimates for end 1973 and end 1978.

HEAVY OILS

The most recent and accessible information about resources which are nearest to being proved reserves is from Canada. The deposits in the Cold Lake area of Northern Alberta have been mentioned in the resources chapter as having perhaps 12-25 Gt of oil-in-place. Although more fluid than the oil in the Athabasca oil sands, yet at temperature and pressure (55°F and 440 psi) of the shallow reservoirs, only 305 m below surface, the 10-14° API oil will only flow into a well bore at a rate of about one barrel per day. The oil must be heated to reduce its 100,000 centipoise viscosity. Various methods have been investigated. Progress has been slow because real-time experiments in the field are essential. Mungen and Nicholls[78] say, 'advancements in technology will continue and should result in large scale commercial development by the early 1980s'. There is a necessary long lead time and heavy capital expenditure in both production and benefication aspects.

OIL SANDS

As noted in the resources chapter, it is only in the Athabasca oil sands area of northern Alberta that any of the very great resource base and great resources can as yet be considered proved reserves. In this area some 3,900 boreholes have delineated the deposits of oil soaked sands of the McMurray formation, the same formation in which are some of the heavy oils of the Cold Lake area to the south. In the part with less than 46 m overburden, it is estimated[iii] that 50% of the sand is recoverable by open pit mining and will be economic to handle; 92-93% of the oil-in-place will be extracted and produce synthetic crude with 75% of the volume of that oil. These estimates are based on the experience of the Great Canadian Oil Sands plant. This plant started in 1967, reached full operation at 6,700 t/day (50,900 bd) of 'synthetic' crude/calendar day in 1972 and is still producing. On this basis 6.1 Gt of bitumen (at 6.23 b/t)

82

will be recovered by surface mining from the area of over 2,000 km² with less than 46 m overburden and produce 3.6 Gt (at 7.36 b/t) of synthetic crude oil. These are proved reserves in the strict sense.

In addition, sufficient work has been done on the deeper deposits with in-situ recovery methods to make estimates possible. The Alberta Energy Resources Board estimate[112] some 88.6 Gt (at 6.23 b/t for 552,000 b) of bitumen-in-place in deposits with between 46 m and 610 m overburden. Extensive field tests began in 1958 and by 1974 the scale of pilot plant tests had increased to about 14 hectares (35 acres) with twelve patterns of 1 to 1.74 hectares of injection and recovery wells. It has been found that an in-situ combustion method can result in 7% of the oil-in-place being burned, 30% heated to 150°F or higher and about 55% of the heated oil being produced. One estimate is that 50% recovery of the oil-in-place should be feasible.[78] Dr G. W. Govier, chairman of the Alberta Energy Resources Board, argued in 1973 that even if only 20% of the bitumen was recovered, then some 18 Gt would be available. Using the same beneficiation factors as for the mined oil, at 75% by volume of the raw bitumen and at 7.36 barrels/tonne, then presumably the 'synthetic' crude oil available would be 11 Gt. These are probable reserves, not proved reserves. This compares with Canada's proved reserves of conventional crude oil of 1.1 Gt at end 1978. No decision has yet been taken to initiate a commercial project and Mungen and Nicholls[78] consider that from that date 'there will be a construction and start-up period of eight to ten years before full production will be achieved'. Hence one can with some justification say that it will be at least the turn of the century before any large inroad is being made in the potential 34 Gt of 'synthetic crude oil' which the Alberta Energy Board considers Alberta might win from the recoverable reserves of 53 Gt of bitumen from the 160 Gt of bitumen-in-place.

The official 'remaining developed established non-conventional oil reserves' at 31 December 1977 were stated in 1979 by the CPA[490] as being slightly over 237 million m³, (ca. 230 Mt), being the amount which could be recovered by the existing oil sands plant and the plant under construction,[491] with each plant calculated on its rated output for 25 years. The initial volume in-place was given as just over 432 million m³, with the note that 'These estimates in no way detract from published estimates of approximately 50 billion cubic metres, which are estimated to be recoverable from Athabasca-type oil sands by mining or in-situ processes'. In previous years this note used the figure of 300 billion barrels.

SHALE OIL

The 1974 WEC Survey,[114] table IV-2, listed the world's reserves and production data for oil from shale. This data is so variable, some being of oil-in-place, some of recoverable oil and the reporting so incomplete that it is impossible to make a total for the world's proved reserves. The Conservation Commission report[424 A] used 30 Gt as exploitable with current technology out of 400 Gt of 'reserves'. They used the US Bureau of Mines Bulletin No. 650, 1970, to quote the world resources in deposits yielding 10 gallons/ton and over of some 477 Gt with the USA having 314 Gt, Brazil 114 Gt, USSR 16 Gt, Zaire 14 Gt, China 4 Gt. Production is known in the USSR, Brazil and China. Another source[537] gives 27 Gt (190 × 10⁹ barrels) as the shale oil reserves of the world land areas recoverable under 1977 conditions, with 441 Gt as economically or technically submarginal reserves, 286 Gt total reserves being in the USA.

The Piceance basin in Colorado, where much research and pilot plant work has been done, might be said to be the deposit which is closest to exploitation. Elsewhere, its recoverable reserves were noted in the resources chapter as 47×10^9 barrels, though the operational problems were also noted. A 1972 study[113] estimated that only 20×10^9 barrels (2.740 Mt at 7.3 b/t) from shales at least nine metres thick with an oil yield of at least 0.156 l/kg could be considered as prospective before 1985. A 1974 study[54] estimated that a maximum production target by 1985 could not realistically exceed 0.5 million barrels of oil per day (25 Mta at 7.3 b/t), involving a capital expenditure of some $3-5 billion in a new and unfamiliar technology. That gives a measure of the effort required before even the richest oil shale patch of the highly industrially developed USA, with a strong energy self-sufficiency incentive, could be termed true proved reserves available for production in the next few years.

FISSILE ENERGY

The resources of uranium and thorium were discussed in the resources section and now discussion will be concentrated on the proved reserves of uranium; communist countries for which information is not available are ignored. Basically these proved reserves are those reported by the Nuclear Energy Agency and the International Atomic Energy Agency as 'reasonably assured resources', recoverable at a stated cost per kilogramme of uranium oxide.

Bowie[31] illustrated the lack of knowledge of reserves by noting the following figures of official reserve estimates and productions, recoverable at up to $20/kg.

Reserves estimated at	KtU	Production period	KtU
End of 1958	820		
End 1961	526	1958–61	82
End 1972	866	1959–72	279

The Joint Report of the Nuclear Energy Agency (NEA) of the Organization for Economic Cooperation and Development (OECD) and the International Atomic Energy Agency (IAEA), on 'Uranium: Resources Production and Demand', dated December 1977, used two cost ranges, (1) up to $30/lb U_3O_8, ($80/kg uranium metal), and (2) $30–$50/lb oxide, ($80–$130/kg U), for both their reasonably assured resources and their estimated additional resources. Those countries with reserves of the first category of more than 5 ktU are, on data available at 1 January 1977, shown in table 25 with the 1975 figures for the reserves first category of a cost of $15/lb oxide, in the first column.

It will be noted that Canada has retained a price criterion which the NEA/IAEA abandoned for other countries in 1975. The $15/lb category in 1975 was then acknowledged as being in line with the 'reserves' of other industries. The 1977 report has broadened the cost ranges, 'to reflect the increased costs of developing resources. Nevertheless, the categories of this report, carefully used, do maintain a basis of comparison with those of previous reports'.[506] The 1975 costs included mining, milling and extraction costs but not exploration costs.

A 1979 study[505] used higher figures except for the reasonably assured resources at less cost than $30/lb U_3O_8. The differences were mainly due to higher figures for the USA, the four columns reading 530, 773, 708 and 1,158 kt U. These were very significant increases. Canada's reserves were also increased, reading, for the same four columns, 167, 392, 182 and 656 kt U. The main differences in the resources allocated to Australia were that they were credited with 296 kt U in the reasonably assured category up to $50/lb U_3O_8.

The USA reserves of the $15/lb category as at 1 January 1973 were reported as 400 kt U; the 1 January 1975 figure was only 320 kt because, although 52 kt had been found and only 20 kt mined, some 108 kt were relegated to the next higher cost bracket owing to increased production costs. This is a good example that the higher price of competitive fuels does not automatically raise the level at which a resource may be considered economic. One of the aims of The Uranium Institute, founded in 1975 by producers, and later joined by some processors and consumers, is to investigate the world's uranium resources and another aim is to provide a forum for exchange of information. Hence it was appropriate that a private study on the USA productive capacity for uranium should be mentioned at the International Symposium on Uranium Supply and Demand organized by the Institute on 16 June 1976, in London.[287] This study suggested that the 320 kt U reserves plus 124 kt reasonably assured resources gives 444 kt, would be reduced to about 246 kt U, when defined in terms of a market price of $30/lb U_3O_8. ERDA released an increased figure of 492 kt U for the reasonably assured resources up to $30/lb oxide cost category, as at 1 January 1976 and retained the estimated additional resource figure at 819 kt.

Table 25
Uranium resources for the world outside communist areas (WOCA)
on data available at 1.1.1977
(kilotonnes uranium metal)

	(1975) (to $15)	Reasonably Assured (1)	(2)	Est. Additional Resources (1)	(2)
Algeria	28	28	0	50	0
Argentina	9.3	17.8	24	0	0
Australia	243	289	7	44	5
Brazil	9.7	18.2	0	8.2	0
Canada (a)	144 (b)	167	15	392	264
France	37	37	14.8	24.1	20
Gabon (1975)	20	20	0	5	5
India	–	29.8	0	23.7	0
Niger	40	160	0	53	0
Portugal	–	6.8	1.5	0.9	0
South Africa	186	306	42	34	38
Spain	10	6.8	0	8.5	0
USA	320	523	120	838	215
Rest of WOCA	33	40.6	315.7 (c)	28.6	43
Total WOCA	**1,080**	**1,650**	**540**	**1,510**	**590**

1. Cost to $30/lb U_3O_8 = $80/kg U
2. Cost $30 – $50/lb U_3O_8, = $130/kg U
a. Canada reported Reserves as minable at prices up to $104/kg U and other Reasonably Assured Resources as minable at prices between $104/kg U and $156/kg U.
b. price and not cost
c. includes 300 kt U from Sweden's black shales.

M. A. Lieberman has attempted[276] to forecast the remaining uranium resources of the USA, using the methods which M. King Hubbert applied to the
petroleum resources. Lieberman drew the analogies that uranium is also a finite
resource, that a few giant deposits in the USA contain the bulk of its reserves,
and that drilling is the definitive exploratory tool. Lieberman calculated a fall
in discovery rates from an early high in 1955–1956 of 18.6 lb/foot drilled, to
about 2.40 lb/foot drilled in 1971–1973; he deduced that, with cumulative
production at 245 kt, reserves at 248 kt, the undiscovered recoverable reserves
in the category of $8/lb U_3O_8, are only 79 kt. He argued that whilst ERDA
admits a shortage of supply to demand in the 1990s, his figures forecast a shortage in the late 1980s. Hence he recommended a limitation to growth of nuclear
power; exploration for intermediate grade ores; production of western lignites
and recovery of uranium from the ash, at a cost of about $25–45/lb U_3O_8;
development of extraction methods for uranium from the Chattanooga shales
and from sea water, with costs for the latter at $50–$1,000/lb U_3O_8; the im-
portation of foreign uranium, even though he recognises that this would be
difficult if there was a world shortage. These extreme recommendations are
based on the findings of a method which suffers from very grave, contentious
items, particularly if considered in any way other than illustrative of general
principles. Knowledge of the origin and occurrence of uranium and the best
exploration technology are little more than primitive. Hence the location of
exploratory boreholes may be based on immature reasoning, liable to considerable improvement; the apparent precision of measurement and the costing of
the lbs discovered per foot drilled is suspect; and the assumption that every
foot drilled has an equal exploratory value is very debateable. The Lieberman
exercise may be useful, within the USA, to urge ERDA to increase its exploration support, but the method would be most misleading if applied elsewhere in
the world, where knowledge is even more primitive, and exploration has been
less and just as spasmodic and sporadic.

The major change between the 1974 and 1976 OECD Reports is in the
reserves of Australia, which have been doubled by the delineation in the Northern Territory of deposits which were discovered in the 1950s. In February 1976
the Deputy Prime Minister set out the Australian government's policy under four
heads: uranium exploration and development would be mainly tasks for private
industry; permission for mining will depend on proof of sales prospects; foreign
majority shares in uranium companies is not favoured; the enrichment of uranium is desirable and the possibility will be explored. This policy reversed the
strongly nationalistic policy of the previous government, but the control of
mining permission emphasises that they wish to ensure orderly development.
The importance of this intention is enhanced by the latest reserves figures[287b]
being even larger than those in table 25, for they show an increase of 67 kt to
312 kt, though estimated additional resources fell by 38.5 kt to 41.5 kt, as some
deposits were upgraded by better definition. Australia's domestic needs for
uranium are comparatively low and the export potential high. Dr S. H. U. Bowie
considered[267] that 'Australia is of major importance as a country with a capability of producing very large tonnages of low-cost uranium, and this could be the
controlling feature in the world market in the 1980s'.

This opinion was reinforced by the later discoveries of copper and uranium
at Roxby Downs in Southern Australia which the Federal Minister for Trade
and Resources said[492] may contain more uranium than the deposits along the

Alligator River, in the Northern Territories, which have been rated the world's best undeveloped uranium ore bodies.

Dr A. J. A. Roux, President of the South African Atomic Energy Board said[485] on 27 June 1979, that South Africa had 355 kt of uranium resources recoverable at a price less than $30 and 625 kt at less than $50/lb U_3O_8. Spot prices at that time were $42–45/lb. Careful note must be taken of the quoted price/cost whenever estimates of uranium resources are compared. Some 40% of the reserves of Southern Africa occur in the Rossing Mine, near Walvis Bay, in Namibia. Rossing production started in 1976 and reached 5 kta, for which the bulk was already contracted in November 1975, including 7.5 kt in a five-year contract to the UK Atomic Energy Authority.[269]

There is the political problem of the eventual government of Namibia, but the general South African view is that, if the transfer of government is peaceful and to people who have the true interests of the indigenous population at heart, there should be no ill effect on the future exploitation of the Rossing Mine. Much of the gold mined in the Transvaal and Orange Free State carries uranium and the grades tend to move together. Uranium production depends therefore on the gold production, and the higher the gold price the lower the grade which is economically recoverable. This has resulted in less uranium being produced, even by plants working at full capacity, because the rise in gold prices in the past few years has allowed lower grade gold ores to be mined. The benefit of the rise in uranium prices has been largely offset by rises in mining costs, and there has been reluctance to invest in new uranium processing plant. However, there are several sources from which production might be increased. There are considerable tonnages of uranium in the tailings dumps of former gold mines. Many of these are now in built-up areas, particularly of Johannesburg, difficult of access and lacking space to handle the large volumes of materials involved. A scheme on the East Witwatersrand started in 1978 to mill 1.5 Mt each month. Other, more recent tailings dumps have been sited with eventual uranium recovery in mind, and so will be easier to develop. To the nine mines producing uranium currently, six more are expected to be added over the next few years. Improved process equipment is designed and should be installed in the near future. Hence uranium production from the goldfields is expected to increase. Some uranium has been produced since 1971 with the copper mined at Palabora in Northeast Transvaal, and this also is expected to increase. The reserves and production systems of the uranium of South Africa are therefore very different from those of Australia and their potentials are subject to different factors.

Canada published some higher resource and reserve figures, revised from those

Table 26
1975 Estimate of Canada's recoverable and prognosticated uranium resources
(kilotonnes uranium metal)

Mineable	Measured	Indicated	Inferred	Prognosticated
Up to $20/lb U_3O_8	63	82	174	129
$20–$40/lb U_3O_8	11	17	85	217
Reasonably Assured	173			
Estimated Additional			605	

Source: *North American Uranium Resources*, G. M. MacNabb, International Uranium Supply & Demand Symposium, Uranium Institute, 16 June 1976, London.

in the 1976 OECD Report, but used different price categories[287a] and their own Canadian resource nomenclature. Once again therefore direct and precise comparisons, even with data on uranium from other countries are difficult, but the figures are summarised in table 26 for illustration.

The OECD Report of 1976 noted that there were no new discoveries in the 1972-1974 period, even though footage drilled for exploration and delineation had increased from 5.0 km in 1972, to 5.4 km in 1973, to 7.4 km in 1974 and some 10.7 km of drilling was said to be planned for 1975.[275] Drilling footage is accepted by OECD as not the best yardstick; for their 1978 Report, they intended to seek data on the number of dollars invested, (*pace* Hubbert & Lieberman!). The discoveries in Northern Australia can be expected to be followed by discoveries elsewhere, for virtually all the uranium deposits reported are in the low cost categories, because it is naturally on such deposits that attention is focussed in the early stages of exploration of a rich area.

Uranium is widespread in the earth's crust but was concentrated in upper rocks of the crust in Precambrian times. Ninety per cent of presently known ore reserves occur in well-defined provinces in Precambrian masses in Australia, Canada and South Africa or in sediments immediately overlying Precambrian rocks, as in the Colorado-Wyoming province of the USA. There are many other situations of both types in the world; the notes of table 25 indicate variations. The present distribution of reserves, with the USA having 33% of the proved reserves, is a measure of the exploration effort applied rather than of the distribution of natural richness, as one can deduce more plausibly for the concentration of crude oil richness in the Middle East. A further example is that the position of both France and Niger is due to the efforts of the French to secure sources of this strategic element under their own control.

PROVED HYDRAULIC RESERVES

Hydraulic energy reserves are those available from existing plant or from authorized plans for the immediate future. They are small in total in comparison with other energy resource reserves. They are often established with other non-energy parallel objectives, particularly for irrigation, navigational improvement and flood control. They are, however, of great local importance. Large schemes are often on the drawing-boards for years, hence in table 27 there have been extracted only the figures for plant currently operating and under construction which are the 'hardest' figures and closest in concept to the proved reserves of other energy sources. Unfortunately there are gaps in the information in the 1978 WEC Survey, and no world totals can be given. Some points which warrant mention in examining the table are:-

1. In order of operating annual generation, on the criterion of the capacity useable under annual average flow conditions, the countries rank as: USA, USSR, Canada, Japan, Norway, Turkey, Brazil, Sweden, France, Italy, India (only G_{95} figure available).
2. The variations in differences between G_{95} and G_{av} data are most marked.
3. Capacities in megawatts have been given where generating data are wholly or partially lacking, for this highlights the differences in reporting.
4. Some of these data are markedly different from those reported in the 1974 WEC Survey, e.g. India was reported as having a probable annual generating capacity G_{95} of 137,180 TJ. Once again one is made aware of the difficulties

in resource assessment, and that the apparent precision of the figures in table 27 should be viewed with caution.

Table 27
Hydraulic energy
(capacity in megawatts; probable annual generation in terajoules)

	Operating			Under construction		
	Capacity	G_{95}	G_{Av}	Capacity	G_{95}	G_{Av}
Ghana		29,536	9,515	—	—	—
Mozambique		1,134	U		64,801	U
Zaire		10,292	U		2,081	U
Japan		44,800	85,234		1,320	3,495
India		35,040	U		27,348	U
Turkey		8,073	9,027		9,643	14,192
Iran		9,439	12,953		12,539	14,832
FR Germany	3,776		—	550	—	—
France		47,040	60,300		2,200	2,800
Austria (1974)		64,440	71,640		23,760	26,442
Italy		139,680	160,560		2,880	3,600
Spain	12,604		35,313	1,061		1,134
Portugal		5,934	8,823		1,750	2,605
UK		3,490	4,149		N	N
Sweden	13,200		U	1,600		U
Finland		7,800	11,600		0	0
Norway		75,000	83,000		13,000	15,000
Romania		18,000	27,720		5,400	8,640
USSR	43,130			13,500		
USA	57,035		271,124	8,200		16,736
Canada	39,475		213,049	17,522		100,975
Mexico		12,470	17,000		5,330	7,300
Brazil		79,289	99,577		129,772	156,466
Venezuela		30,895	3,456		10,544	15,401
Argentina	1,720		5,806	5,872		16,267
Chile		18,000	20,376		13,644	18,500
Australia		12,000	13,000	1,660		U
New Zealand		14,863	17,486		3,711	4,366

G_{95} = Maximum generating capacity usable 95% of the year.
G_{Av} = Capacity usable under average annual flow conditions.

Source: *Survey of Energy Resources, 1978,* World Energy Conference, London.[422]

The WEC Conservation Commission Report[424a] used other standards and estimated that the world total installed and installable hydraulic capability, based on the 1976 WEC Survey, in terawatts (TW) in terms of generating capacity at a 50% factor was made up as follows:

Developing Countries   1.06 TW (48%) of which approx. 7% was
installed = 74 GW

OECD Countries   0.52 TW (24%) of which approx. 46% was
installed = 240 GW

China, N. Korea, Vietnam   0.36 TW (26%) of which approx. 3% was
installed = 11 GW

USSR, E. Europe   0.27 TW (12%) of which approx. 12% was
installed = 30 GW

TOTAL WORLD   2.2 TW (100%) of which approx. 16% was
installed = 355 GW

The proportion of installed and currently operating capacity emphasises the great possibilities for further development in the developing countries and the Communist countries, in contrast to the current high figure in the OECD countries of installed capacity and less scope for expansion.

5
Production

Production is obtained from proved reserves but the determinants of the scale of production in the industry and country components of the world total are many and complex with some unique to the individual component.

Cumulative production to last year's end is fact. This year's production is part fact, part surmise. Next year's production is all surmise and as one moves to the future one moves further from fact into fiction.

None of the figures, fact or fiction, stand in a vacuum; they are interrelated and interdependent on other figures. The scale of production of any one resource has a significance which varies in the context of other resources, of its and their economics and of the stated area of local district or country, region or world.

A simple table or graph of annual production of a resource in some unit, of a country, is therefore only a start and not the end of the discussion. Information is also required of the other energy resources and the relationship of each to past annual productions, rates of growth or decline, to proved reserves, individual and aggregate, to the past and present energy mix, to imports, exports (and bunkers for coal and oil), to gross national product, to use for electricity generation or as direct heat, etc – all expressed in meaningful geographic and/or economic/political aggregations.

The UN Statistical Papers, J Series, are becoming increasingly comprehensive and go a long way towards supplying annually much of the basic information, at

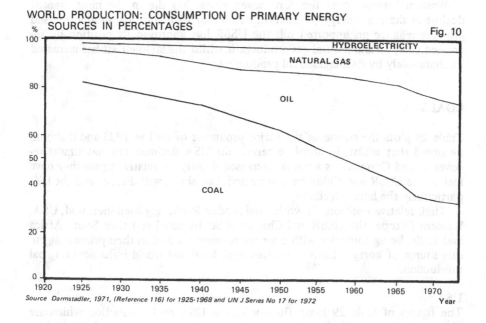

WORLD PRODUCTION: CONSUMPTION OF PRIMARY ENERGY
SOURCES IN PERCENTAGES   Fig. 10

Source Darmstadler, 1971, (Reference 116) for 1925-1968 and UN J Series No 17 for 1972

Table 28
Energy production by major regions
(gigatonnes of coal equivalent, rounded)

	1925	%	1967	%	1972	%	1976	%
USA	0.77	49	1.97	32	2.07	27	2.05	23
W Europe	0.53	34	0.54	9	0.59	8	0.68	7.6
USSR	0.02	1.7	0.12	18	1.3	17	1.67	18.7
Middle East	0.01	0.5	0.76	12	1.2	16	1.69	19
North Africa	—	—	0.2	3	0.24	3	0.27	3
Remainder	0.23	15	1.55	26	2.14	28	2.59	20
World	1.57	100	6.15	100	7.57	100	8.95	100

Sources: 1925 & 1967 Darmstadter et al. 1971 Table 9 p. 27.
1972 & 1976 UN Series J.

least since 1949, with some data available for 1929 and 1937. J. Darmstadter, 1971,[116] compiled a mass of information, primarily on coal, petroleum and hydroelectricity, basically for 1925–1965, winnowed the mass and related the results to other economic variables. 1925 can be a useful benchmark because it is in the second quarter of this century that the big changes have occurred with the explosion in total growth, the decline in importance of coal and the rise of petroleum. The Industrial Revolution of the 18th/19th century was founded on coal; the Mobility and Comfort Revolution of the mid-20th century was founded on oil.

The graph, figure 10, illustrates the changing role of the primary energy sources from 1925 to 1972. Table 28, indicates the change in production source.

Western Europe, over the forty-seven years, has shown the most marked decline in share of energy production because of its loss in coal production and growing reliance on imported oil; the USSR has improved its position by increased coal, oil and natural gas production, whilst the Middle East has increased its share solely by its increased oil production.

COALS

Table 29 plots the course of the major producers of coal in 1925 and it should be noted that whilst, in absolute terms, the USA declined but has almost recovered, and Germany as a whole increased slightly, in relative terms they have lost as the USSR and China have increased their share, with France and the UK, particularly the latter, declining.

Their relative positions in world coal production having been indicated, USA, Western Europe, the USSR and China will be discussed and then South Africa and India, being countries with a strong reliance on coal as their prime indigenous source of energy. Each illustrates both local and world influences on coal production.

USA

The figures of table 29 mask fluctuations in USA coal production which are illustrated in figure 11. The figures in the graph do not agree with those used by

Table 29
World coal production — seven countries
(megatonnes coal equivalent, rounded)

	1925			1965			1977		
	Total	Hard	% *	Total	Hard	%	Total	Hard	%
USA	556	555	43	505	504	22	623	608	22
Germany (E & W)	189	146	15						
West Germany				168	135	7	121	85	4
East Germany				78	2	3	71	1	3
France	48	47	4	53	51	2	24	22	1
UK	247	247	19	190	190	8	124	124	4
USSR	16	13	1	449	362	20	597	500	21
China	23	23	2	305	304	13	500	500	18
Remainder	201	175	16	545	424	24	798	635	28
World	1,280	1,206	100	2,293	1,972	100	2,858	2,475	100

Notes: France peaked at 58 Mt in 1957, West Germany peaked at 182 Mt in 1957, East Germany peaked at 80 Mt in 1964
UK declined from 262 Mt in 1929 and in 1933 had a production of 211 Mt

Sources: Darmstädter et al. 1971, Part III, Table II & Profile 21
1972 & 1976 UN Series J.

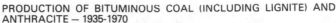

PRODUCTION OF BITUMINOUS COAL (INCLUDING LIGNITE) AND ANTHRACITE — 1935-1970   Fig. 11

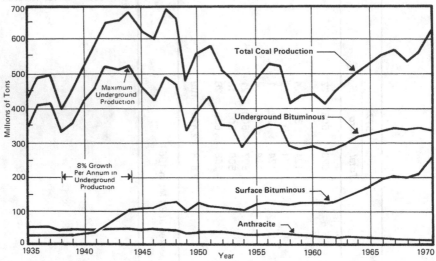

Source. U.S. Energy Outlook Coal Availability. Nat. Petr. Council p25

either Darmstadter or the UN J Series but make the point that as complete a series as possible of figures must be used. The table suggests a decline from 1925 to 1965 and an increase to 1972. The graph tells of a rise from 1935 to twin peaks in 1943 and 1947, a bumpy decline to 1961 and a smoother rise to 1972. The graph also shows that much of the increase has come from open-cast mining in the west — a rise from 31.4% in 1960 to 43% in 1970.[117] Indeed some 33.4% of USA proved reserves are said to be recoverable by surface, open-cast mining.

The 1974 WEC Survey[118] gave 182 Gt for total known recoverable reserves, all ranks, for the USA, with total resources of 2.9 Tt. The National Petroleum Council (NPC) 1973 study[61] uses 150 Gt of recoverable total resources. Noting the differences in total resources and recoverable reserves merely as examples of problems of measurement and definition, the NPC study (p. 7) is interesting as listing the following factors as affecting future coal supply in the USA:

— Developments in improved mining technology must be substantially accelerated to offset the severe impact of the Coal Mine Health and Safety Act of 1969 on production capacity.
— A programme for rapid development of manpower, both mine workers and mining engineers, must be rigorously pursued.
— The number of hopper rail cars and utilization efficiency must be increased and some bottlenecks on river systems and at coal export ports must be removed.

The NPC study splits future production by use into supply of conventional markets from existing conventional mines, and into supply through synthetic gas and liquids, with the latter predominantly coming from the cheaper open-cast coals of the Rocky Mountain area. Using various parameters and cases, the future coal supplies are estimated as varying:

a. from presently used reserves for domestic conventional supply:
   1975, 665-603 Mt; **1980**, 851-705 Mt; **1985**, 1,093-819 Mt
   noting, however, that if there were a complete ban on all surface mining there would be minima of 351 Mt, 413 Mt and 480 Mt.
b. annual requirements for synthetic gas plants in 1985 from 23.2-46.6 Mt.
c. annual requirements for synthetic liquids in 1985 from 106.8 Mt to nil.
This NPC study therefore offers a wide choice for future production.

A study by a task force of the National Academy of Engineering[54] takes a more pragmatic approach. It suggests that the 1973 output of 600 Mt could be doubled to at least 1,200 Mt by 1985. Some 700 Mt would be used for direct firing for electricity generation, 310 Mt for producing synthetic gas and oil and the remainder for industrial uses and export. Western surface mining would be providing 560 Mt, eastern surface mining maintaining its 1973 rate and eastern underground mining expanding to its 1940 capacity, which from fig. 4 of the NPC study would be about 410 Mt. They estimate that, by 1985, twenty syngas plants could be producing an aggregate of 5×10^9 cu.ft/day = 0.8 million barrels/day which is of the same order of production as case 1 of the NPC study with thirty plants producing 6.8×10^9 cu.ft/day. These projections of the possible production of coal for synthetic gas are an important indication of coal industry thinking.

A summary in the Engineering Academy study estimates that, at a cost of US $ 16-22,000 million, coal-based synthetic fuel production in 1985, using western coal (8,500 Btu/lb, 4076 kJ/kg) could be as in table 30. They note that methane from coal technology is ready for immediate application. Methanol technology is almost ready, but uses on a commercial scale have to be developed; liquefaction by hydrogenation requires more research and development effort, but could be significant in the late 1980s and coal processing for power use has further possibilities. However it must be appreciated that for an input of 310 Mt coal there is an output of energy equivalent only to 1.7 mbd when their estimate of 1985 domestic crude oil production is 12.5 mbd and natural gas the equivalent of 14.15 mbd. In brief, synthetic gas and oil production are for the longer term future than the next ten years. Furthermore, it is important to consider the government actions which the task force considered were necessary to ensure the starting goals of table 30.

These were:
(a) For the expansion of the domestic coal mining industry
   1. market forces: the price must be set to encourage private investments,
   2. stability: a long-term US energy policy must be promulgated,
   3. prospects: Western federal lands opened for surface mining in 1974 to allow for lead times (2-4 years) for exploitation,
   4. standards: lessening of some standards and criteria for design and mine operation.
(b) For expansion of transportation system to allow 800 Mta more coal to be carried in 1985 than in 1974
   1. financial assistance to improve main-line railway tracks and signalling systems and highway crossings,
   2. improvement of waterways,
   3. expediting of rights of way for new railways tracks and pipelines,
(c) For establishment of domestic coal-based synthetic fuels industry
   1. approval of gasification projects,
   2. sponsoring of liquefaction plants,

3. safeguarding of proprietary data, patent rights and expertise when acquired during the research and development period.

Table 30
Possible coal-based synthetic fuel production, USA, 1985

	Input, Mt coal	Production		New plants required
		TJ	mbd crude oil equivalent	
High Btu gas	150	1.7 TJ	0.8	20 x 250M cfd methane
Methanol	60	0.64 TJ	0.3	8 x 40,000 barrel / d.
Syncrude	50	0.64 TJ	0.3	10 x 30,000 barrel / d.
Medium Btu gas	50	0.64 TJ	0.3	

Source: *US Energy Prospects: an Engineering viewpoint,* Task Force, Ch. K. W. Davis, Nat. Acad, Eng., 1974.

These ten points are from a recent study by a scientific/technical, multi-industry task force in the USA with its pride in free enterprise. There could scarcely be a more striking illustration that other factors than reserves capacity or production technology will determine the production, the resource availability in the future. One of these other factors is the influence of the struggle to conserve high environmental standards with implications on the industry structure.

The US National Coal Association in 1974[88] argued that the Clean Air Act of 1970 had 'gone beyond the point where the benefits derived are worth the cost'. As noted earlier, a recent US Bureau of Mines report is quoted[88] as suggesting a shortfall in coal supply in 1980 of 202 Mt in the utility sector alone, equal to 38% of the coal utility demand. They point out that if this is not provided then major shifts in consumption pattern will be necessary. The shifts to oil, which are already occurring, lead to development of an oil facilities infrastructure, working inland from the coast. This development results, in their opinion, in an inexorable inertia to increase oil consumption. An alternative is low sulphur coal from both the Southern Appalachians and the far west. Both areas have increased their supplies to utilities but mining costs are high in the east and transportation costs high from the west. Natural gas is the third substitute and 20% of its use is as a boiler fuel. However, growing scarcity gives no hope for long-term use. Import of foreign low sulphur coal is the fourth substitute, as mentioned earlier. The authors showed most concern for the erosion of the infrastructure of the railway system and, particularly, of manpower, etc, and the lack of investment in the coal industry. They say, 'coal mining investment is recoverable only over the long term and the inherent risks from internal and external forces, to which have been added environment regulations, have made investors very wary, and a target of 1.4 Gt in 1985 seems impossible.' These arguments and opinions reinforce the need for government action to attain a 1.2 Gt target for 1985 previously quoted from the US National Academy of Engineering report.

Table 31
USA coal production and exports
(megatonnes)

	1977	1978	1979 est	1983 likely	1983 range
Production					
Eastern Mines	478	425	476	528	500–500
Western Mines	149	168	195	277	236–303
Total	627	593	671	805	736–853
Exports					
To Canada	16	14	15	17	14–19
To elsewhere	33	22	34	36	27–39

Source: Energy Review; US Coal, P. Cheesewright, *Financial Times*, 12 Oct. 1979, London.

The difficulties facing the American coal industry and the need for government help were no nearer being resolved in late 1979, despite the years of growing dependence on oil imports. The USA hard coal productions in megatonnes coal equivalent for 1972-1977[421] were:

1972	1973	1974	1975	1976	1977
537	530	539	575	586	608

Exports in 1973 were about 50 Mt, having fallen from 67 Mt in 1970 but rose again[298] to almost 65 Mt in 1975 but were estimated[430] in late 1978 to be only 49 Mt in 1977. A study by the Bankers Trust Company projected a steady increase in demand for coal to 998 Mt in 1985 and 1,360 Mt in 1990. These figures compare with those of earlier studies which forecast[54] a production of about 1,200 Mt by 1985. Other suggestions were that 1977 production will only double to 1,200 Mt by the end of the century.[264] In May 1979 estimates of 700 Mt for 1979 were made by the chairman of one coal company and the Carter Administration were said to have lowered their target for 1985 from the original 1,200 Mt to 900-1,000 Mt. The coal industry was reported to consider the lower figure attainable, if controls were changed. Another figure suggested for 1985 was 842 Mt.[422] The Carter Plan of July 1979 included a new Energy Mobilization Board to expedite critical energy facilities, which might be assumed to be concerned with excessive regulatory controls.

In brief, the US coal industry entered the 1980s with a great many estimates of future production. The actual production will depend less on new technology and more on the differential between the cost of coal, including the cost of ensuring environmental acceptance in all phases from mining to end-use, and the cost of oil and gas, if decontrolled to rise to world prices. Western, low-sulphur coal production was expected to rise more rapidly than that for eastern coal. The long haul transportation problem for the west had been lessened by power plants being sited in the south and west, but the availability of land for open-cast mining was still a constraint in 1980, nevertheless one estimate[505] was for 400-500 Mt in 1985, with another, more optimistic one[504] by the US National Coal Association being given in table 31.

Western Europe

On the figures of table 29, it is the UK coal production which has suffered the greatest decline, for the aggregation of the Federal and Democratic Republics of Germany show increases. The realistic reduction of working proved reserves of coal in the UK was mentioned in chapter 4. In the 1960–1970 decade pits were closed at the rate of forty per year, the number of faces in the remaining pits was reduced fourfold, output was reduced by 25% but productivity rose 50%. This was effected by development of new coal-faces, transportation and coal-handling equipment.[119] Even so, particularly in the underground sector, new mines must be continually opened for each mine has a finite life and the existing collieries in the UK have an average life of eighty years. Furthermore, there is a lead time of about ten years from location to production. The recent addition to proved reserves of about one gigatonne in Yorkshire, mentioned earlier, should be producing ten megatonnes by the early 1980s and so continue for many years. Other new areas should be producing possibly 30 Mt from the middle 1980s. In Britain, therefore, one anticipates a reversal of the closure of old pits of the 1960s, the maintenance of present production from existing pits and the additions of new pits. The total should be of the order of 150 Mta.[101] It is interesting to remember than in the 1960s, when the coal industry was fighting for survival, the absolute minimum of 200 Mt was spoken of as a last ditch production figure before catastrophe. One of a number of forecasts made by the UK Department of Energy, used in evidence to a Select Committee of Parliament, in March, 1975, and taken from the Secret List in April was published in June 1979. The possibility suggested was that from a low of 115 Mt in 1974 production would rise to 130 Mt in 1980, 135 Mt in 1985 and reach 150 Mt by 1990. Later a target was set of 170 Mt in 2000.[417]

The USSR

Table 29 also shows the great increase in coal production as the Soviet Union moved into place as a major industrial power. Production rose from 1925 to 1972, relatively from 1.3% to 20% of the world's total and, absolutely, from 16 to 480 Mta. In contrast, a 1974 Soviet report[108] used figures of 655.2 Mt for 1972 and 668 Mt for 1973. Between 1960 and 1970 this report notes the relative decline of coal production from 53.5% to 33.7% of total primary energy production, with the hydrocarbon share rising from 37.9% to 60.3%. In that same period the increase in production was 30% but 400% from 1940 to 1970. It might be noted that the 1974 WEC Survey[120] gave a solid fuel production of 694 Mt for 1971, whereas the UN official figure[121] gives 461 Mt for coal and lignite, presumably excluding peat which only had 57 Mt in the former reference. There can be major discrepancies in reporting of production. However, the distribution of the producing centres, as well as the total production is important in a country of the size of the USSR. Between 1940 and 1972[122] the share of coal production from the regions east of the Urals rose from 28.7% to 45.6% of the total energy production, but the share of the hydrocarbons rose more spectacularly, being for oil 6.3% to 25.3% and for natural gas from 0.5% to 33.5%. This same source estimates that 80% of the geological gas reserves of fuel are in the eastern regions and that future production is expected increasingly to reflect this distribution. Coal is expected to maintain its share of energy production after 1980. The increase will come from the opencast, high productivity, Asian fields, with some processing of the coal to effect competitive value nationwide, either as a primary source or a secondary source by conversion to electricity. However, there are major scientific, technological and economic problems which will need to be solved to provide the infrastructure in these remote areas with a harsh climate, before mass flows of fuel and energy resources of the equivalent of over 1,000 million tonnes of coal equivalent are moved annually from 2,000 to 3,000 km.

The USSR coal production objective in the tenth Five Year Plan of December 1975, is for 800 Mt ± 10 Mt, in 1980, from a 1975 production of 700 Mt,[259] and, again, there is talk of doubling by the end of the century, to about 1,400 Mt.[264] There are no apparent changes in policy which would suggest that coal will be used as a major export and source of hard currency, even though the 1979 adverse trade balance with the West would encourage consideration of exports of any practical kind.

Poland
Poland is the major coal producer in Eastern Europe and in 1974 produced 162 Mt coal with almost 14 Mtce as lignite (39.8 Mt original tonnage).[262] Almost all the coal came from the Upper Silesia Basin with some coking coal from Lower Silesia, but the main future hope is the recently discovered Lublin Basin in the east, from which the first production is expected in 1980. Lignite deposits, at depths between 20 m and 200 m, occur in several regions, but are currently mined in the Lower Silesian and Poznan Basins, and they essentially supply power stations, with little export.[297] Exports of hard coal in 1973 to both Western Europe and Eastern Europe/USSR, were 18 Mt nett each; in 1976 total exports were 40 Mt, estimated.[421] Total production in 1975 was 170 Mt, but the 1976–1980 plan calls for a production by 1980 of 200–210 Mt of deep-mined coal; the long term, more tentative objective is for 250 Mt in 2000. This plan is ambitious, more particularly in the short term. The total share of coal in the total energy consumption is now 80%, but was planned to drop to 70.5% in 1980 and to 60% in 2000. However, industrial expansion has run Poland into big deficits in foreign trade balances and has necessitated curtailment of planned imports of oil, gas and heavy machinery. There is, therefore, a probability of a slowdown in industrial activity and a greater reliance on indigenous coal supplies. On the other hand, Poland and FR Germany were reported[303] to be planning new economic agreements aimed at increasing exports in order to redress the Polish adverse position with FR Germany. The deals being discussed in mid-1976 were for two coal gasification plants, a coal-based chemical plant, equipment for a new large coal mining operation in Upper Silesia, and a copper export scheme. Poland is also suffering from the increase of its oil imports from the Soviet Union; in 1973 10.5 Mt of oil and 1.7 km^3 of natural gas were imported from USSR.[262] The Soviets had raised their export prices in 1975 to 33 roubles/t for Comecon countries, but to 60 roubles/t to other customers; Poland still had a favourable trade balance with the USSR in 1976.[298]

The People's Republic of China
The 1974 WEC Survey[120] gives China's 1971 coal production as 410 Mt. The UN official figure[121] is 390 Mt for 1971 and 400 Mt for 1972. Another source[63] gives an estimated total mine capacity of only 330 Mt with an output in 1970 of 244 Mt, being an 86% rate of utilization. Some 300 plus Mt is given as the capacity in modern mines and 30 Mt as the indigenous pit capacity. It is said that production was 425 in Mt in 1960, but led to excessive wear and tear as normal capacity was over-stretched. No official statistics have been published since 1960

but some observers have estimated that production fell to 220-300 Mt in the mid-1960s. The same source reports that 'according to one estimate, seven large coal producing centres in north and east China, each producing 10–20 Mta, supplied 100 Mt in 1969 (30% of the total). A further 20–25 centres each produced 1–10 Mt that year; small and medium mines, producing less than 1 Mt made up the remainder'. During the third Five Year Plan it is reported that two hundred and sixty-seven new mines were constructed in 1970 with an aggregate capacity of 43.75 Mta. The paucity of hard facts make any analysis difficult. However, as stressed earlier in chapter 2, production is being encouraged in the south to obviate long hauls from the north and east, small local mines are locally important, low grade coals can thus become economic, and there is emphasis on industrial use. Production has fluctuated with political events,[433] but UN statistics[421] show a steady rise from 400 Mtce in 1972 to 500 Mtce of coal and lignite in 1977. The 450 Mt for 1974 was in line with a report[361] of November 1976, with coal providing 63% of total primary energy needs in 1974. The constraints facing the industry are the low quality of much of the coal, limited rail transportation and high exploration and production costs. These are said to be higher than for oil, but this is perhaps a Western-type judgement which does not properly assess the position in China. An Australian Coal Mission to China in 1976, as reported in June 1977, noted[362] that one third of production is in small mines for local consumption, 10% is from open-cast mines, and about half the underground production is by mechanized mining. The overall impression of the Australian mission was that there was every probability that the Chinese coal industry will expand, but that its export capability seem unlikely to be significant before 1985. Even then, they suggested, trade might be confined to steam coal and high-volatile coking coal to nearby countries, Japan, Korea and Vietnam. There seems little to challenge this view. It is difficult to see how any massive export capability could be developed before the early 2000s, and then only if that was Government policy, which would be more obvious in the late 1980s than in the late 1970s. Nevertheless coal must be a very important, if not the most important primary domestic energy source on which industrial expansion must be based. This was appreciated by Chairman Hua, who in February 1978 set a target[430] of doubling production by 1987, say to 1 Gt, and doubling again by 2000 to say 2 Gt. The plan was to tap the potential of existing mines, develop and update the 20,000 small mines, construct more large coal mining centres and complexes, increase productivity and complete the mechanization of all mines, (presumably except the very small ones) within ten years. Mechanization could be achieved most quickly with imported equipment and technology. Coal industry missions, particularly from the USA, FR Germany and the UK were encouraged during 1978 to investigate and discuss cooperation. By the beginning of 1979, however, the euphoria of early 1978 seemed to be lessening as the problems of financing simultaneously all the great plans for agriculture, industry, defence, and science and technology became apparent. Thus in February the Chinese proposal to pay for British mining equipment and mining consultancy services with coal from the mines which these would develop, slowed down the negotiations with the British National Coal Board. Some equipment was bought and 60 Mt for sale on the 1985 world market was still a target in February 1979.[432] In June 1979, however, Chairman Hua slowed down the general expansion plans to a more steady pace, somewhat vaguely defined, but stressed that China's economy was flourishing and gave a figure of 618 Mt for 1978 coal production.[434]

South Africa

South Africa is overwhelmingly dependent on coal for its energy. Coal is almost its sole primary energy source for electricity generation, railway transportation and for the metallurgical industries. In 1972 coal provided 76.1% of the gross energy needs of the country, oil 23.3% and hydro-power 0.6%. D. J. Kotze, in a neat energy balance exercise in 1974[123] forecast coal use in 1980 and 2000 from a 1972 base. The figures are simplified and summarized in table 32.

Table 32
Solid fuel production and use in South Africa
(megatonnes)

	1972			1980			2000		
	Coal	Coke	Gas	Coal	Coke	Gas	Coal	Coke	Gas
Direct uses	16.22	3.86	64	23.25	6.95	64	18.77	18.27	64
Conversion uses									
Electricity generation	32.8			38.6			94.9		
Coal to oil	1.7			1.7			1.7		
Gas & coke	6.5			12.7			32.0		
Mines power	1.3			1.6			3.0		
Total production	58.4			77.8			150.4		

Source: 'A forecast of energy supply and demand in South Africa', D. J. Kotze,
IX WEC Preprint, 1.2.—14, 1974

This table of end-uses is included here to emphasize that production, past, present and future, has, does and will depend on sector demand and not just on total demand. It will be noted that the coal to oil conversion has been kept static. No expansion of the SASOL plant has been allowed, though the study was published after the quadrupling of oil prices in 1973/4. However, more recently, plans for a new plant of double the size have been announced. (The coal to oil process is discussed later in the non-conventional hydrocarbon section.) The drop in the direct use of coal after 1980 is due to the elimination of sale of coal as bunkers, which were 8 Mt in the 1980 balance.

In considering whether South Africa has adequate coal reserves, Kotze takes 16,431 Mt as the saleable coal reserves[124] whereas the 1974 WEC Survey reports 10,584 Mt. Kotze uses a bell-shaped forecast curve and concludes that, 'for the next approximately sixty years, the availability of coal will place no restriction on consumption of the mineral.'

Southern Africa has very large coal deposits, which are imperfectly known or assessed. In South Africa, whilst mining companies are required to report certain data to the Government, there has been no systematic collation or publication of all the data so acquired. The Coal Advisory Board reported in 1969, (the van Rensburg Report), that reserves were not as great as had been thought, that recovery was inadequate, and that there was urgent need to conserve coal, as it was their only indigenous energy resource. The Petrick Commission was set up to investigate the situation, and, after considerable delay, a report was published in 1976.[304] From this detailed report there are three basic and important figures, from the resource standpoint, but they do not provide as clear a picture nor guidance as to potential production, as it would seem might have been possible if the authors had not been apparently inhibited by past estimates and procedures. The qualifications and assumptions surrounding the figures are vital to their understanding – this is a further example of the care with which resource figures must be considered, and loose comparisons avoided. The report confines itself to areas where coal is indicated, and does not extrapolate into areas where coals may be likely but have not been seen. The first important figure is that of 92 Gt (91,767 Mt), of raw bituminous coal-in-place, which it is hoped could be mined after taking into account certain restrictions. A widespread problem in Southern Africa is that caused by dolerite sills and dykes, which are of post-Karoo age, younger than the coals, and have penetrated the coal seams, and in some places are interstratified with or overlie the seams; having come up from below as hot molten liquid, the dolerite has often burned or devolatilized the coal, rendering it unmarketable. A geological loss factor has therefore been deducted before arriving at the 92 Gt figure, for even the Electricity Supply Commission (ESCOM), which takes normal run-of-the-mine coals for its power stations, has a volatility criterion in its contracts to avoid burnt coal. Most coals at a depth below 300 m and all coals below 500 m are considered unmineable and have been excluded. Thin seams of less than 0.7 m for good quality and 1.2 m thickness for poorer quality coals are excluded. In thick seams only 6 m of coal is considered as mineable, being the maximum possible thickness which can be mined under present operational conditions. Under such restrictions but only down to 300 m for all ranks of coal, the mineable amount of bituminous coals is reduced from 92 Gt to 81 Gt. There are, therefore, physical, technological and economic factors peculiar to South Africa which are involved in the selection of the coals which total to 92 Gt, and any resemblance between that resource figure and, for instance, the 99 Gt reported as the known coal-in-place in the UK is purely coincidental. Within the 92 Gt of mineable raw bituminous coal-in-place in South Africa, the Petrick Report makes two separate and distinct calculations or collations from the data which were collected. The first calculation is of 25 Gt (24,915 Mt), as the amount which would be extractable under current conditions, if all the coal was mined by underground methods, using the bord-and-pillar (room-and-pillar) technique, with recoveries based on a formula which allows for depth and thickness of seam, and giving full recovery only to the higher grades of coal when at or near maximum depth. An estimate is then made that of these 25 Gt some 21 Gt would be marketable; this would seem to be reasonable, for, on the estimates to 2000, ESCOM might be expected to take 60% as run-of-the-mine coal, and the losses would be incurred only in the washing, grading etc. of the 40% remaining. A second and separate calculation of the amount which would be extractable down to 100 m from surface by open cast methods gave figures of 11 Gt, using the ratio of rock thickness to coal thickness of 5:1, 24 Gt on a 10:1 ratio, and 28 Gt on a 15:1 ratio. Perhaps the figure of 24 Gt, (23,680 Mt), is the most meaningful figure under the current economics and techniques available in South Africa. If one now assumes that to the 24 Gt obtainable by mining in open cast pits down to 100 m were to be added the coals which could be mined by underground, bord-and-pillar methods down to 300–400 m, according to quality, a guess might suggest that a total of about 40 Gt might be extractable under current conditions. The Petrick Report does not specifically make this calculation, yet some such figure is necessary if one is to consider the potential future production even in round terms.

This discussion of the coal resources of South Africa, as they have been reported, should indeed be a lesson in the care with which any resource estimates should be considered, the need to dig as deep as possible into the background to the figures, and to take account of the local circumstances which may make the selection of data subjective and not directly comparable with, or equally affected by conditions in other places. A review of production can now illustrate other considerations which affect availability.

South African 1974 production was 65 Mt, of which under 2 Mt were exported.[262] In 1977 production was 85.4 Mt.[430] In 1971 an agreement had been made for 27 Mt of coal to be exported to Japan over the period 1976-1987. Since then firm contracts have been made with France, Italy, FR Germany, Japan and the USA to a total of 9 Mt. A protectionist school in South Africa has argued against any exports, but the coal industry is convinced that it can both safeguard domestic interests and export.[305] The South African Government published plans to export 20 Mta in the late 1970s, rising to 40 Mta in the 1980s. To provide facilities for increased exports, it was decided in 1971 to construct a coal port at Richards Bay, separate from the existing raw minerals exporting port. Facilities are now planned to load two ships simultaneously, at a maximum rate of 6,500 t/hour, with adjustments possible to accommodate ships from 10,000 tons to 250,000 tons.[297] It is said that Phase I of the construction should allow export of 15 Mta, though in 1975 Government approval was only for 12 Mta. A new rail connection, from Witbank Colliery in Eastern Transvaal, was complete and traffic began in 4,000 t unit trains.[297] It is said that in order to double the export capacity to about 40 Mta both the rail link and the port facilities would have to be doubled. It is these facilities which might be the restriction on exports in the late 1980s, even if the political attitude favours exports. It may be noteworthy that the 1975 estimates of the Department of Planning for internal coal requirements in 1980 were almost 74 Mt, with 10 Mt exports, and in 2000 were 180 Mt, but no exports were included.[297]

Botswana and Swaziland are known to have considerable coal resources. Botswana has very large deposits flat-lying, with seams up to 10 m thickness in places, and with a smaller dolerite problem than in the Republic of South Africa. Total resources might be, perhaps, of the order of those of South Africa, but much is under the Kalahari Desert and much is of low grade, which would not be suitable for grading into export coal but might be accepted as run-of-the-mine coal for power stations. Hence, on present evidence, it might be that the proved reserves and the export potential might both be small. Swaziland is more accessible and the coals include some of high quality with low sulphur content and some anthracites with high calorific value. However, the dolerite problem is serious, the seams are less flat and much faulted; boreholes at 200 m intervals may be necessary in some parts to define the structure for mining purposes. Hence one estimate of a potential of 5 Mta[265] might be reasonable, despite what may be very considerable tonnages of coal-in-place. Mozambique reported to the 1974 WEC Survey 100 Mt of known coal-in-place, with 80 Mt recoverable reserves, of which 80% could be open cast mined. On these figures a potential of 10 Mta[265] would seem optimistic, but very little is known about these resources.

In summary, it would seem unlikely, on present knowledge, that Southern Africa, as a whole, would have more than 20-40 Mta for export from the mid-

1980s, even if South African production increased from its 1975, production of 69 Mt to over 200–220 Mta in 2000, and export facilities were constructed; such a programme would need a very big effort in every aspect.

South America

In South America, Mexico may double its 1973 production of 3.5 Mta by 1980, and produced 6 Mt in 1977[430], but only Brazil, with 1977 production of 3.0 Mta, would seem to have a real chance of developing a production of more than a few million tons in the medium term. Venezuela had a 1977 production of only about 800 kta, but is said to have 12.8 Gt of producible coal; the bulk, perhaps 10 Gt, is in the as yet unexploited Perija foothills, due west of Lake Maracaibo; some 2 Gt may be in the Lobatera deposits which have 20 Mt proved reserves and provide the present production; maybe 46 Mt proved reserves and 500 Mt of inferred reserves are in Naricual, which had a peak production of 30,000 tons in 1921, but closed down in 1963, when production was only 12,000 tons. Despite these small numbers and the large petroleum reserves, the Venezuelan Government intend[292] to conserve oil and gas, diversify their energy resources and develop the steel industry. If these intentions hold, then there may be a good possibility that Venezuela will try to exploit their coal resources, but they also have the heavy oil resources of the Orinoco Tar Belt, and big hydro-electricity potential, and these may have priority.

Asia

In Southern Asia, India is a good example of high reliance on coal for energy. Coal constitutes over 90% of its commercial energy resources, and contributes 60% of the total value of mineral production.[125] Since the first systematic production in the 1850s output has risen consistently to 76.8 Mt in 1972. The Fifth Plan of the Government set a production target of 143 Mt in 1978. This was over-ambitious. In 1976 production, all ranks, was 102 Mtce, about 84% of total energy production[421] but production was about the same in 1978. Underground mining provides about 80% of production and in 1970 the average seam thickness was stated to be 4.1 m and average working depth 138 m, but the average depth to which reserves were calculated in 1977 was 600 m and the minimum seam thickness 1.2 m. Coal India's 'Project Black Diamond', 1977/78 to 1987/88, aimed[455] at improvement in all aspects of the industry and the implementation of new technology. The big increase was expected to be in open-cast mining, from almost 24 Mt in 1977/78 to over 90 Mt in 1987/88, contributing to targets in 1982/83 of 153 Mt and over 210 Mt by 1987/88. If these targets are achieved then the target suggested by the Conservation Commission of the World Energy Conference[424] of 235 Mtce, all ranks, in 2000 should be attainable but the jump to 500 Mtce in 2020 would be a very impressive feat.

Elsewhere in Asia, outside China and the USSR, there seem to be few chances of large productions. The hopes which Indonesia had around 1975 seem to have wilted. South Korea has climbed slowly to a 17 Mtce production in 1977[421] but Japan's production has been declining slowly from 22 Mtce in 1973 to 18.135 Mtce in 1977.[421]

Australia

Australia, with technically and economically recoverable hard coal resources of 14–22 Gtce and brown coal of 9–40 Gtce[425] had a production of almost 82 Mtce in 1977[421] and is expected to have a production in 1985 of between

139 Mtce[329] and 150 Mtce[424] with productions in 2000 and 2020 at about 300 Mtce.[424] Exports have been estimated to rise from the net 31.34 Mtce in 1976[421] to 41-57 Mtce[329] in 1985 to an export availability in 2020 of some 240 Mtce.[424] This last figure would seem to presuppose both strong growth in other energy sources for domestic use and also the political will to export. Australia has strong nationalist feelings. In January 1976, Australia demanded that some of the coal exported to Japan must be carried in Australian ships at rates about 50% higher than the current market rates. The Japanese were reported to have refused, because they were already over-committed in a surplus shipping market.[265]

Canada

Output in 1976 was 25.4 Mt and in 1977 was about 28.5 Mt with imports from the USA into eastern Canada at 15.6 Mt but exports at 12 Mt.[430] The Canadian Government was reported in April 1979[431] as suggesting that production could be 155 Mt in 2000 and 310 Mt in 2024, if transportation, social acceptability and mining economic problems were solved. These figures are much higher than the 115 Mt in 2000 and 200 Mt in 2020 which were used by the Conservation Commission of the World Energy Conference.[424]

Summary

The resources of coal are very great, the proved reserves large but the practical capability of digging out or utilizing these underground reserves to the extent required by most conventional forecasts of energy demand, are still very doubtful. In the UK the exploration programme of the National Coal Board, using modern methods, has continued to find new large reserves, but the environmentalist lobby is very strong, saying that the coal is not needed. At the same time the productivity of the miners is falling. In November 1977, a miners' ballot refused to negotiate a new productivity deal with the Coal Board which was aimed at improving the amount of coal won and paying the men for the extra effort. In the USA, the regulations restricting operations have been strengthened by Congress. The coal industry has a tough fight ahead if it is to play its proper role.

There are some signs that the position is slowly becoming apparent. One example is the doubt which Dr Hans Lansberg cast in an article[358] in July 1977, at the possibility of US coal production reaching a 1985 target of around 1200 Mt. This is the production which the Carter Administration requires if coal is to be the 'swing fuel' instead of oil. Dr Lansberg was on the NEPSG[320] which produced the report which showed the very complacency to which criticism has been directed. A report[54] had pointed out in 1974 that such a production was impossible unless certain actions were taken, and none of them were taken in the interim. Despite this and other evidence, the Leontieff report[319] suggested in its central scenario that world coal consumption would rise from just over 2 Gt in 1970 to almost 8 Gt in the year 2000. The Executive Summary of the draft report[328] on demand, prepared for the Conservation Commission of the World Energy Conference, suggested on their energy constrained high growth scenario that coal consumption might be 142 EJ (4.7 Gt) in 2000, and 361 EJ (12 Gt) in 2020.

The Executive Summary of the draft coal resources report for the WEC Conservation Commission was prepared by a German team, using published information and their own extrapolations to augment the replies which they had received from a world-wide questionnaire. Both the 'geological resources' and the 'technically and economically recoverable reserves' which they compiled were higher than those reported in the 1974 World Energy Conference Survey of World Energy Resources. The revised figures, with the 1974 figures in parentheses, were geological resources 10.1 Ttce, (9 Ttce), and technically and economically recoverable reserves 636 Gtce, (560 Gtce). The upward revision is the result of increased exploration and closer definition of reserves stimulated by the higher prices for fuels, after the jump in oil prices in late 1973. The maximum potential world productions given in the report were, however, 5.65 Gtce in 2000 and 8.7 Gtce in 2020, with export potentials in those years of 520 Mtce and 666 Mtce. These forecasts should be noted as covering the demand noted above for 2000 but falling well short of the 12 Gt 'demand' for 2020. The potential production of 5.65 Gt in 2000 is higher than the estimate given in the First Supplement to the first edition of this book (1976) (p. 21), when it was stated:

> In summary, these speculations would seem to suggest that the total coal production could rise from 2.5 Gt in 1975 to 4 Gt, perhaps 4.5 Gt in 2000, but that the world movement of coal, which was some 200 Mt in 1973 would need a massive and early effort to double to 400 Mt. in 2000.

The authors of the draft CoCo report stressed that 'action must be taken now if the maximum use of the potential offered by coal is to be made'. They illustrate that coal reserves (not resources) of 1,200 Gt are needed to support the production of 8.7 Gt in 2020. In other words a further 600 Gt must be added to the existing 600 Gtce proved reserves and that this is technically possible, but they then say, 'Nevertheless, it is doubtful that these reserves will be available in time.' This view is confirmed by the personally expressed view of one coal authority,[371] that a maximum of only 6.5 Gt is possible for 2020. This would suggest that a production of 4.5 Gt in 2000 might be more reasonable than 5.65 Gt. The report certainly therefore strongly supports the opinions that the present efforts to increase the production of coal are not sufficient to meet the projected figures of demand. A 1980 report[540] added another set of estimates.

Finally it may be wise to emphasize that:

— despite the great resource base, the proved reserves of coal, in the strict sense, are less than the combined proved reserves of crude oil and natural gas, and a proved reserves/production ratio only about 2½ times that of crude oil,

— over half the proved coal reserves and the less firm, probable and possible reserves are in the USSR and China,

— there has been a revolution in the exploration and mining fields, which has become more obvious and widespread with the acceleration of effort since the interruption of oil supplies in October 1973, and has brought new men and ideas into the industry,

— the major coal producing countries, the USA, USSR, China, and both West and East Europe, will need their coals for domestic consumption, with little chance of large surpluses for export to the end of the century,

— even where the indicated resources far exceed domestic demand, as in Southern Africa and Australia, there are a number of factors inhibiting exports,

— hence there are few indications of any major international trade in coal, even by the 1990s, which would satisfy the demand for energy in the world, and in particular in the industrial but energy deficient countries, if petroleum supplies do not develop parallel to energy demand on the scale which is normally forecast,

— the appreciation of the need for coal has, however, stimulated not only exploration and mining, but also the revival and improvements of known methods of utilization, and conversion to hydrocarbons, and in particular to substitute natural gas; SNG should be available to replace first the declining natural gas supplies of the USA in their gas distribution system, and then be ready to supplement European natural gas supplies, as they decline, possibly towards the end of the century.

— the use of coal-based synthetic feedstocks for the chemical industry will develop slowly world-wide, except in countries with coal as the only indigenous source,

— international collaboration is growing rapidly in the coal industry and should help the industry in its major task, which is to:

— convert its resources into reserves and maintain momentum, particularly through the next few years of apparent surplus, e.g. in the United Kingdom, in order to be in a position to take up a bigger share of the energy market when petroleum supplies become less available.

CRUDE OIL

Statistics on crude oil production are generally available and reliable. Statistics on natural gas production are reliable only where marketed. Measurements are made both for operational control and for taxation collection control tor crude oil, natural gas and petroleum products.

The objective of this section is not to catalogue annual production figures of all countries, but to use some as examples of problems which must be acknowledged when looking at the past and present figures as guides to the future. The main points for consideration are:

— the smooth growth of aggregated crude oil production masks many irregularities in its components. (Examples are taken from Mexico in contrast to North America, and then from the US Petroleum Administration Districts in contrast to the aggregated USA figures.)

— the production curve of any area depends on many factors, some international and some local, and not just on size of reserves or national demand. (California is used as the example.)

— giant fields and secondary recovery are very important. (California.)

— over 17,000 Mt of crude oil, a third of the total world production, has since 1931 been produced under control by local demand. Hence prorationing must be understood. In North America rationing later led to conservation concepts and production of fields as units which had been accepted in the 1930s in Iran.

— production is controlled in different countries for different reasons. (Examples are taken from Venezuela, Kuwait, and Libya.)

— Romanian production offers many illustrations of many factors.

To some readers this presentation will seem unstructured. In a desire for apparent clarity and directed education they may feel a lack of bold, crisp generalizations. The author considers that any such generalizations would, at best, be only half-truths and would have as spurious and misleading effect as have many of the crisp economic theories from which the world suffers.

The curve of the aggregated total world crude oil annual production since 1930 is a remarkably smooth one, fig. 12. It shows also that the spectacular growth has been since 1949 and only possible because of the growth in the Middle East; North American production, in contrast, seems to have increased steadily, despite some broad variations. North American production, plotted on different time and quantity scales in fig. 13, shows vigorous continued growth with only minor aberrations. Inspect then the USA and Mexican components of fig. 13. The noticeable feature of the USA curve is the drop from the 1970 peak. The Mexican curve at the bottom is very different. It shows a slow decline from 1921 to 1930 and a slow recovery from 1943. This is not the way that the Mexicans view their petroleum history, nor should any student of oil production take this view either. Increase the vertical scale of production ten times and the real story becomes more obvious.[126] Between 1918 and 1920 production rose from 9 to 22 Mt (conversion factor 7.1 bl/tonne) because of the development of flush production in the Tampico/Golden Lane areas. The decline to the low of 4.6 Mt in 1932 was partly due to the depletion of the known fields, but also to many other external and internal factors, many rooted in history, and with an importance varying according to the different protagonists in the argument. The unsettled conditions in Mexico in the early 1930s, coincided with growing awareness of the richness of Venezuela, to which, therefore, exploration effort was preferentially directed. It was not until 1938 that the Mexican government expropriated the oil industry for reasons which were complex and often mutually incompatible. Since 1945 the Mexican oil

WORLD CRUDE OIL PRODUCTION 1930-1978 Fig. 12

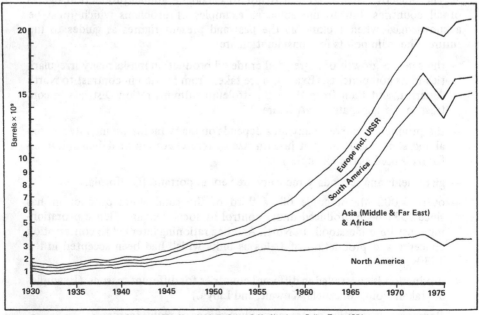

Source: Twentieth Century Petroleum Statistics, 1974, Chart No.4. De Golyer & MacNaughton, Dallas, Texas 1974, c.

ANNUAL PRODUCTIONS OF CRUDE OIL — NORTH AMERICA   Fig. 13

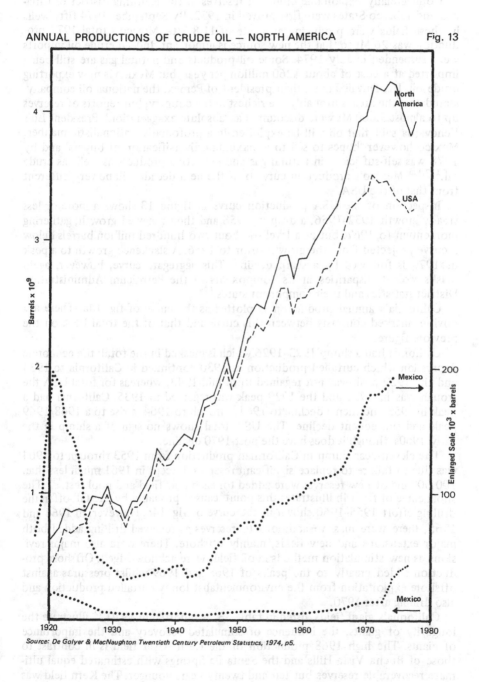

Source: De Golyer & MacNaughton Twentieth Century Petroleum Statistics, 1974, p5.

industry has shown steady growth, and recently reported discoveries in southern Mexico give promise of continued growth.

Commercially exploitable crude oil reserves in the Reforma district of Chiapas and Tabasco States were first proved in 1972. By September 1974 fifty wells in these fields were producing at a rate of 11.8 Mta. Mexico's total 1972 production was 26 Mt so that the new source is important. Indeed crude oil imports were suspended in July 1974. Some oil products and natural gas are still being imported at a cost of about $260 million per year, but Mexico is now exporting crude oil. Dr. Duvali Jaime, then president of Pemex, the national oil company, stated that the area is probably the richest in the country, but reports of reserves up to almost 3,000 Mt were discounted as 'absolute exaggeration'. President Luis Echevarria said that oil will be exploited in a profoundly nationalistic manner. Mexico however 'hopes to sell to a maximum diversification of buyers' and by 1978 was self-sufficient in natural gas and exporting products as well as crude oil.[127,128] Mexico's production curve over the next decade will be very different from that of the USA.

Inspection of the USA production curve in figure 13 shows a more or less steady growth 1932-1956, a drop in 1958 and then renewed growth, gathering momentum to 1967, but at a level of about two hundred million barrels below a curve projected from the growth prior to 1956. A slackened growth to a peak in 1970 is followed by a sharp decline. This aggregate curve, however, again masks violent disparities in its components of the Petroleum Administration District statistics and their component states.[129]

California's annual production is plotted as the curve of fig. 14a. There are obvious marked contrasts between this curve and that of the total USA on the previous figure.

California had a slump 1923-1926 which is masked in the total; the economic depression which curtailed production in 1930 continued in California to 1934 and the 1929 peak was not regained until mid-1943, whereas for total USA the trough was in 1932 and the 1929 peak was regained in 1935. California had a peak in 1953 and then a decline to 1961, through to 1964, a rise to a 1968/1969 peak and subsequent decline. The USA total shows no sign of a slump in the early 1960s, though it does have the post-1970 decline.

The eleven-year slump in Californian production from 1953 through to 1964 was due to failure to replace significant reserves. Except in 1961 much less than 100,000 bls of new reserves were added for each new field and pool test.[130] The short curve of fig. 14b illustrates this point, caused probably by the fall-off in the drilling effort 1955-1960 shown in the curve of fig. 14c. However, in 1964 and 1965, there were massive additions to reserves as renewed drilling added both major extensions and new fields, mainly offshore. There were also major revisions as new stimulation methods in oil fields were acknowledged. Offshore production added greatly to the peaks of 1968 and 1969 whilst pressures against offshore exploitation from the environmentalist lobby curtailed production and also drilling after 1970.

California's giant fields, listed in table 33, as reported in 1970,[72] illustrate the longevity of giants, the influence of stimulated recovery and the importance of giants. The high 1968 production of the Kern River field is in contrast to those of Buena Vista Hills and the Santa Fe Springs with estimated equal ultimate recoverable reserves but ten and twenty years younger. The Kern field was subjected to successful steam stimulation which increased its production markedly. These eleven giants produced over 51% of California's production in 1968

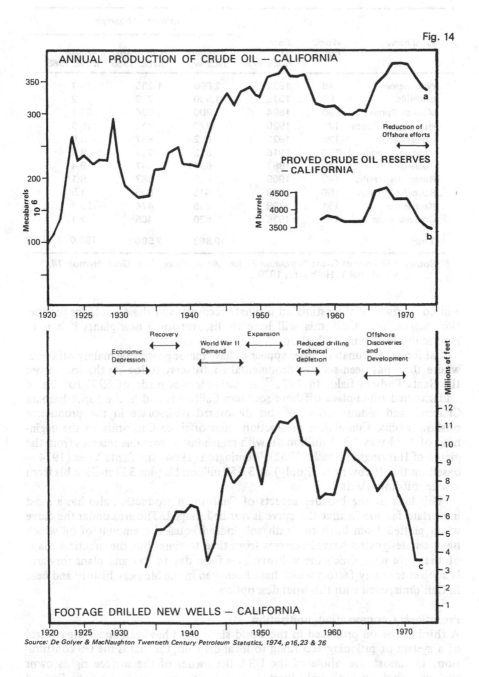

Fig. 14

ANNUAL PRODUCTION OF CRUDE OIL — CALIFORNIA

a

Reduction of
Offshore efforts

PROVED CRUDE OIL RESERVES
— CALIFORNIA

b

Recovery
Economic
Depression
World War II
Demand
Expansion
Reduced drilling
Technical
depletion
Offshore
Discoveries
and
Development

c

FOOTAGE DRILLED NEW WELLS — CALIFORNIA

Source: De Golyer & MacNaughton Twentieth Century Petroleum Statistics, 1974, p16,23 & 36

111

Table 33
Giant oil fields of California

Field name	World rank	Date of discovery	million US barrels		
			Ultimate recovery	Cumulative prod. 1968	Annual prod. 1968
Wilmington	48	1932	2,600	1,236	78.7
Elk Hills	83	1919	1,300	279	1.2
Midway Sunset	90	1894	1,200	1,024	33.1
Huntington Beach	121	1920	970	812	18.3
Long Beach	125	1921	892	857	3.9
Ventura Avenue	128	1916	812	732	8.4
Coalinga	140	1887	664	587	9.4
Buena Vista Hills	149	1909	615	587	6.1
Santa Fe Springs	150	1919	615	595	1.5
Kern River	151	1899	615	474	25.3
Coalinga Nose	170	1938	520	405	7.1
Total			**10,803**	**7,590**	**193.0**

Source: 'Geology of Giant Petroleum Fields', *Amer. Assoc. Petr. Geol. Memoir 14*, ed. Michel T. Halbouty, 1970.

and comparison of the estimated ultimate recovery and the cumulative production suggest that California will have to discover some new giants if it is to reverse the decline in production.

California, fortunately, does appear to have major prospects, mainly offshore, where there has been such environmental controversy following the spills from the Santa Barbara field. In 1971[37] an estimate was made of 89×10^9 bls of undiscovered oil-in-place offshore southern California and in the Santa Barbara Channel, and almost 26×10^9 bls discovered oil-in-place in the productive onshore basins. Cumulative production from offshore California to the beginning of 1973 was 1,371 million bls with remaining recoverable reserves from the giants of Huntington Beach (1962), Wilmington (1964) and Santa Yuez (1974 — based on the discovery well only) at 3,652 million bls plus 337 million bls from smaller offshore fields.

This look at the broader aspects of Californian production also has a most important feature in that the curve is *not* bell-shaped. The area under the curve when plotted from birth to death will indeed equal the amount of oil which has been designated proved reserves from time to time. But the practical shape of the curve with which the industry lives from day to day and plans forward, is subject to many factors which have been seen in the Mexican history and here in California, even with this brief description.

Prorationing, conservation, unitization.
A third of the oil produced in the world since 1931 has been under the control of a system of rationing according to local demand, for that is the US contribution. In almost the whole of the USA the owner of the surface rights owns the mineral rights with only limited number of areas acquired by the Federal Government and designated Navy Reserves. In certain states the Federal Government owns the shale oil rights. By the Rule of Capture, a courts-made law, the

owner of a tract of land acquires title to the oil or gas which he produces from wells drilled thereon, although part of such oil or gas may have migrated from adjoining land.[131] Such conditions inevitably led to petroleum being produced as fast as possible with wells as close as possible to each other, and to property boundaries. The waste, resulting from exploitation of this system, of the east Texas field, discovered in 1931, was so great and prices dropped so disastrously, that regulation became essential. The regulations in Texas included a provision known as Rule 37, which set minimum distances between wells and from property lines. The objective was to prevent physical waste of the resource. From the minimum well-spacing concept grew the optimum well-spacing concept, based on the optimum exploitation of each reservoir according to its production engineering characteristics. Gradually there grew the concept in the USA, and adopted more thoroughly and widely in Canada in the 1950s, that each reservoir, no matter how many owners, should be developed and produced as a unit. In Canada, the extreme was the state-enforced mandatory unitization, if voluntary unitization was impossible because the owners could not mutually agree as to equitable shares. Over the period of the 1940s and 1950s as reservoir engineering knowledge improved and became more widespread, the concept of an optimum production rate for each well became established but not uniformly defined. Sometimes a maximum production rate (MPR) is established per well or per field. Sometimes a maximum efficient rate of production (MER) is defined, which ideally, takes account of the reservoir pressures, permeability, porosity etc with some economic factor concerned with anticipated costs and revenues over the life of the reservoir.

Despite well spacing regulations and appreciation of optimum withdrawal rates, there could still be over-production. This could lower prices, make some wells uneconomic and some operators bankrupt. Therefore, to prevent both physical and economic waste, prorationing to market demand was adopted in some, but not all, states in the USA, and in all provinces in Canada. The rules, however, vary; most states use a depth-acreage formula to determine relative production rates between wells within their state. An onshore Louisiana well, 5,000-6,000 feet deep, would have an 'allowable', if there were no market restriction, of 159 barrels/day if on a field with a designated 40 acre spacing, but 239 bd if there was 80 acre spacing. A market demand factor may then be applied to all wells, for example an allocation of 30% of the full allowable for all wells for a stated period, or, in Texas, an allocation of so many producing days per month. Commonly excluded from prorationing are wells with very low production capacity of a few barrels/day. These so called 'stripper wells' can be providing a significant part of total production in many states. The USA average production per well in 1973 was eighteen barrels/day, and in 1977 was 16.[482] In the state with a depth-acreage/market demand formula, prorationing is not tied to MERs nor to reserves; wells of similar depth and spacing will have the same allowables no matter what the differences in reservoir characteristics. In the Rocky Mountain states, with significant production starting later, and even more markedly in Canada, advantage could be taken of the mistakes of others. There the well-spacing laws, the unitization and prorationing regulation are much closer to sound engineering and energy/resource conservation.

USA production in 1950 was 52% of the world total production and still 14% in 1978. Hence it is important not only to realize that that production was not just a total USA figure to meet a total USA demand, but an aggregation of individual wells producing at a rate decided in different ways in different states.

In 1973 some 9,500 producing wells were drilled, some 13,000 shut-in or abandoned and almost 500,000 were producing on 31 December 1973, in the USA.[12] Furthermore it should now be appreciated that increased demand may not be able to increase the flow rates from most stripper wells, though higher prices might resuscitate some which had been shut-in as uneconomic. Even in those wells or fields with shut-in capacity, the new 'allowables' will be different in different places according to the ruling system, none of them simple.

Control of production

Elsewhere than in the USA there are other forms of control. In Iran all fields were produced as units from the early 1930s, well-spacing and capacities determined by calculated MERs. This was possible because of single ownership and advanced energy/economics conservation thinking and practice. Total field production was prorated according to market demand. However, because of single, integrated control of oilfield and refinery, surplus products, whether light or heavy, could be returned from the refinery to the oldest oilfield. In this way there was greater flexibility and the ultimate recovery increased by the recycling of the 'surplus' products, maintaining the pressure of the field recovery mechanism.

In Venezuela production has been controlled since 9 April 1959, when Resolution 557 of the Ministry of Mines and Hydrocarbons set up a Co-ordinating Commission for Conservation and Commerce of Hydrocarbons. It was established to be 'capable of studying and recommending the regulations on the commerce of hydrocarbons and of co-ordinating them with the conservation policy advised by the supreme interests of the nation'.[132] It can be noted that the oil consultation commission, later to be known as the Organization of Oil Producing Countries, OPEC, grew from meetings on 20 April 1959 between delegates of Iran, Kuwait, Egypt and Venezuela. On 3 December 1959, Venezuela made compulsory the unitization of reservoirs exploited by more than one producer. From 1960 onwards Venezuela controlled production mainly by restricting the granting of new permits to produce and, since 1963, substituting service contracts for the older form of concessions with less restrictions. The Government also aimed at a general 4%/year growth, varying somewhat with estimates of reserves. Demand for Venezuelan oil is also closely tied to demand for heavy fuel oil on the eastern seaboard of the USA, which is its principal market.

Kuwait, in the 1960s, began to appreciate that its petroleum reserves were finite, even though very large, and started to control production. Libya cut back its production in 1971, officially on conservation grounds, but there were also political reasons. The drastic reduction of production in Iran in 1951 was due to the nationalization of the oil industry. Even after operations were organized in 1954 by a consortium of international oil companies, it was mid-1956 before production regained the 1950 level. Production rose[482] from almost 39 Mt in 1957 to 296 in 1976, then a slight drop to 284 Mt in 1977, owing to a weak market. However by the end of 1978 a new factor of political troubles began to affect production, so that the 1978 output was only 260.4 Mt, culminating in a cessation of exports and then a restricted resumption so that 1979 output may be under 200 Mt. These political problems in Iran, leading to the overthrow of the Shah, affected all the OPEC members, as will be discussed later.

In Romania almost from the drilling of the first well in 1851 bitter commercial competition, political machinations and complicated market forces interacted with considerable operational problems in both exploration and production from the many small, multi-horizon reservoirs found there. The story before nationalization after the Second World War is fascinating.[134] Annual production rose from very low levels in 1885 to almost 1.7 Mt in 1913, fell to 1 Mt by 1920, rose to 8 Mt in 1934–36, then fell to about 4 Mt, 1944–51. The First World War caused the decline from 1913, but during the German occupation in the Second World War the basis was laid for the rapid rise to over 10 Mt in 1955 and then the slower rise to 14 Mt in 1973 and almost 15.5 Mt in 1976. Production then fell[493] to below 14 Mt in 1978, despite a programme of enhanced oil recovery (EOR). This programme was started in 1952 with water and gas injection and increased annually with other methods being researched. The programme was aimed to add over the period 1976–1980 12.5 Mt to an estimated 57 Mt, which would be produced without EOR, to achieve a planned average annual production of about 14 Mta. The full planned EOR is aimed at obtaining a final average recovery factor of 42% over the whole country in 1990. Such high recovery may well be necessary for Romanian reserves are reported to be declining.

Enhanced oil recovery
The concern about the cost of finding new oil has increased worldwide interest in improving the recovery from known fields. The USSR first applied massive waterflooding to new fields in the late 1940s and 260 fields had been treated by 1978 with 87% of oil produced in the USSR being obtained with waterflooding.[496] There are dangers of water encroachment unless flooding is very carefully controlled and the case in the USSR is discussed later. Elsewhere, particularly in the USA, increasingly sophisticated methods of reservoir stimulation are being developed and employed. But EOR remains a high-cost technology requiring threshold prices ranging from $10 to $25 per barrel. Some estimates of costs were given[497] in 1979 as being, per stock tank barrel.

Steam drive	$11–16
In-situ combustion	$13–20
Carbon dioxide	$13–23
Surfactant/polymer	$20–32

Costs of this order should be a challenge towards improved technology because the need is obvious when the average recovery factor, worldwide, is only about 30% of initial oil-in-place.

Forward oil supplies
Methods of forecasting oil supply have acquired greater prominence since the 1973 politically imposed oil shortages and the resulting increased involvement of governments and of economists in academic, industry and banking circles. The fashion for computer models and systems analysis has led to a proliferation of forecasts. Some results have been misleading when biased input has led to biased output.[387] Others, such as the reports[388] of the Supply Analysis Group of the Massachusetts Institute of Technology, can show hopeful signs that there can be fruitful cooperation between the many disciplines which *should* be involved, as long as wide oil industry experience is given its full due.

The Delphi Poll organised by Dr. Desprairies for the draft report prepared for the Conservation Commission of the World Energy Conference[376] was, in effect, an attempt to tap that experience directly. WAES tapped the oil industry perhaps less comprehensively and more indirectly. Oil companies and associations

Table 34
Some possible world crude oil production capacities — 1975–2025
(million US barrels per day)

	1975	1980	1985	1990	1995	2000	2005	2010	2015	2020	2025	Curve shown in figure 15
OPEC: 1(a)	27	32	39	45	45	45	45	42	33	28	22	C
Non-OPEC, WAES 1(b)	16.3	(20)	24.7	(25)	(25)	25	23	21	19	16	13	D
SUB-TOTAL — WOCA	43.3	(52)	64	(70)	(70)	70	68	63	52	44	35	
CPE 2(c)	(11.5)	20	20.5	21	24	24	20	15	13	10	8	E
TOTAL WORLD	55.2	72	80.5	91	94	94	88	78	65	54	43	
Exxon, 3(d)												
OPEC, mid-east	19	28	32	37								F
OPEC, non-mid-east	9	10	10	9								G
WOCA	46	59	68	76								H
MAXIMUM CAPACITIES (Ult.Rec.Res.239 Gt)												
2(e)	90	90	98	100	97	90	80	68	59	48	36	A
2(f)	68	68	77	80	79	75	68	60	52	46	35	B

ACTUAL PRODUCTIONS (4)	1970	1971	1972	1973	1974	1975	1976	1977(E)[5]
WOCA	39.95	41.98	44.10	48.20	47.61	43.76	47.34	48.18
WORLD	47.80	50.32	52.96	57.84	58.17	55.21	59.56	61.67

Sources: 1. WAES[322] 2. Desprairies[376] 3. Exxon[390] 4. BP Statistical Review, 1976 5. *Petroleum Economist* Jan 1978, Table 11, p. 6. See List of References for full source details.

Notes: a. Tables 3–8, 3–9, Case C1   b. Tables 3–5 & p. 134 Case 1   c. Fig G p. 21, estimated
d. Chart 8   e. Fig. D Curve 11 estimated   f. Fig. D Curve 1 estimated

publish forecasts which may not receive appropriate credit for impartiality in all quarters. It should be remembered, however, that not only is the competition between oil companies real and tough, but each company has a different base, structure and profit centre network. Joint ventures in one area do not lessen competition in another. Agreement on supply forecasts indicates coincidence more often than collusion. Some companies are more inclined to publish than others; some have the self-confidence in their own ability to make better use of the figures than their competitors, as illustrated by the continuing early publication of data by one of the most successful explorers; all know that any 'phoney' figures will be immediately suspect by the industry. In general, oil company information and forecasts of world oil production can therefore be an indication of their real opinion at the time of publication; their commercial secret is what they are going to do with them. Hence much of the suspicion with which some view such forecasts is misplaced.

Some crude oil annual production possibilities are given in table 34 and graphically presented in figure 15. The complications of the table and figure are both due to the dissimilarities in the source data and to the desire to explain how those dissimilarities have had to be reconciled, without distorting the source data, in order to make meaningful comparisons.

The CoCo draft oil resources Executive Summary by Dr. Desprairies[376] dealt with maximum productive capacity in gigatonnes for the whole world for the period 1985 to 2020. The WAES study[322] dealt with possible actual production in mbdoe (million barrels per day of oil equivalent), for only WOCA (world outside Communist areas) for the period 1985 to 2000. The Exxon study[390] dealt with estimated production in mbdoe, but with no arbitrary restriction on production such as WAES postulated in some scenarios, for WOCA for the period 1975 to 1990. The zones into which the reports divided their 'worlds' differ and are reported differently.

The reconciliation may best be explained by taking the curves of figure 15 in turn. The CoCo draft oil resources summary provided curves A and B. Curve A is the median of a three curve family based on a 'medium' depletion rate of ultimate recoverable reserves of 238 Gt. Curve B was the lower of the same set of curves using the same ultimate reserves but the lower depletion rates which, according to published data, were current in 1975 in the zones into which the world had been divided. In the Executive Summary two other families of curves were shown which were based on ultimate reserves of 174 Gt and 347 Gt, depleted at similar rates to those of the median reserves. The form of all the curves was derived from cumulative discoveries by zones, assuming that, 'petroleum discoveries in the world's oil-bearing zones as a whole will follow the pattern of the United States . . .' Dr. Desprairies specifically offered the two curves reproduced as curves A and B in figure 15, 'for consideration by political and industrial leaders'. He stressed six conditions for achieving these maximum capacities for production: sufficient funds; deployment of exploration effort to the most prospective areas world-wide irrespective of politics; substantial government aid in developing improved techniques and training specialized labour; overcoming the reluctance to invest; wise resource management by government and industry, and education of the public to appreciate the need for such management. These are valid conditions, but the problem with curve A is that it seems unreasonable to suggest that these conditions would be accepted so fast that a capacity of 4.5 Gt would be available in 1980. The Exxon study suggested that there was an OPEC surplus capacity in 1977 of 5 mbd. It seems

117

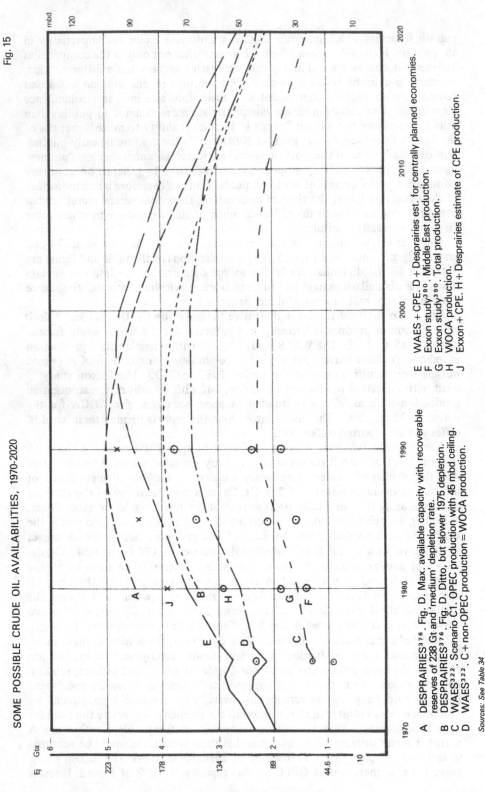

SOME POSSIBLE CRUDE OIL AVAILABILITIES, 1970-2020

Fig. 15

A DESPRAIRIES[376]. Fig. D. Max. available capacity with recoverable reserves of 238 Gt and 'medium' depletion rate.
B DESPRAIRIES[376]. Fig. D. Ditto, but slower 1975 depletion.
C WAES[322]. Scenario C1. OPEC production with 45 mbd ceiling.
D WAES[322]. C + non-OPEC production = WOCA production.
E WAES + CPE. D + Desprairies est. for centrally planned economies.
F Exxon study[390]. Middle East production.
G Exxon study[390]. Total production.
H WOCA production.
J Exxon + CPE. H + Desprairies estimate of CPE production.

Sources: See Table 34

improbable that non-OPEC producers can have a surplus of more than 1 mbd. Oil production in 1976 was 59.5 mbd. The total capacity therefore in 1977 must be about $60 + 5 + 1 = 66$ mbd or 3.3 Gt. It seems unlikely that the extra facilities would be financed in a time of surplus nor constructed within three years. Therefore the starting point of curve B would seem more possible at about 3.5 Gt.

WAES juggled with various discovery rates, recovery factors, depletion rates and, most importantly, with constrained maxima in OPEC production rates at 33 mbd, 40 mbd, and 45 mbd. A number of scenarios of possible productions were calculated for WOCA but no firm preference was given. Their thinking is illustrated from their Scenario C1 in which the proved reserves for WOCA at the end of 1975 were taken as 555 bb (billion barrels), with gross additions to reserves at 20 bb/year to the year 2000 and then declining to 4 bb/year by 2025. The WAES numbers are given at the top of table 34 and the relevant curves in figure 15 are labelled C, D and E. Curve C plots OPEC production with a ceiling of 45 mbd. To the values of this curve were then added their non-OPEC production estimates to give a total WOCA production as curve D. Dr. Desprairies in figure G of his Executive Summary gave estimated maximum capacity values for the centrally planned economies (CPE); these are unlikely to include any spare capacity, for the reasons discussed later. It therefore seems reasonable to accept these values as production values. On this assumption, the composite total for the world, based on the WAES Scenario C1, is shown as curve E. The encircled points in figure 15 indicate curves from the Exxon study.[390] Curve F gives possible Middle East productions which are so obviously the bulk of the total OPEC production which is shown as curve H. The addition of the same figures for CPE from the Desprairies report then constructs an Exxon-based world view. This curve J obviously does not have the 45 mbd restriction on OPEC production which is part of the WAES Scenario C1, and does lie between the Desprairies curves A and B. It must be stressed again that the WAES production values and the Desprairies CoCo capacity values, have been selected as the more plausible ones, in the writer's opinion, from the many different curves offered by these two reports.

It is not surprising therefore that when in figure 16 a personal choice is presented of a possible world crude oil production curve that this curve should be somewhat similar. This curve assumes ultimate recoverable reserves of conventional crude oil around 300 Gt, which is somewhat more optimistic than the weighted mean of the experts consulted by Dr. Desprairies in his Delphic Poll. Gross discoveries averaging around 29 bb/year into the 1990s are assumed to decline after 2000 but to be partially compensated by improving recovery factors. Assumptions on the Middle East are argued later in this chapter. Most of the work for this figure 16 was done in early March 1977, and was the latest in a personal series of such curves of which only few have been published. One curve was published[391] in 1956 and is interesting in that the USA and world totals for 1975 were close to the USA and world actuals; on the other hand, the spectacular contributions of Libya and Nigeria were not foreseen, but in the event other productions moved over to accommodate the African production. Such internal modifications to the individual components of the curve 16 must be expected, but the aggregates may not be far away from that curve. As has been said many times before, any 'forecast' such as this single curve is out-of-date even before it is printed as new facts and new opinions are incorporated into one's working papers in the continuing attempt to keep abreast of the dynamic scene. A later opinion is given in the Supply and Demand Chapter.

Mexico

Mexico was used earlier as an illustration of the need for care and understanding when considering past production. Mexico can also illustrate the care with which future production potential must be viewed. The promise of new oil mentioned earlier has been improved by new oil discoveries in the Gulf or Sound of Campeche, west of the Yucatan Peninsula. There have also been gas discoveries on and off the long peninsula of Baja Mexico on the Pacific coast. Pemex, the state oil company, announced an ambitious six-year development programme in December 1976 and by November 1977, was already indicating that this was probably conservative. The plan[394] was for oil production to increase to 2.24 mbd (114 Mta) in 1982, of which 1.1 mbd (56 Mta), would be for export. Pemex, has shown caution in declaring reserves. However, in mid-1977 they increased[393] their proved reserves from 11 bb to 14 bb (almost 2 Gt), and confirmed a figure of about 60 bb (8.45 Gt) as the probable total oil and gas reserves. Grossling[389] argued that on the declared policy of maintaining their proved reserves at a level equivalent to a minimum of 20 years of current production, then production from the Reforma area of 440,000 bd would argue a proved reserve of 3.2 bb for those fields, as proved; if five times as many structures remain to be tested, and there are now thought to be four different trend lines in the Reforma area, then five times 3.2 bb should be possible for the Reforma area alone.

In the light of such figures, the statement of May 1976, by the then Minister of Natural Resources, that he hoped to see 7 mbd of Mexican oil exports in the late 1990s may not be beyond all reason. Mexican domestic demand was estimated[329] to be 1 mbd in 1980. If the 1965–1975 annual growth rate of 7%

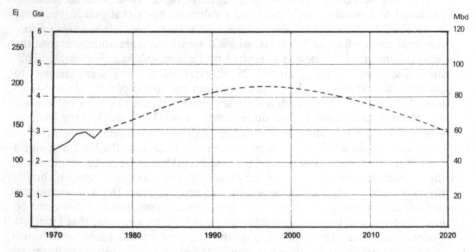

WORLD 1970-2020
ONE PROBABLE CRUDE OIL AVAILABILITY FORECAST   Fig. 16

Source: D. C. Ion. 1 Nov 1977.

MEXICO — OIL GAS DEVELOPMENTS   Fig. 17

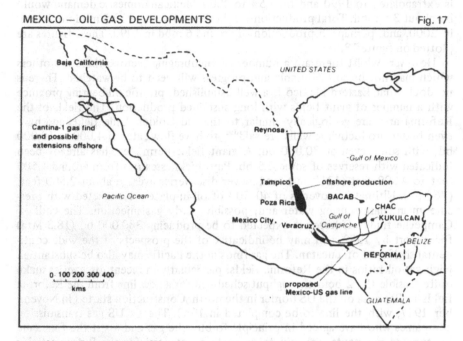

Source: Offshore Engineer. August 1977

MEXICO — CRUDE OIL PRODUCTION & EXPORTS 1970-2000
SOME ESTIMATES   Fig. 18
Million barrels per day

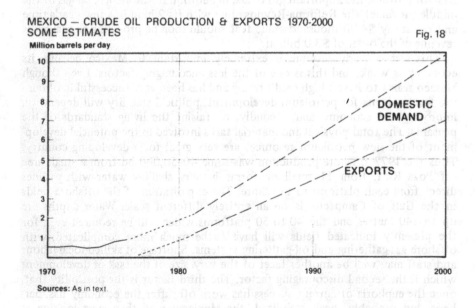

Sources: As in text.

is extrapolated to 1990 and then 5% to 2000, Mexican domestic demand would be about 3.5 mbd. Total production would then have to be 10.5 mbd minimum, in 2000, and, perhaps, a production of around 6 mbd in 1990. These figures are plotted on figure 18.

However, whilst there are a number of encouraging factors, there are others which may not be so favourable and progress will need to be watched. There is no doubt that Eastern Mexico is a well established, prolific oil-bearing province with a number of giant fields with long sustained production. The fields of the Reforma area are geologically similar to the old Golden Lane fields and have even better productivity; wells are said[393] to have flow rates of 4,000 to 18,000 bd, with some even at 30,000 bd. A giant field, Bermudez, has already been indicated with reserves of some 2.5 bb. Pay thicknesses run from around 1,500 feet to 4,000 feet (1219 m.), and newer discoveries average about 2,500 feet, (760 m.). Ultimate recoveries of 40-50% of oil-in-place are expected with pressure maintenance using water and, possibly, early gas injection. The Gulf of Campeche finds offshore are expected to be producing 360,000 bd (18.3 Mta) for export by 1982, and may be indicative of the prospects of the wide continental shelf west of Yucatan. The gas finds in the Pacific may also be substantial. High gas/oil ratios in the Reforma fields, particularly in recent discoveries make quite feasible the 2 bcfd throughput scheduled for a new line from the Reforma fields to Reynosa on the US border in the north. Construction started in November 1977, with the line to be completed in 1982. The six US gas transmission companies who have agreed in principle to buy the gas and assist the financing by forward payments, are said to estimate a potential for the Reforma fields of some 20 tcf (trillion cubic feet = 566 km^3), with 10 tcf already proved as probable. Mexico's total gas production is already at 4 bcfd, because the North-Eastern province has important gas fields, in addition to the associated gas of the middle province. The Reforma-Reynosa line with its 2 bcfd in a few year's time and gas at say $4.00/thousand cubic feet should soon be providing annual gross revenue of the order of $3.0 billion.

This sort of early revenue is extremely important to Mexico because its economy is weak, and this is one of the less encouraging factors. Even though Mexico seems to have a high credit rating and has been very successful in obtaining foreign loans for petroleum development, political stability will depend on improving the economy and, secondly, on raising the living standards of the peasants. The total physical and material tasks involved in the potential development of the new petroleum resources are very great for a developing country. Thus the 1977 offshore production was some 40-50,000 bd from a small area off Poza Rica, from 11 small platforms in very shallow water with gaslines direct from each platform to the shore. The exploitation of the offshore fields in the Gulf of Campeche is on an entirely different scale. Water depths are up to 140 metres and the 40 to 50 platforms which will be required even for the presently indicated fields will have to be much more complicated, with offshore gas-gathering and oil-gathering systems. Shortage of skilled local labour and staff may well be another facet of the very size of the task of development which is the second discouraging factor. The third factor is the possibility that, once the euphoria of current success has worn off, after the economy has, perhaps, but probably, been stabilized, the Mexicans may then start looking at their petroleum resources as a wasting asset, and adopt a conservation attitude. They have the example of nearby Venezuela which has been pursuing such a policy successfully for about twenty years. The WAES report[322] suggested that

Table 35
Middle East – an exercise in possible future crude oil reserves – the assumptions
(gigatonnes)

	Proved reserves	Additional gross reserves 1977–2020	Estimated average annual additional reserves	Approx ultimate recoverable reserves at 1.1.2021	Reserves/production ratio 2021
Saudi Arabia	(1) 20.6	20.6	0.48 (3.5 bb)	41.2 (300 bb)	12 : 1
Iran	(2) 8.6	5.0	0.12 (0.88 bb)	13.6 (100 bb)	12 : 1
Iraq	(2) 4.6	6.2	0.14 (1.02 bb)	10.8 (80 bb)	12 : 1
Kuwait	(2) 9.2	0.8	0.02 (0.15 bb)	10.0 (73 bb)	12 : 1
Others	(2) 6.08	3.8	0.09 (0.7 bb)	9.88 (72 bb)	12 : 1
Totals	49.08	36.4	0.85 (6.2 bb)	85.48	

Cumulative production to 1.1.77 12.6 Gt (3)
Total proved reserves + cumulative production = 62 Gt.
Conversions from US barrels to metric tons have been made at country average crude gravities.

Sources: (1) *Oil & Gas Journal*, 15 Feb 1977[398] for proved reserves only
(2) *Oil & Gas Journal*, 27 Dec 1976[399] for proved reserves only
(3) *20th Century Statistics*, 1976, DeGolyer & MacNaughton, Dallas, USA

OPEC might well put a ceiling on their production in the near future. Mexico, once it has improved its economy and the standard of living of its poorer people, might well be chary of too rapid development. Mexico's leaders have repeatedly said in the past few years that Mexico's petroleum resources will be exploited for the good of Mexico only. On the other hand, at least in the 1980s, the policy might be to earn money with which to develop its other energy resources for the long term, particularly if this becomes the popular trend, whether led by OPEC or not. Mexico has:

- coal – 1.7 Gt positive, 6-10 Gt potential,
- hydropower – present 15 TWh/year but probably 60 TWh/year developable,
- uranium – 6,122 t positive but only 1% of the country explored and prospects good,
- geothermal – potential considerable
- solar energy – possibilities obviously greatest in the desert areas of the north.

In brief, towards the end of 1979 it would seem that Mexico could be considered to have a possible petroleum potential of several millions of barrels per day of crude oil and several billions of cubic feet a day of natural gas, with the major market of the increasingly petroleum-deficient USA on its northern border. Whether, however, crude oil exports will rise to Middle East scale of the order of 7 mbd in the 1990s, will be dependent on many factors which at this time are all in a very dynamic stage.

Other areas outside the Middle East

The 1980s will see increasing production from Mexico, the North Slope of Alaska, South-East Asia, the USSR, China, parts of Africa and most countries of Latin America. I have discussed elsewhere some aspects of the impact of indigenous offshore petroleum on the energy policies of the North Sea countries.[337] Grossling has argued strongly that Latin America is underdeveloped.[389] South-East Asia, and particularly Indonesia, has not yet perhaps lived up to its potential for a number of reasons, mainly political. The USSR and China will be discussed later.

The Middle East

Although there have been many estimates of crude oil supply in the 1977 reports, most give figures for OPEC, or OPEC in parts, or the Arabian Peninsula producers and few for the Middle East, however defined, as a whole or for the countries of the Middle East separately. This is unfortunate. In present proved reserves and in current expectations it is the Middle East, and in particular Saudi Arabia, Iran, Iraq, Kuwait and the United Arab Emirates, (UAE), which do dominate world trade and which will dominate it for the next ten or fifteen years. In proved reserves, these five Middle Eastern States have 56% of the world's proved reserves and 65% of those of WOCA; and of the 1976 production they gave 36% of the world's production of crude oil and 46% of WOCA's.

A thorough study of the potential of the Middle East is not to be found in the literature, although a 1977 study[406] is the closest. An exercise was therefore constructed as one way in which to feel the scale of the problem. The assumptions used are given in table 35 and the resulting annual average productions in table 36. The OECD estimate[329] of capacities are also given in this table. Figure 19 displays the suggested figures as production curves.

WAES in their estimate of OPEC production made four assumptions about supplies from the Middle East:

Table 36
Middle East — an exercise in possible future crude oil production by countries
(megatonnes)

A	1975	1976	(1980) OECD Capacity	1980	1985	1990	1995	2000	2005	2010	2015	2020	2025	Curve in Figure 19
Iran	267.7	295.0	(350)	335	360	385	400	390	370	340	290	250	200	
Iraq	110.9	112.2	(300)	160	200	240	250	235	220	180	160	130	120	
Kuwait	92.4	98.4	(175)	130	160	190	200	200	200	200	200	175	150	
Others	145.7	160.4		150	150	150	150	150	150	150	150	150	150	
Sub-Total	614.7	666.0		775	870	965	1000	975	940	870	800	705	620	
Saudi Arabia Unrestrained	343.9	421.6	(750)	615	800	920	980	1000	965	880	770	640	480	
MIDDLE EAST Total A	958.5	1087.8		1390	1670	1885	1980	1975	1905	1750	1570	1345	1100	A
Saudi Arabia Restrained				525	615	685	735	750	750	750	750	750	750	
MIDDLE EAST Total B				1300	1485	1650	1735	1725	1690	1620	1550	1455	1370	B

Sources: 1975, 1976, *BP Statistical Review* 1976
1980–2025, D. C. Ion, based on assumptions of Table 35
(1980 Capacities, *World Energy Outlook*, OECD, 1977)

B	1985	2000	2020
Inland oil consumption	150	300	650
Export potential on Total B	1335	1425	805

Source: D. C. Ion, on assumptions described in text.

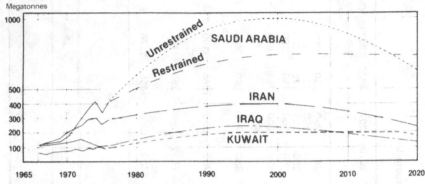

MIDDLE EAST – MAJOR COUNTRY CRUDE OIL PRODUCTION
PROJECTIONS, 1966-2020, AN EXERCISE   Fig. 19

Sources & Assumption: See Table 36. Figures see Table 37.

— countries outside the Arabian Peninsula would roughly maintain current production at 20 mbd (1,000 Mta). These countries include Iran and Iraq.
— for the countries of the Arabian Peninsula the assumptions were that:
   — governments would set limits at near present production, about 13 mbd, or Saudi Arabia would allow its production to rise to (a) 15 mbd, or (b) 20 mbd, which they considered technically possible.
   — other producers in the Peninsula would produce 5 mbd in total in both cases,
   — production capabilities were based on OPEC average discovery rates of 10 bb/year or 5 bb/year.

Current production of the other OPEC countries seems somewhat arbitrary, because of Iraq's potential, as noted above, and the prospects in the other major OPEC producers, where exploration is, in my view, very far from complete. Even in Iran there may be possibilities of other discoveries which may have been inhibited by what might be called the 'Khuzistan complex' – the over-reliance on the 'established', well-worn, perhaps out-worn, rationale of current reserves. Such opinions are behind the assumptions of table 36 and the curves of figure 19.

However, in 1978 and 1979 the OPEC trend hardened towards controlling production at that level which generated in their opinion adequate funds for their own development. OPEC found that the market would bear increased prices and, as a corollary, sufficient funds could be raised whilst holding down production. Higher prices were originally justified because inflation increased the cost of manufactured goods and services from the consuming countries, aggravated by the fall in value of the US dollar in which oil payments were made. The drop in Iranian production and exports turned the apparent surplus of 1978 into a shortage, though partially alleviated by Saudi Arabia increasing by 1 mbd the target they were setting of 8.5 mbd as their requirement. It was stressed that this was merely a temporary measure. Other countries profited from this unexpected short supply situation to raise prices unilaterally, so that

126

Table 37
Middle East — a revised exercise in possible future crude oil production
(megatonnes)

	1976	1978	1980	1985	1990	1995	2000	2005	2010	2015	2020	2025
Iran	295.0	260.4	200	220	240	260	260	260	250	250	250	250
Iraq	118.2	127.6	140	160	180	200	200	190	180	170	160	150
Kuwait*	110.6	105.8	105	100	100	100	125	125	125	125	125	125
UAE & Qatar	117.6	112.2	115	115	120	110	100	100	90	90	85	80
Saudi Arabia*	433.8	420.2	425	450	450	500	550	600	600	600	600	600
TOTAL	1075.8	1026.2	985	1045	1090	1170	1235	1275	1245	1235	1220	1205
Other OPEC	464.1	464.7	470	480	480	490	500	500	490	480	480	470
TOTAL OPEC	1539.9	1490	1455	1525	1570	1660	1735	1775	1735	1715	1700	1675

* includes half of Neutral Zone production.

Sources: 1976 & 1978, *BP Statistical Review*, 1980–2025 DCI estimate on restricted OPEC 1985 production.

Sheikh Yamani of Saudi Arabia said in October 1979, that the price system was out of control.[495]

In mid-1977 there was a general feeling within the oil industry that OPEC had a sustainable production capacity which would be about 40-43 mbd in 1985, (2.0-2.15 Gta). In September 1979 it was as firmly held that 35 mbd (1.75 Gta) was a more likely figure and, if the OPEC governments adhered to the full conservation policy about which they were talking, actual output might be limited to 29-30 mbd (1.45-1.5 Gta). The OECD 1977 estimates for 1980 capacity in table 36, which were some 1.575 Gt for the four major countries alone, were therefore over optimistic.

An alternative to table 36 was constructed in October 1979 to take account of both the suggested ceiling for OPEC in 1985 and the drop in Iranian production with a probable slow recovery, and a continuing very restricted production system for the Middle East OPEC. Table 37 can therefore be taken as another scenario to contrast with the 1977 table 36. An important feature of such a low production regime would be the effect on the reserves position. Table 38 therefore reviews the cumulative productions and the reserves involved.

It can be seen that the Middle East OPEC would enter the next century with 22 Gt remaining from their 1977 proved reserves plus whatever new oil had been found. In table 35 a reasonable rate was given for new discoveries as 850 Mta. Even if reduced exploration and development activity halved that rate, 1977-2000, then almost 10 Gt would have been added, making a total of 32 Gt. A production declining slightly, to indicate lower demand as energy systems had been adapted to the tight supplies, from 1235 Mta to 1205 Mta in 2020, would by that time use up some 24 Gt, leaving 8 Gt, again plus whatever new discoveries were made, which might be greater than in the earlier period because falling reserves might stimulate greater effort. Hence a rate of 500 Mta has been assumed, so leaving 18 Gt as the remaining proved reserves in 2021.

Table 38
Middle East OPEC and other OPEC, cumulative productions
& proved reserves 1908—2020
(megatonnes)

	Time period	Middle East OPEC:	Other OPEC
Cumulative production	to 1.1.1976	11,500	
Proved reserves	1.1.1976	49,295	12,331
Cumulative production	1976—2000 inclusive	27,344	12,015
Remaining 1977 reserves	1.1.2001	22,000	300
Additional discoveries @ average 425 Mta	1977—2000	10,000	
Cumulative production	2001—2020	24,000	9,800
Remaining 2001 reserves	1.1.2021	8,000	
Additional discoveries @ average 500 Mta	2001—2020	10,000	
Total remaining reserves	1.1.2021	**18,000**	

Note: A 1979 source [536] uses 50.8 Gt for Middle East OPEC proved reserves and 11.14 Gt for non-Middle East OPEC.

Sources: Cumulative production to 1.1.1977; *Twentieth Century Petroleum Statistics.* All others: calculated from Table 37.

However, the low OPEC production to 2000 might have forced the rest of the world to find not just alternative energy sources but to devise alternative energy systems. Maybe even the 22 Gt remaining from the 1977 reserves might be sufficient to satisfy the restricted demand for crude oil from the Middle East and the less known resources which have been called forward into reserves may never be exploited.

Table 39
Estimated commercial production of natural gas
1974 report and in 1978
(cubic kilometres)

	1974	1978	Regional totals as percentage of total production %	Regional proved reserves as percentage of world total after Adams & Kirkby, 1975. %
AFRICA				
Algeria	3.05	14.1		
Nigeria	15.71	0.5		
Libya	14.05	5.0		
Total Africa	**38.3**	**21.0**	3 1.4	8%
ASIA				
China	21.0	49.5		
Pakistan	3.9	5.2		
Brunei	3.4	8.8		
Indonesia	4.2	10.0		
Iran	40.8	18.7		
Kuwait	17.1	6.3		
Saudi Arabia	31.7	6.0		
UAE	11.4	3.6		
Total Asia		**91.5**	11 6	37%
EUROPE				
France	10.6	7.9		
West Germany	17.7	20.31		
Netherlands	58.6	90.2		
Italy	14.2	12.5		
UK	26.0	39.2		
Total Europe	**179.5**	**185.9**	13 13	9%
USSR & E. Europe	**212.0**	**428.4**	15 29	27%
N AMERICA				
Canada	73.1	71.6		
USA	637.5	556.7		
Mexico	18.7	26.5		
Total N America	**732.8**	**654.8**	53 45	15%
S AMERICA				
Venezuela	46.0	15.0		
Argentina	8.0	7.8		
Total S America	**70.6**	**60.5**	6 4	2%
OCEANIA				
Australia	3.4	6.9		
New Zealand	0.4	2.2		
Total Oceania	**3.73**	**9.0**	0 0.6	2%
TOTAL WORLD	**1389.0**	**1459.0**	**100** **100**	**100%**

Source: 1974 *WEC Survey*, Table IV-S, p. 111.
1978 *Petr. Economist*, Vol. XLVI, No. 8, August 1978.

NATURAL GAS

Although the piped gas industry based on coal is over a hundred and fifty years old, only in the past thirty years has the natural gas industry become a significant factor in world energy terms. In that time, its growth rate has far outshone that of other energy forms because of its four inherent advantages as a fuel source.[5] First is the efficiency of pipeline transmission, which is economic in terms of both large and small energy quantities. Second is the ease of storage in pressure vessels or in natural or artificial underground reservoirs or as liquefied natural gas (LNG). Third is the flexibility of control and efficiency in use. Fourth is that its combustion is non-polluting, except for that of sour gases from which the sulphur must be removed prior to use. This is why in 1971 domestic gas consumption in Britain was one half of total use, in the USA one third and in USSR one tenth.[5] Transportation of a gas, except by pipeline, is difficult but marine transportation, as LNG, has now started to solve some problems. However, the technology is still young as the following dates emphasize.[135]

1951	Major Canadian gas line to USA
1964	Export from the Netherlands of gas from the Groningen field
1964	Algerian LNG shipped to UK
1968	Soviet gas piped to Austria
1969	Alaskan LPG shipped to Japan
1970	Iranian gas piped to USSR
1973	Soviet gas pipeline extended to West Germany.

It is the transportation problem and the youth of its solution which has left most consumption local on indigenous supplies. There is a big disparity between production and proved reserves as shown in table 39 by comparison between the regional production and reserves as a percentage of world figures. Some 90% of the world's consumption is in the USSR, USA and Canada, and Europe, perhaps 4% in China and Eastern Europe and the remainder in the many other countries of small local consumption.[135]

The USSR

Production in the USSR is increasing fast. In 1973 the production at 236 km^3 was 14.6 km^3 more than in 1972. Growth in comparison to other energy sources can be seen in table 40. 1977 production[421] was 350 km^3.

Table 40
Primary energy production, USSR, 1940—1973
(rounded)

	1940	1960	1970	1972	1973
Oil (Mt)	31.0	148.0	353.0	400.0	421.0
Natural Gas (Km^3)	3.0	45.0	198.0	221.0	236.0
Coal (Mt)	166.0	510.0	624.0	655.0	668.0
Hydropower (Mt) (coal equiv)*	0.6	6.3	15.3	15.1	

* standard coal equivalent at 7,000 k cal/kg

Source: 'Fuel and Power Economy of USSR', P. S. Neporozhiny *et al*, *IX WEC Preprint* 1.2—4, Table 1, 1974.

The USSR is fortunate in having a number of giant fields. The Medrezhie field, credited with 1 T km^3 in 1970[72] and 1.55 T km^3 in 1974 of proved commercial reserves, has an average gas flow rate per well of 1.5 M m^3/day. Much of the Soviet gas reserves are in the far north in permafrost areas. The middle Asian fields are under abnormally high pressures and the gas has some highly corrosive elements, e.g. Orenburg field has 4.5% hydrogen sulphide. These problems, though great, are being solved. Pipelines of over 1.6 m diameter, and 2,000–3,000 km in length are planned and will operate at high pressures.

There is every expectation that production will continue to grow and with it the shift of energy supply eastwards as table 41 indicates.

Table 41
Region east of Urals, share of USSR fuel resource production
(rounded percentages)

	1940	1960	1970	1972
Natural gas	0.5	2.0	30.0	33.5
Oil	6.0	7.0	18.0	25.0
Coal	29.0	36.0	43.0	45.5

Source: 'Fuel and Power Economy of USSR', P. S. Neporozhny *et al*,
IX WEC Preprint, 1.2–4, Table 5, 1974.

North America
The USA was the first to exploit natural gas on a large scale and produces now almost half of the world's gas. It has however, reached a scarcity point despite imports from Canada and plans for LNG imports from Africa. A 1974 Study[54] will be used to highlight the problem and the many-sided effort which will be required if that scarcity is to be reduced. Total US 1973 consumption of natural gas was 723 km^3/y* supplied as to 5.5 km^3/y from onshore, 14 km^3/y from offshore fields and 3 km^3/y from imports. The known gasfields are expected to decline from 692 km^3/y** in 1973 to 407 km^3/y in 1985. By then new fields, they say, including 31 km^3/y from Alaska's North Slope, should provide 378 km^3/y, so offsetting decline. An additional 145 km^3/y would be available, granted adequate incentives, including higher prices and more leases made available. Yet a further 31 km^3/y are suggested as possible from low permeability sands by nuclear explosion stimulation. The total supply in 1985 of 843 km^3/y can be considered the result of a hard, practical, engineering study but its attainment depends not just on incentives but the effort indicated as necessary by the Task Force on the crude oil front would be to raise the 1973 production of 490 Mta*** to 605 Mta in 1985 with new oil at 266 Mta, partially offsetting a decline in known fields of 290 Mta. The authors judge that the effort and results are the upper limit reasonably attainable 'with existing technical capability, existing natural resources, early access to new natural resources, removal of administrative and regulatory impediments, and efficient operation of the nation's economic system by providing incentives to attain the desired

---

*   Source conversion factor, 1 million barrel oil equivalent/day: 2.06×10^{12} cu.ft/day
** 1×10^9 cubic feet/day = 10.34 cubic kilometres/year = 10.34 k^3/y
***   7.3 US barrels = 1 metric ton; 1 million barrels/day = 48.4 t/year

end results'.[54] The effort which they suggest is illustrated in the table 42. This table and this argument emphasize the massive effort required at a capital cost for both oil and gas from 1973 to 1985 of about $200,000 million.

There is a gap between gross production and marketable production of 21.1 km^3/y (21% of total) in 1972 because of prior use of that gas in re-injection into gas fields after stripping off the natural gas liquids, in flaring, in loss of gas processing plants during recovery of the butanes etc or in use as process fuel. The rise in marketable production is more marked if the start is 1945 production of 1.6 km^3 to 82 km^3 in 1973.[136] In 1945 about 83% was sold locally in Alberta and most of the rest in Ontario. In 1972 45% of production was exported to the USA and the rest supplied all provinces from British Columbia to Quebec.

Canada can illustrate three important points about natural gas production, namely the rigidity due to long-term contracts, the gas industry as an entity and control of export availability.

Table 42
Supporting activities to USA oil and gas production — 1973—1985

	1973	1985
Offshore employees	49,000	165,000
Total employees (excl. refineries)	265,000	330,000
Mobile offshore drilling rigs	50	180
Platform drilling rigs	40	570
Numbers of platforms in place	2,000	3,700
Tubular goods (million tons/year)	1.5	2.8
Pipeline goods (million tons/year)	1.9	2.4
Platform steel (million tons/year)	0.3	1.9
Other steel (million tons/year)	0.4	0.4
Total steel required (million tons/year)	4.1	7.5
Wells drilled per year	30,000	58,000
Average depth per year (feet)	5,040	5,400

Source: *US Energy Prospects — an engineering viewpoint,*
U.S.Nat.Acad.Sci. p. 84, 1974.

Table 43
Canadian natural gas production, consumption and exports
(cubic kilometres, rounded)

	1960	1966	1970	1972
Proven remaining reserves of marketable gas	—	1,526	1,875	1,868
Marketable production	15.6	39.1	65	79.5
Canadian consumption	11.7	24.8	36.9	44.4
Exports to USA	3.9	15.3	27.5	35.7
Imports from USA	0.2	1.6	0.4	0.5

Source: *An Energy Policy for Canada: Phase 1*, Vol. II, p. 308, 1973

Transportation facilities require a heavy initial capital cost for plant and pipelines (or, indeed, for tankers for liquefied gas) which necessitate amortization over many years. Hence long-term, fixed quantity contracts are commonplace, often with price subject to escalator clauses. There are no large 'spot' gas sales or a brokerage business as with crude oil. There is far less flexibility or variation in gas production than with oil. Associated gas whose production rate is tied to that of crude oil is therefore far less valuable to the gas industry than non-associated gas which is not subject to such possible fluctuations. The rigidity of the system with its contractual base load does not favour interruptible supply contracts unless there is a large and flexible cheap storage facility.

The second point is the nature of the natural gas industry as an entity. Natural gas in the USA, in the beginning, was a by-product, a waste product of crude oil production. It was therefore cheap and handled by others than those in the crude oil producing industry. The US gas industry started from the market end. The US gas industry therefore grew up separate from the oil industry with the oil man happy to have the gas taken away. Natural gas therefore had a low price and there has been a continuing tendency for natural gas to be underpriced in relation to alternative energy sources. In the USA federal control of inter-state trade has kept prices low and so reduced incentive and inhibited exploration and production. With basically fixed contracts for both supply and market and a fixed investment, with little risk capital involved, it was easy to calculate a rate of return on investment. Federal control tended therefore to set profit limits as a public service or utility industry. The US natural gas industry was staffed and financed as such; it never took an active operating part in the exploration and producing end of the petroleum industry. US supplies are now scarce yet the large US natural gas companies tend to loan funds to producers to finance exploration and/or production rather than integrate backwards into production. This is in contrast to the habit of oil companies. Even those, like the Shell Group, who started as marketers have moved backwards into production and exploration, whereas the initial explorers, like the British Petroleum Group, have moved forward to become marketers. In Canada the natural gas company habit is similar to the USA but the industry is younger and more oil money has gone further towards the market by investment in the major trunklines and there are some close ties between the gas producer and the pipeline company as in British Columbia. Yet, as in the USA, the essential rigidity of the system has made availability susceptible to government control.

In Canada the Government objectives have been stated[80] as:

a) to assure Canadian consumers of the availability of adequate natural gas supplies at economic prices.
b) to encourage the development of the abundant natural resources.
c) to encourage continued expansion of exploration by allowing the sale of surplus natural gas in export markets, thus improving Canada's international trade accounts under price conditions which:
  (i) recover the appropriate share of the costs incurred,
  (ii) are not less than the price to Canadians for similar deliveries in the same area and,
  (iii) do not result in prices in the US market area materially less than the least cost alternative for energy from indigenous sources.
d) to safeguard national security, both in terms of physically available resources and of capacities to provide and use them.

One aspect is that applications to export gas have therefore to prove to the provincial and federal Energy Boards that the quantities proposed are covered, for the period proposed, by proved reserves sufficient to ensure the continued cover of Canada's expanding needs. A yardstick used by the Government has often been a term of thirty years rolling demand. This system has entailed far greater involvement by the government agencies in forecasting demand and supply than in most other countries and consequently the long-term control of production far greater than that over oil under normal prorationing. In Canada control of gas is under double control because the provincial government have full rights to the administration and operation of their natural resources. They therefore deal with conservation aimed at the elimination of waste, including prorationing of gas to market demand from field to pipeline, but also control of the distribution and prices through regulation of the gas utility companies. These provinces, Alberta, Saskatchewan and Ontario have statutes to prevent the removal of gas without a provincial permit. The federal government, in addition to having jurisdiction over the Yukon and Northwest Territories, claims jurisdiction over all offshore areas and also has the right to regulate interprovincial and external trade. This double system can obviously give rise to problems. Alberta is contributing about 81% of Canadian gross natural gas production and wishes to increase the price to reflect its commodity value relative to alternative fuels whilst also setting a lower price for Albertan consumers by a rebate system. This raising of gas prices to competitive level is caught up with both the increase in world oil price in 1973/74 and in the disagreements between the federal and provincial governments over an export levy, and royalty and income tax expensing. There can be no doubt that these political problems discouraged exploration for new sources and the plans for gas exploitation in 1974 and 1975.

Alaskan Supplies

Progress has also been made, and at least the basic choice has been taken as to the general route of a pipeline from the North Slope of Alaska through Canada to the 'lower 48'. The formal agreement between Canada and the USA was signed on 20 September 1977. The line will run from Prudhoe Bay south to Fairbanks and then along the line of the Alaska Canada Highway through Southern Yukon and across British Columbia and Alberta to the US border – a distance of 4,425 km. The cost may reach $13 billion, with construction 1980–1982 and completion in 1983. Access will thus be gained to the 680 km^3 proved reserves and 1,133 km^3 potential resources of Prudhoe Bay.[381] These are of the same order as the current reserves of the UK North Sea. If a spur line is built to the Mackenzie Delta, access would be given to additional resources which were reported[382] as being 137 km^3 proved and 476 km^3 probable at 31 December 1976. The main line will carry 25.5 km^3/year, which compares with the earlier UK North Sea line capacity of 10–16 km^3/year. The main line capacity will rise to 36.5 km^3/year if the Canadian gas is brought in. This line will be of benefit to both Canada and the USA and, in mid-October 1977, there was talk[383] of the USA being willing to discuss additional supplies of Canadian gas. The need of the USA for imported gas is illustrated by some estimates given in August 1977 and reported in November[392] that the USA (East and West Coasts) might well be importing 63.3 km^3 in 1985, and 90 km^3 by 1990. It may also be noted that LNG costs are rising. The 1977 costs for handling 6 km^3/year include $700 million for a liquefaction plant, $200 million for a re-vaporizing plant, and $120 million for each 125,000 m^3 LNG tanker. Big money is involved and the methods of financing such schemes are various.[415]

Middle East Gas

Some more recent estimates[381,385] differ somewhat from those given for individual countries by Adams and Kirkby[75] but current and future availability is much more important than relatively minor differences in reserve estimates. An estimate[381] of current availability gave the figures of table 44.

Table 44
Major Middle East countries: natural gas availabilities[1]

	km³/year	billion cubic feet/year
Iran[2]	48	1,700
Saudi Arabia[3]	43	1,500
Abu Dhabi[3]	14	500
Kuwait	10	350
Iraq	9.6	340
Qatar	5	180

[1] based on estimated 1977 crude oil production.
[2] National Iranian Oil Co. suggested that the country's ultimate gas reserves might be as high as 830 trillion cu.ft.
[3] each include half of the production from the Neutral Zone.
Note: These availabilities are potential availabilities in that the bulk is associated with the production of crude oil and even in Iran, which has the greatest utilization, roughly half this gross production of 48 km³ will be flared.

Source: 'Larger Role for Natural Gas', E.S. Tucker, *Pet. Econ.* pp. 334–347, Sept. 1977.

Tucker quotes a recent estimate of costs of transportation from the Persian Gulf to Japan for liquefied natural gas (LNG) as $44/t, liquefied petroleum gas (LPG) as $28.2/t, and crude oil as $6.4/t. With LNG tankers costing some $120 million for a vessel of 125,000 m³ capacity and the costs of the plants at each end as costly as noted earlier, the ocean transport of natural gas is expensive. One can note that the ambitious LNG project from Iran to the USA and Europe, involving El Paso of the USA and Distrigaz of Belgium, was shelved in 1977.

Abu Dhabi started exporting LNG from the Persian Gulf in April 1977, the first in the Middle East, and there are plans for further development. Iran was actively pursuing a number of plans. Kuwait started exporting LPG in 1962, but requires a new non-associated gas source if its future plans are not to be restricted by the declining oil production which is expected. Deeper drilling is therefore planned through the Burgan oilfield to tap horizons which are gas-bearing elsewhere. Qatar is now hopeful that the recent discovery in the 'North-West Dome' area, some 40 miles offshore, may be sufficiently large to put gas into the international LNG trade. Iraq is still only utilizing about 15% of its 9.6 km³ current gross availability.

Saudi Arabia appears to be content to export LPG and natural gas liquids, (NGL), and not enter the LNG trade. Dr. Taher, Governor of the state oil company, Petromin, is reported[381] as saying that on a production of 12 mbd (almost 600 Mta) in the mid-1980s (itself an interesting forecast), virtually all the associated gas at 57 km³/year will be processed to provide 16.5 km³ of methane for inland consumption, and 15 Mta LPG and 5–6 Mta NGL plus substantial tonnages of sulphur for export. This implies both a scaling down and slippage of the original massive scheme of a few years ago. Costs have risen from $5 billion in 1975 to $16 billion in 1977 for processing some 41 km^3/year (4 bcfd), of wet associated gas. It is also realised that impossible tasks were being set in the simultaneous construction of all the industrial projects to utilize the gas. Two new NGL lines, (28 inch and 30 inch) are to be constructed over the 1,200 km from the eastern fields to a new terminal at Yanbu on the Red Sea, as well as a parallel 48 inch crude line.

These plans for the utilization of their natural gas resources, particularly in Iran and Saudi Arabia, are important. Although situated so far from the big industrial markets of the USA and Japan, and Europe, it is only from these great sources that the bulk of the expected natural gas can come in the future, supplementing North African supplies. The availability of associated gas, however, depends on the rate of crude oil production. Saudi Arabia was producing crude oil at the rate of 500 Mta in April 1977. If production in the mid-1980s is limited to 450 Mta, then the increase in availability of LPG and NGL will depend mainly on increased recovery.

The natural gas resources Summary of the draft CoCo report[377] suggested that the production capability of their OPEC Group 1, (Kuwait, Libya, Qatar, Saudi Arabia, and the United Arab Emirates), would rise to a plateau peak around the year 2010 at 595 km^3/year, falling to between 438 and 467 km^3/year by 2020. Their OPEC Group 2 (Algeria, Ecuador, Gabon, Indonesia, Iran, Iraq, Nigeria, and Venezuela), were given a production capability curve moving upwards rapidly from 1995 at some 170 km^3/year to a sharper peak about 2015 and only falling from 1,274 km^3 to 1,189–1,133 km^3 by 2020. In contrast both North America and Western Europe plateau before 2000 at about 790 km^3/year and 248 km^3/year respectively; the USSR and Eastern Europe peak around 2005 at 1,540 km^3/year. The only countries which it was suggested might have a production which was still rising in 2020 were the small producers to a total of 280 km^3 in 2020. WAES[322] calculated that by 2000 the production of the OPEC countries would need to be about 855 km^3/year, (15 mbdoe in their units), to satisfy their own needs and the import needs of other countries. The OPEC total in the draft CoCo report[377] is some 1,027 km^3/year, but it must be remembered that this gas production is dependent to a very large extent on the assumed rates of crude oil production.

In brief, North American and European gas production capability may well peak before the year 2000 and so would be increasing thereafter their demands on outside sources, in competition also with the presumably increasing need of Japan. These outside sources, mainly in the USSR and the Middle East would themselves be peaking within the first quarter of the next century. The world aggregate production, according to the draft CoCo report would peak at around 3720 km^3 (143 EJ), about the year 2000. The gas position in the next century would seem therefore to depend on the natural gas supplies associated with a declining crude oil production, on the amount of new non-associated natural gas sources, perhaps from gas hydrates and geopressured zones, and on the production of synthetic natural gas (SNG) from coal and from very heavy oils.

Far East and Australia
LNG exports from Brunei to Japan started in 1972 and were planned to be of the order of 7 km^3 by the mid-1970s. Indonesia also was very eager to get its

share of the expanding Japanese market for LNG and exploit its proved reserves of about 1,000 km^3, and in 1978 was exporting 5 km^3 by LNG tanker.

The gas discoveries in Australia, particularly offshore, first in the Bass Strait and then on the north-west shelf, are important to Australia. Production had risen to 6.9 km^3 by 1978. Delivery of gas from the earliest discoveries in the Gippsland basin, offshore Victoria, started in 1969 and provided 10% of the state's total energy needs in 1974 and was expected to provide 20% by 1980. One source[141] in 1974 estimated the recoverable reserves off Victoria as 209 km^3 and that by the year 2000 cumulative consumption of natural gas in Victoria for purposes other than electricity generation will have risen to 170 km^3 and the annual rate of consumption to 14 km^3. In general with total Australian proved reserves at the end of 1978 estimated as possibly of the order of 1,000 km^3 it seemed in 1979 that Australia would be inclined to reserve its natural gas resources for its own domestic use.

Northwest Europe
In Europe the gas industry is separate from the crude oil industry for different reasons than in North America. In Europe the manufactured gas industry from coal was well established before natural gas was exploited in significant quantities. In Britain the manufactured gas industry, established by private enterprise for over 150 years, was nationalized in 1949 and has a monopoly right to purchase all gas produced. The British Gas Corporation had an exploration partnership on land with the British Petroleum Company in the early 1950s. The Corporation went into offshore exploration in the early 1960s, mainly in partnership with Standard Oil of Indiana. Offshore success, particularly working without partners off Liverpool and in Morecambe Bay in the Irish Sea, has given the nationalized company valuable experience in all phases of the industry. In the British water and land areas the Corporation has the right of purchase for all gas found. In the British sector of the southern gas-bearing basin of the North Sea the price given for the first gas discovered, the West Sole field of British Petroleum, undoubtedly encouraged investment in exploration and production; however, lower prices were offered for gas discovered later. This just as surely inhibited further investment and encouraged the explorers to move into the northern basin where oil was considered the more likely prospect, as indeed it proved to be. However, increasing potential supplies of associated gas in the northern basin, from which the Corporation started taking gas under a contract signed in 1975 for the gas from the Brent oilfield, made them negotiate contracts in 1978/1979 in the southern basin[439] aimed at greater flexibility in production control to handle a possible surplus over anticipated demand. Nonetheless, rising oil prices and energy shortages caused by the reduction in Iranian crude exports, again tended to increase the demand for natural gas, which the other nationalized British energy industries, coal and electricity, already accused the Corporation of selling at too low a price.

The development of the petroleum resources offshore northwest Europe is still at a very young and dynamic stage, but there are three aspects of this development which are important as examples of current trends affecting world availability. The first is the offshore legal framework within which exploration and development are taking place. The second is the deep involvement of governments in the energy field, and, on the continental shelf of northwest Europe, there are variations between countries which emphasize the importance of differing government attitudes on current and potential production. Thirdly, the natural gas situation, in particular, illustrates the dubiety in reserve estimation, the control on production by the available facilities, the importance of resource management, the growing international character of the natural gas trade and continuing influence of the Middle East, and, finally, the turn full circle to the importance of considering all energy sources at the same time, with the potential importance of coal as the source for substitute natural gas.

Problems of international jurisdiction

The Geneva Convention of 1958 is the legal basis for the exploitation of the continental shelf of northwest Europe. The Convention became international law when Britain became the twenty-eighth signatory in 1964. The guiding principle is the drawing of an imaginary line from land boundaries in such a way as to maintain equal distance between the land on either side outward to meet another median line, as in the North Sea, or to the edge of the Shelf, which is defined as that contiguous part of the seabed which is under less than 200 m of water or is capable of exploitation. These so-called median lines in the North Sea were agreed between Britain and its neighbours and between the other states with no problems, except a temporary one between FR Germany and its neighbours, which will be mentioned later. The agreement in the North Sea is extremely important, because the most prolific oil and gasfields of the northern basin lies close to or even straddle the median line between the UK and Norway; the giant gasfield of Frigg, and the giant oilfield of Statfjord straddle the line and will be produced by agreement between the two countries. There are however areas of dispute elsewhere, which might inhibit exploration and cause problems.

In the far north, Norway administers the Spitzbergen Islands (Svalbard) under a Treaty which was ratified in 1925 by the Great Powers, including the USSR. This treaty did not give Norway exclusive rights to the exploitation of the natural resources. The Soviet Union operates coal mines which produce some 450,000 ta; this is matched by Norwegian-operated mines. The Treaty covers the sea area for four miles offshore, and it is within these limits that in 1975/1976 both the USSR and Norway each drilled an exploratory hole for petroleum; Norway tested 1.34 km^3/year of natural gas, (130 mcfd). The problem becomes complicated outside the Treaty limits into what Norway considers Norwegian waters but no agreement has yet been reached with its mainland and islands neighbour, the USSR. Starting from the mainland mutual boundary, which is well east of North Cape, a median line would run north across the Barents Sea, midway between Bear Island (Norway) and the Novaya Zemlya Island group (USSR), and then east of Svalbard (Norway) and west of Franz Joseph Land, (USSR). The Soviet Union, however, argue that a wedgeshaped allocation south from the North Pole should be adopted in arctic waters similar to that adopted for Antarctica. The Russians are sensitive, because they fear that any offshore structures or operating craft on the sea or in the air could be used for surveillance of their naval units, based on Murmansk, which use the all-weather route in what would be under Norwegian jurisdiction if areas were designated. Both suggested lines would run east of the geological basement high, which separates two potential petroliferous basins. The Soviet sensitivity also applies further southwest, and the Norwegians decided[527] in October 1979 to delay until March 1980 the decision as to whether work could start north of 62° N. latitude. There would seem to be no reason why either country should wish to explore in the Barents Sea, for both have more than sufficient known resources elsewhere to

NORTH SEA ESTIMATED OIL PRODUCTION   Fig. 20
(in 10,000 bd)

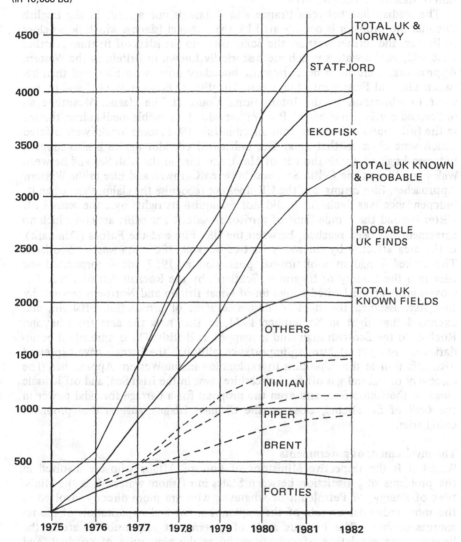

TOTAL UK & NORWAY

STATFJORD

EKOFISK

TOTAL UK KNOWN & PROBABLE

PROBABLE UK FINDS

TOTAL UK KNOWN FIELDS

OTHERS

NINIAN

PIPER

BRENT

FORTIES

1975   1976   1977   1978   1979   1980   1981   1982

Source: Wood Mackenzie reported Petroleum Economist Vol. XLIII, No. 5, p. 176, May 1976

keep them busy for many years. However, southwest of North Cape and off Tromsoe there are two basins which are potentially petroliferous, and it is the declared policy of the Norwegian Government to help their northern regions. It is therefore obvious that Norway, a founder member of the North Atlantic Treaty Organization, (NATO), has a political problem which affects the exploitation of its shelf in the far north.

The median line between France and Britain is not agreed. In the English Channel the problem is complicated by the Channel Islands, which lie so close to France, and further west are the Scilly Isles and the islets off Brittany; further west still, in the waters which are historically known in Britain as the Western Approaches, there must be a tripartite boundary point with Eire, and then between Eire and France into the ocean. The dispute between the UK and France went to arbitration to the International Court at The Hague. Meantime an unlicensed corridor has been left on either side of a possible median line, though in the fifth round of licences, announced in July 1976, some blocks were offered which were close to that area. The unlicensed corridor device is also adopted between England, (with the Isle of Man), and Eire in the Irish Sea and between Wales and Eire in the Celtic Sea, and between Cornwall and Eire in the Western Approaches. Eire claims, but the UK does not recognise the claim, that, when its independence was declared, it did not relinquish its rights over the seabed off Ulster beyond the 3 mile limit of territorial waters. The other area on which no agreement has been reached, between the UK, Eire and the Faroes (Denmark), is the area affected by the rocky islet of Rockall, 300 miles west of Scotland. The United Kingdom took formal possession in 1957 and incorporated the islet into the County of Inverness, Scotland, by the Rockall Act of 1972. The Government of the United Kingdom of Great Britain and Northern Ireland, by this Act, assumed the right to designate areas of jurisdiction offshore, and exercised that right in September 1974, so that now the area from beyond Rockall to the Scottish mainland is mapped as British. These unresolved boundaries are not yet of burning importance, although the French have expressed dissatisfaction at the inhibition to exploration in the Western Approaches. The discovery of natural gas off the Lancashire coast in the Irish Sea, and off Kinsale Head in the Celtic Sea, and even the proposal for a barrage for tidal power in the Gulf of St. Michel, close to the Channel Islands, suggest that problems could arise.

The involvement of governments

Whilst it is the respective Ministries of Foreign Affairs who are involved in the problems of jurisdiction between States in offshore waters, it is the Ministries of Energy, or Petroleum or whatever, who are more directly involved in the more industrial aspects of the exploration for and exploitation of the resources offshore. The obvious areas of government responsibility are in the licensing and regulation of operations by establishing rules of conduct, and policing activity to ensure conformity, particularly in matters of safety. Governments may also encourage their own industries to participate in the activity and to benefit from the opportunities, whether in the service, supply, or relevant manufacturing and energy industries, if necessary by promoting training schemes of various kinds for their own nationals. There are many similarities in the ways in which the different governments handle their role, but many dissimilarities. It is not possible to note even all the latter. One example is that the permit areas in the North Sea are on a grid system, which in British waters have a unit area of 210-250 km², in Dutch waters 390–420 km², and in Norwegian waters 500–570 km². There are also differences in the terms and duration of licences and their relinquishment, not only between countries but within countries, from time to time; work obligations are expressed in terms of seismic programmes and required wells in the British and FR German sectors, but in fixed expenditure levels in the Dutch sector. There was no original plan agreed by the interested countries; each country developed its own rules, erratic step by erratic step, as it has seen its own needs, and has sought ways to satisfy them. This point can be seen best by looking in turn at Britain, Norway, Sweden, Denmark and the Federal Republic of Germany, which will emphasize that the availability of the petroleum from the northwest continental shelf of Europe, depends on many other complexities than the geological ones, and these are only being unravelled slowly.

The United Kingdom
Britain was the first to encourage competitive interest in the exploitation of its offshore waters, which it did by soliciting bids in certain licence areas for which work plans had to be submitted. Selection of successful applicants was then made on the strength of these work plans; in cases of overlapping applications, the general, technical, and financial resources of the applicants were weighed by the Government Department, together with, if necessary, an assessment of the contributions made by competing companies to the British economy and, in the case of foreign-owned companies, the comparable treatment being given to British companies operating in their countries of origin. Such criteria are, therefore, wider than economic or technical. The system of sealed cash bids without work obligations, which is the system common in the USA and Canada, was tried for a few permits in 1971, but has not been repeated. The first applicants were experienced, well-established oil companies, but, with time, and particularly after prospecting moved from the gas-bearing southern basin to the oil-bearing northern basin, many smaller, younger oil companies, and non-oil companies participated, often in groups combining petroleum expertise and risk capital. The British Gas Corporation (BGC), has the legal right to the first refusal to buy any gas discovered on land or at sea under British jurisdiction, but it also decided to enter the offshore industry directly and first in partnership with Amoco (UK), the local subsidiary of the major American oil company, Standard Oil Company of Indiana, which had a lot of foreign and offshore experience world-wide. Later the National Coal Board took up some interests, first with the American oil company, Continental Oil, with whom we have seen it has mutual coal interests. British Gas, as the monopoly buyer, and encouraged by the Government, paid a price which was considered reasonable by the oil industry for deliveries from the first discovery of gas from the southern basin. Prices for gas from later discoveries were not considered to be so favourable. This opinion both inhibited further exploration in the south and encouraged the move northward into what materialised as a predominantly oil basin but is also an important source of gas. The freedom to trade in any oil discovered attracted the small oil companies and the non-oil companies. The result is that there is a very heterogeneous collection of interests now represented in the British sector of the North Sea, including a number of foreign nationalised companies, e.g. from Iran and FR Germany, and, most recently, the British National Oil Corporation (BNOC).

The Government White Paper, *United Kingdom Offshore Oil and Gas Policy*[248] announced, in July 1974, the proposal to form a national oil company, and proposed new taxation on offshore oil profits, changes in the terms of existing licences, majority participation by the state in existing and future licences etc. These proposals created an atmosphere of uncertainty which was enhanced by the slackening in demand for petroleum and energy, because of world economic recession, higher prices, conservation and economy measures, and mild winters. The Government enacted the Petroleum and Submarine Pipelines Act in 1975 and continued discussions with the industry about majority participation. Uncertainty, however, continued, partly because many of the clauses of the Act had an element of discretion by the relevant Government authority, and the impact of those clauses was difficult to pre-judge until precedents had been set. BNOC became a legal entity on 1 January 1976, and assumed all the oil interests of the National Coal Board, its relevant, small staff, and its considerable forward development cost obligations. The chairman of BNOC stressed that he wished to work with the private sector as a partner but did not disguise the fact that he wished to take advantage of their knowledge and expertise. BNOC was committed by mid-1976 to spending £375–425 million in 1976, including £90 million out of a total possible £500 million development costs of the National Coal Board; Burmah Oil Company's share in the Ninian oilfield was bought for about £900 million, and 65% of its share in the Thistle oilfield for £87 million, with a further future development obligations of some £300 million. Other deals with other companies have or are being made and BNOC is fast moving into the industry. A deal with British Petroleum will provide training and experience in the refining and marketing management aspects, for BNOC has the right to take up interests in these aspects as it may consider appropriate, although it had not made any firm commitments in its first six months of existence. The Government, in a different area, has made it compulsory for all employers to accept statutory trades union recognition on offshore structures and grant general trades union recognition under the Employment Protection Act. The Government affirmed that these measures merely extend offshore the normal industrial relations in force on land. These examples of the involvement of the British Government in the offshore oil industry are not comprehensive, other aspects are considered later in dealing with natural gas, but they should provide sufficient to allow comparison with other countries.

Norway

The Norwegian Government had little knowledge of the oil industry in 1963 and Norwegian industry, unlike British industry, had only been involved directly in the tanker sector. The first licences in Norwegian waters were therefore given in 1965 to large international oil companies who had the expertise and funds to explore, and the state retained interests of 5% to 17.5%. However, as it became obvious by 1970 that resources were being discovered which were far beyond Norway's modest needs, the Government changed its attitude. The State acquired an interest, eventually 51%, in Norsk Hydro, which had become a major chemical company and had shown an interest in the developing petroleum scene. Statoil was established as a state company on 1 January 1973, and given the responsibility of acquiring between 20% and 50% participation in the companies operating in the Norwegian sector. In February 1973, Government policy was expressed as following three major principles:

i. Government control of exploration and production by majority participation, and of refining, petrochemical and marketing aspects by a lesser involvement,

ii. a moderate scale of production of oil and gas,

iii. a slow introduction of the resulting revenues into the economy.

Then a new state-owned refining and marketing company was formed in 1976, with the state owning 71% directly, Statoil 15% and Norsk Hydro 6.7%. The structure of this company illustrates that the Norwegian distrust of monopolies, of big companies, and of a specialist elite which might be difficult to control democratically, applies to their own people as well as to outsiders. (This is in contrast to the way in which the British Labour Party aimed to concentrate all aspects into one integrated company in competition with the major international oil companies.) The second objective of slow growth in production may create problems for Statoil which is supposed to be a self-sufficient and profitable entity, and may therefore wish to increase production to earn revenue to finance forward operations, at a time when the Government wishes to hold back. The restraint of the second and third policy objectives, is based on a deep rooted conviction that the newfound wealth must not destroy the Norwegian way of life. Hydro-electricity was developed from the early 1900s and gave Norway an indigenous energy source which fully satisfied its needs until petroleum products grew in importance from the 1950s. However, when oil equalled hydro-electricity as an energy source in 1972, the amount was only about 6 Mtoe, and indigenous oil production was already starting from the North Sea at about 1.3 Mtoe. Hence Norway had no import burden of 100 Mtoe of foreign oil as has Britain, and therefore no cause to press for rapid production to replace imports. Furthermore the Norwegians have developed, since the catastrophe of their country being occupied during the Second World War, a strong economy, a high level of culture and technology, and a work system which permits them to have leisure to enjoy the beauties of their country at all seasons. They wish to preserve their system. They see no reason why they should produce their oil for the benefit of others, if, in so doing, they destroy themselves. This is the driving force behind their restraint, and why they would like to have full control over their oil and gas, and land it direct to Norway. Geography has hitherto prevented this.

The Norwegian Trough or Trench is a deep steepsided submarine feature, close offshore the southwest angle of Norway, which has acted as a moat and prevented the laying of submarine lines to carry oil or gas to the mainland. Only thickwalled, small diameter pipe can normally withstand the pressure of 450 m of water; these would not permit a throughput which might be considered economic and, more particularly, would have serious maintenance and repair problems over the long distances involved. Nevertheless the Norwegians are making considerable efforts to overcome the disadvantage of the Trench. Perhaps their easiest route will be from the more northerly fields, like Statfjord, to Bergen, where the Trench is shallower. A similar deep sea crossing also faces the Algeria–Italy gas line with the Straits of Messina 650 m maximum, or an Algeria–Spain line via the Straits of Gibraltar with about 600 m of water. Meantime however, because of the Trench, oil from the Ekofisk group of fields is going west to England, the gas from these fields is to go south to Emden in Germany, and the gas from the Frigg gasfield further north will be piped to Scotland.

In brief, Norway, as will be seen, has large quantities of oil and gas in its sector of the North Sea, especially in large fields close to the median line with Britain, but has a political boundary demarcation problem in the north, with the Soviet Union, a social attitude which affects the speed of development of these resources, and a geographical problem, which has prevented absolute control of the disposal of production and will continue to do so, in that once pipelines have been laid and contracts signed, those supplies become a permanent and almost inflexible feature.

Sweden
Sweden, on the other hand, has little prospect of finding oil or gas in its offshore areas, either in the Baltic, or in the narrow waters towards Denmark and FR Germany, but has an energy consumption some 2½ times that of Norway, and relies on imported oil for 80% of its energy. Coal production is only 10,000 ta so that its main indigenous energy is hydro-electricity, which, with the help of some nuclear power, provides 16% of the total energy demand of 50 Mtce. One 50% state-owned company was established in 1969 to prospect for domestic supplies of petroleum, and a similar company in 1973 to explore overseas. Sweden talked to Britain in late 1975 about the possibility of participating in the North Sea activity, but its obvious natural links are with its fellow Nordic countries. In March 1976 Sweden made an agreement with Norway to co-ordinate their petroleum and petrochemical industries, studies were initiated as to how the new petroleum supplies could be used to mutual benefit. One suggestion has been that, if a gas supply could be piped to Bergen, a joint petro-chemical plant might be constructed on nearby Sotra Island. The Norwegian Trench is a moat to Sweden as to Norway.

Denmark
Denmark sketched out its offshore territory on the lines of the Geneva Convention and awarded an offshore exploration licence in 1962 to the Danish/American consortium which had been exploring for petroleum on land for some years. This permission was given for the whole area, and was given before the Convention became law with the twenty-eighth signature. The Federal Republic of Germany objected that the small wedge-shaped area which the strict delineation by the Convention would allocate to them was not commensurate with their ability to explore for and exploit whatever resources were under the seabed. They applied to the International Court at the Hague, who expressed sympathy with their case but suggested that they agree on some adjustment with their neighbours. This they did, with both Denmark and the Netherlands ceding some areas. However FR Germany agreed to honour concessions in these areas which had been already granted. Although the American representation in the Dansk Undergrunds Consortium (DUC), has changed, Denmark is still relying on their expertise and has not formed a national company. However, a report, in late 1975, of a Committee set up by the Minister of Commerce, expressed concern at the slow progress and particularly the overcaution with which DUC appeared to have estimated the size, (26–42 km^3), and economics of the recoverable gas reserves in the five structures indicated by drilling after the original discovery by DUC in its second hole in 1967; an independent estimate made for the Committee by the Dallas, Texas, consultants deGolyer and MacNaughton, suggested that recoverable gas reserves might be 48 m^3 plus 20 km^3 probable reserves. In May 1978 deGolyer and MacNaughton estimated that Danish gas finds might supply 2.5 km^3 in 1985, rising to 4 km^3 in 1995 and be sustained at that level for 12 years from the four major fields of Dan, Cora, Vern and Bent, which in mid-1979 were taken to have 40 km^3 proved and 40 km^3 of probable reserves. DUC estimated that between 53 km^3 and 75 km^3 might be recoverable from the total reserves which would include at least 5 km^3

of proved reserves from three minor fields. In February 1979 the Dansk Olie & Natur Gas A/S (DONG) contracted to import 1.7 km^3 from 1982 to 1986 from the German Ruhrgas and in March 1929 some 55 km^3 between 1984 and 2009 from DUC, so allowing a gradual build-up of the necessary infrastructure and markets.[469]

The Federal Republic of Germany

FR Germany has also been disappointed with its sector of the North Sea. The North Sea Consortium of German companies was given sole rights in 1964 over the area originally allocated to FR Germany, and later, over the areas ceded to Germany by Denmark and the Netherlands, which had not already been allocated. The last of twelve dry holes drilled by the Consortium was abandoned in 1967 and large areas were returned to the Government. Exploration has not been vigorous and one reason is said to be that there is a lack of definition in the jurisdiction between the States, (the *Länder*), and the Federal Government. The lack of success in its own sector spurred on the German Government's determination to seek oil elsewhere, through its chosen instrument, the Deminex company which was set up in 1969, with the Federal Government contributing DM 575 million and private enterprise DM 191 million; DM 700 million had been spent by 1975 and the 1976-1979 budget was said to be for a further DM 1,000 million. In January 1976, Deminex had interests in the North Sea, including a share in the Thistle oilfield, in the Irish and Celtic Seas and off eight other countries around the world. FR Germany has thus become involved in yet another different way, through a mixed company, in areas outside its own sector of the continental shelf, because, though it has a broad-based energy supply pattern, yet oil in 1975 did contribute 129 Mt and natural gas 33 Mtoe out of a total of 242 Mtoe, and its own onshore production only 5.7 Mt of oil and about 17 Mtoe of natural gas, most of it from the extension into Germany of the Dutch Groningen gas field.

The Netherlands

Exploration in Dutch waters has not been very successful, and the giant gasfields of the English sector of the southern basin of the North Sea appear to be separated by small structures from the giant gasfield of Groningen on land. The first offshore licences were granted in 1968, and an independent oil company, Placid International Oil of Dallas, Texas, found gas in January 1970, in a wildcat well some 44 miles offshore and in 26 m of water. A production licence was granted and the Government declined to take up the 40% interest to which it had the right by the terms of the licence which it had been granted. However, when FR Germany took up an option in December 1972 to buy the gas, the Dutch authorities blocked the deal. The Germans objected that such action contravened the Treaty of Rome which established the European Economic Community of which both countries are members. In 1973 an agreement was signed allocating half the proposed gas supply to FR Germany with half to remain in Holland. Placid's supply contract is for 6.5 km^3 per annum, for 24 years, with the reserves estimated to a maximum of 150 km^3. The gas is piped to landfall at Uithuizen where the line divides, one half going to the Gasunie system for distribution in Holland and the other line to Germany near Emden. The main line has a capacity of 12 km^3 per annum and therefore has spare capacity for collection from future fields. A second line of smaller capacity is planned from a small group of gasfields further west to landfall further south, near Den Helder, for Dutch consumption. The Government in late 1974 announced that it would take up its share in all future discoveries. The change in attitude from that of three years earlier was due to the realization that the Groningen field was not as big as originally estimated, that most of the gas was committed to export and supplies to the domestic market might be starved in the 1980s. In this situation, it is perhaps surprising that, despite warnings from the oil companies that any increased burden could inhibit investment in further offshore exploration, the Government announced in April 1976, new terms for future licences which would increase the overall take by the State.[308] This action does, however, fittingly emphasize the involvement of Government in the exploitation of the continental shelf.

The availability of indigenous natural gas resources

Britain. Natural gas was first discovered in the British sector of the North Sea in 1963 in what eventually became recognised as the gas-bearing southern basin. Natural gas was later found, associated and non-associated with crude oil, in the northern, predominantly oil basin in both British and Norwegian waters. More recently, gas had been found in commercial quantities in the Irish and Celtic Seas to the west.

Denmark, as already noted, may have reserves of between 26 and 49 km^3 proved. The Netherlands authorities say that there are about 322 km^3 offshore. However, again, such a figure can only be an indicator, for exploration is at a very early stage but with little hope of giants and yet 322 km^3 is more than half the proven reserves of the British sector of the southern basin of the North Sea, hence, once again, it is doubtful if these two figures are based on the same criteria.

Production — Britain. Table 45 sets out an estimate of the annual productions from 1975 to 1982 of the known fields, with known production plans, for both Britain and Norway.* In the southern basin, the maximum of the annual production plateau is 1979 production at 4.355 mcfd, with the larger fields having individual plateaux of from 3 to 6 years duration, before, as the West Sole field, the oldest, implies, a long gradual decline. This is itself instructive. In the northern basin only Frigg and Brent are named, but, as will be seen later, there is some gas to be taken by BGC from the Forties field, whilst some is being used in the refinery at Grangemouth already. Hence one may expect a little but not much more gas in this period.

Britain had an intensive pipeline distribution system for the domestic supply of town gas manufactured from coal. The advent of natural gas was easily handled by substitution and adaptation of appliances. Natural gas now supplies 97% of the domestic market to over 14 million customers. British Gas came under some criticism in February 1976, that reserves were being depleted too rapidly, and gas was under-priced and so in unfair competition with coal and electricity. The BGC policy, as restated in June 1976,[310] is 'to conserve gas by giving priority to the premium markets, i.e. a domestic usage, other space heating, and industrial processes where a high grade fuel is required'; supplies to non-premium customers will be on an interruptible basis and restricted to the amounts necessary to allow operational flexibility, without incurring provision

---

* Actual productions in 1976, 1977 and 1978 were somewhat lower than the estimates being,[442] respectively, for the UK, 3841, 4030 and 3823.

Table 45
North Sea: estimated gas production
(million cubic feet/day)

Name of field	Estimated recoverable reserves tcf	1975	1976	1977	1978	1979	1980	1981	1982
UNITED KINGDOM									
"Big Dotty"	0.4	–	60	130	190	230	230	230	230
Hewett	3.0	730	750	750	750	750	750	750	680
Indefatigable	4.5	615	700	700	700	700	700	700	665
Leman	10.5	1,510	1,640	1,640	1,640	1,640	1,560	1,480	1,410
Rough	0.4	2	130	150	150	150	150	140	130
Viking	4.5	550	650	700	700	700	665	630	600
West Sole	2.2	180	230	230	210	185	170	150	130
Total, S. Basin	**25.5**	**3,587**	**4,160**	**4,300**	**4,340**	**4,355**	**4,255**	**4,080**	**3,845**
Brent	3.0	–	–	–	–	200	600	600	600
Frigg	3.4	–	–	72	432	682	682	682	682
Total both Basins	**31.9**	**3,587**	**4,160**	**4,372**	**4,772**	**5,237**	**5,507**	**5,362**	**5,127**
Probable Finds	4.0	–	–	–	–	70	400	750	1,050
Total United Kingdom	**35.9**	**3,587**	**4,160**	**4,372**	**4,772**	**5,307**	**5,907**	**6,112**	**6,177**
NORWAY									
Ekofisk	12.6	–	400	800	1,200	1,200	1,200	1,200	1,200
Frigg	3.7	–	–	78	468	738	738	738	738
Heimdal	1.8	–	–	–	75	200	350	400	400
Total Norway	**18.1**	**–**	**400**	**878**	**1,743**	**2,138**	**2,288**	**2,338**	**2,338**
TOTAL, UK + NORWAY	**54.0**	**3,587**	**4,560**	**5,250**	**6,515**	**7,445**	**8,195**	**8,450**	**8,515**
Oil equivalent in million barrels per day		0.6	0.8	0.9	1.1	1.3	1.4	1.5	1.5

Source: Wood Mackenzie, Edinburgh, reported in *Petroleum Economist*, vol. XLIII, no. 5, May 1976, p. 177.

of uneconomic storage or under-utilization. Industrial use has increased sixfold in the past decade, and almost 6 out of the 13×10^9 therms consumed in 1975 went to industry, with more than half going to large customers using more than 100,000 therms. Some critics fear that this trend is liable to accelerate when supplies arrive from Brent and Frigg, and may rigidly mortgage the future through being tied to long-term contracts, which is the Dutch position, as will be seen later. On the other hand, some companies entered the North Sea in the hope of acquiring access to feedstocks for their chemical plants, and became concerned lest new regulations might cut them off from such supplies. In July 1976, Government assured industry that natural gas supplies (ethane, propane and butane) would be available to contracts for the next 20 years, and that methane supplies might be available for the next ten years.[309] The fluctuations in demand for gas are difficult to handle, and British Gas negotiate, for example, with a power station near the St. Fergus terminal of North Sea gas pipelines from the northern fields, in such a way as to help their flexibility, even though power stations cannot be classed as premium users. British Gas consider that, given their policy, and with anticipated future discoveries, 'there is every expectation that natural gas supplies will hold up past the end of the century'. This is a most important opinion, because natural gas may be expected to supply from 5.5%–13.6% of the total UK energy needs, according to different scenarios as to growth. In the six scenarios which were the basis for a 1976 discussion on the Energy R & D in the United Kingdom.[295] Those which had, in turn, the dominant features of low-growth, limit on nuclear power, high-cost energy, and self sufficiency, all assumed that natural gas annual productions would peak around 1990 and be small in 2015, as the following figures in Mtce show:

1975	1990	2000	2010	2025
52–53	65–80	45–60	20–35	5–10

whilst two scenarios, one based on a multifold increase in world energy prices in the early 1990s, and the other on a high-growth rate, assumed a higher peak in about 1990 and a sharper decline, though the same figure for 2025, viz.

1975	1990	2000	2010	2025
52–53	75–95	40–50	15–25	5–10

Other opinions, also tabled at the National Energy Conference, June 1976, were varied; the Energy Research Group, Cambridge, for both high and low supply projections of indigenous energy, gave 33 Mtce, (22 Mtoe) on their high demand base and 37.5 Mtce on their low demand base for 2000; the Fuel and Power Industries Committee of the Trades Union Council, recommended 60 Mtce in 1985, which compares with the 90 Mtce in the early 1980s of British Gas, the 72–66 Mtce of the Cambridge paper, the 65–95 Mtce of the Department of Energy, and the 58 Mtce in 1982 from table 45, adding Frigg production to the UK total, and converting at 7.64 barrels/t for North Sea UK oil. This range between estimates for only nine years ahead is somewhat surprising, but they will become closer as the size and producibility of the new reserves become better known; but perhaps much more important is the effect which decisions on the possible pipeline collecting systems will have, and these decisions will have to be taken very soon if there is not to be a great waste of hydrocarbon resources as the oilfields start coming on to production; but first a word about production in the Norwegian sector.

Production – Norway. The general attitude of the Norwegian Government and people favouring a slow development of their offshore petroleum resources has already been described, but, as was stressed earlier, the high investments in pipelines necessitates a rigidity, which is an important feature of the industry.

The natural gas productions from Ekofisk and Heimdal, in table 45 at 16 km^3 in 1982, are greater than the 6.7 km^3 per annum contracted to FR Germany etc. for 20 years, see table 46, through the Ekofisk-Emden line of 11 km^3 per annum capacity. Hence further contracts are implied to fill the line. The Norwegian Trench has hitherto sent Norwegian gas to the south and west, and there are three important considerations inhibiting gas going eastward in the quantities which could be available. Neither Norway nor any other Nordic country has the individual market size to merit a line and none have an existing gas pipeline distribution system which has made the transition from manufactured gas to natural gas quick and cheap in the other northwestern European countries. Furthermore, the rugged topography of Norway's coasts makes onward land transportation difficult. A line to southern Sweden via Denmark could benefit from flattish land, but Norway would find it difficult even to pipe large volumes to a few industrial users. Possibly the only alternatives are to construct at landfall chemical plants, as proposed near Bergen, or plants to convert, for onward sea transport, to LNG using special tankers, or methanol using normal tankers. Hence the bulk of Norway's gas, unless production of both oil and gas are severely restricted, would seem to be forced into export, at least on conventional ideas, and a gas gathering system becomes essential.

North Sea gas-gathering systems. The gathering systems in the southern basin are well established and need no further comment. In the northern basin the gas lines from Frigg were completed in 1977 and that from Brent in 1979. The gas from the Forties oilfield, which BGC have contracted for the period 1977–1980, is that which is carried in solution in the oil line and will be separated on land. The major problem is with the smaller deposits of non-associated gas which may be found, or the associated gas in small oilfields which would not themselves merit a gas line or might only be economic if both the gas and oil could reach market. All the gas will not be flared, even if it cannot be got to market from producing oilfields because some may be reinjected in order to maintain the oil-producing mechanism. The optimum amount may be difficult to pre-determine, and the gas/oil ratio is very variable between existing fields, so that it is difficult to assess how much associated gas could be available. However, both the British and Norwegian Governments have had preliminary studies made. The British Report was reviewed in June 1976.[286] The scheme was for a four pipeline system of 800 miles, with 550 miles needing pipe of 610 mm or more diameter. The scheme suggested that if a 6.2 km^3 per annum base load was available from adjacent fields then quantities of associated gas down to 0.5 km^3 per annum could be economically collected, but only down to 1 km^3 per annum of non-associated gas, which would have to carry the cost of all drilling etc. alone. The system was expected to have full energy peak in 1990. There were however some major, simplifying assumptions made and the Government recognise that the £1,590 million cost estimate could be far below the real cost. They said that some 22.7 km^3 per annum (2,200 mcfd), might be available for the new scheme and would be additional to the Brent gas which would go into the system, and the Frigg gas which would have its own line. This scheme is obviously in a very early stage but the Government have called for industry comment and, if further studies were favourable, would be considering completion in 1981–1982; this may be optimistic. The Norwegian Government issued a White Paper in May 1976, outlining, as the most economical for Norwegian waters, a gas-gathering spinal system running north to south with short connecting lines to individual fields and a capacity of about 20 km^3 per annum from total reserves of the order of 300 km^3.

Some form of collection of the associated gas and small deposits of non-associated gas in the northern basin of the North Sea is obviously essential if this natural resource is to be optimised. The cost, the delayed return, the intricacies of the political and commercial problems involved, and the long-term resource investment aspect, are all facts which private companies may not find economically attractive, and would seem to call for a joint UK/Norway venture. Such a venture would probably be without precedent, but one which it should not be beyond the wit of man to devise for mutual benefit. A total of around 43 km³ per annum of natural gas is involved, which is slightly less than the peak export sales from the giant Groningen field in 1978, see figure 21.

In 1979 the Norwegian Government seemed to adopt a somewhat more aggressive attitude to the exploitation of their North Sea oil and gas and a policy paper was promised for the autumn. France and Germany both seemed willing to consider financing pipelines which would bring more gas to them, particularly from the 50 km³ associated with oil in the giant Statfjord oilfield on which development was proceeding. British Gas had offered to take this gas as it was already taking the non-associated gas from the Frigg gasfield, but the Norwegians seemed reluctant to be tied to only one buyer but British Gas seemed just as reluctant to agree to the gas being brought to Britain in transit to continental Europe.

Production – The Netherlands. The management of the gas supplies from Groningen can offer a number of lessons of importance, particularly to UK and Norway. When the large size of the field was appreciated it was over-estimated, and the early thinking also was that its gas might capture 30% of the domestic market, leaving the remainder for export. In fact, the natural gas share of domestic energy consumption rose from 1.8% in 1963 to 47.5% in 1973, when total energy consumption was 61.3 Mtoe.[283] By 1976 gas had captured over 90% of the Dutch space-heating market, provided the fuel for 80% of the electricity generated, and met 75% of that part of the industrial energy market which was not tied to a particular fuel. The Arab embargo, in October 1973, on all

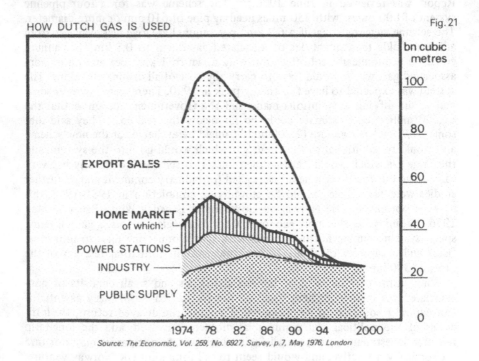

HOW DUTCH GAS IS USED   Fig. 21

Source: The Economist, Vol. 259, No. 6927, Survey, p.7, May 1976, London

oil to the Netherlands, caused a reappraisal of their only significant indigenous source. The Dutch became increasingly conscious that though this resource was great, it was both limited and over-committed, and the revenues from its export, and the savings from replacing imports, would start dwindling from about 1980. Figure 21 illustrates this.

The Gasunie organization is responsible for gas distribution, is owned 10% by the State, 40% by Dutch State Mines, 25% by Shell Group and 25% by the local Esso subsidiary, and put forward a Gas Marketing Plan in 1975 with lessening priorities of:-

— maintenance of supplies to small household consumers to 2005,
— supply to 1990 of high-grade industrial users to whom gas has specific value,
— run down, without renewal, contracts to power stations, with possibly no supplies from 1978,
— do not renew export contracts which peak in 1978, fall rapidly after 1985, as most were 20-year contracts, and virtually cease by 1994.

This policy is similar to that of British Gas, but the Dutch have had to learn the hard way. The remaining proved reserves of Groningen and other onshore and offshore fields will be 350 km^3 short of the 1,855 km^3 cumulative sales to 2000, including 840 km^3 of committed exports.[311] The Centralplan Bureau calculated that the surplus of 18 Mtoe on the energy trade balance in 1974, will be only 3 Mtoe in 1980 and become a deficit of 24 Mtoe in 1985. In money terms, the Algemene Bank Nederland calculated that gas export sales, plus energy savings, minus reduced export sales of manufactured goods because of the high value of the guilder, resulted in a net gain in the current trade balance of fl 5,000 million. Loss of gas exports will reduce this and manufactured goods will still be at an export disadvantage.[311] In brief, the Netherlands enjoyed a fast economic growth and put the money direct into the economy with a rapid build-up of social welfare — a pattern that Norway specifically wishes to avoid — and now faces rough times because the miracle of the new natural energy resource was not well managed.

The Finance Minister on 1 March 1979 argued[426] that though the increase in public spending had begun before the increase in revenue from gas exports, the cost of improved social conditions had been masked by that revenue; its decline would increase the balance of payments deficit by the late 1980s by fl. 15-20 billion (£3.75-£5 billion, 1979 money), unless drastic measures were taken to reduce public spending. The position seemed to be improved somewhat by July 1979, for Gasunie upgraded the Netherlands reserves.[427] The reporting of the Dutch gas reserves during 1979 did not give a clear picture of future availability. The Gasunie[427] and the Geological Service[507] appeared to revise the Groningen reserves upwards, the latter by 150 km^3 to almost 1800 km^3. The additions were made as to 56 km^3 to proved reserves with 90% probability of recovery, and 94 km^3 to probable reserves with 50% recovery probability, with a total proved plus probable for both onshore (1,905 km^3) and offshore (322 km^3) reserves of 2,227 km^3. Gasunie noted that the current 25 Year Plan required 1,691 km^3 of gas to the year 2003. However they then calculated that demand might be reduced, by only moderate economic growth and active energy conservation, to only 1,496 km^3, (891 km^3 domestic and 605 km^3 export), so easing the situation. The trend therefore, as the Netherlands goes into the 1980s, is probably that the picture is perhaps not as gloomy as figure 21 might imply.

NORTH SEA OIL & GAS DEVELOPMENTS   Fig 22

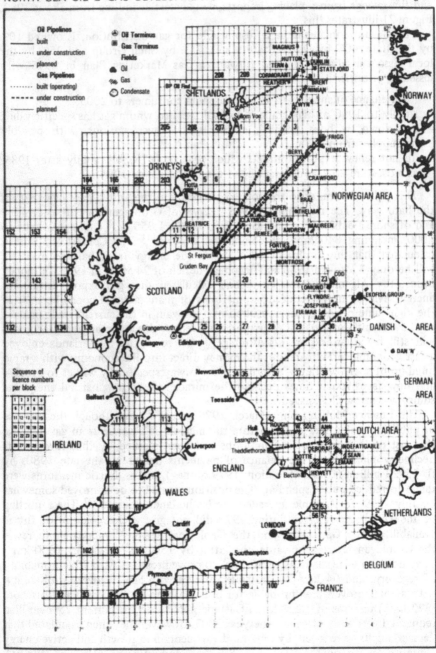

Source: North Sea and Europe Offshore Year Book and Buyer's Guide 1978

The availability of supplementary external natural gas resources

Table 46 illustrates the growing dependence of Northwest Europe on imported natural gas and its sources. One estimate was that the EEC would be importing gas at 115 Mtoe in 1985, being 25% of primary energy demand; a more recent estimate by the Commission scales this down to 94 Mtoe, or 110 km^3 of natural gas.[308] There would seem little likelihood of British Gas exporting, unless gas-from-coal was developed to substitute for natural gas in some areas and allow export or SNG to be exported. Economics and politics would both be involved and the outcome is doubtful. The difficulties of forecasting Norwegian production has been mentioned, as have the Dutch offshore gas problems. Production from the Irish and Celtic Seas is probably certain to start by 1980 but the amount is uncertain. Hence one must consider the non-European sources and their potential.

The proved reserves in these countries are very large, and the figures of Adams and Kirkby, still hold within the general limits of error in the aggregation of imperfect data. Estimates for the USSR have risen from their 17,900 km^3, through an estimate[311] of 23,000 km^3, to a recent Soviet figure of 27,000 km^3

Table 46
Schemes for natural gas imports to Northwest Continental Europe
(cubic kilometres)

		1975	1980	1985
A. Existing Schemes				
From the USSR	— to France	—	4.0	4.0
	— to FR Germany	3.1	8.5	8.5
From Algeria	— to France	4.0	5.1	7.0
	— to Belgium	—	1.1	3.0
	— to FR Germany	—	2.2	6.5
From Iran, via USSR	— to FR Germany	—	—	5.5
	— to France	—	—	3.66
	Sub Total	**6.2**	**20.9**	**38.16**
B. Additional Schemes under Discussion				
From the USSR	— to Belgium	—	—	4.0
	— to Sweden	—	—	4.0
	— to FR Germany	—	—	4.0
From Algeria	— to Holland & FR Germany	—	5.5	12.0
From Iran, as LNG*	—	—	—	30.0*
	Total	**6.2**	**26.4**	**62.16**
C. North Sea Schemes				
From UK & Norway	— to United Kingdom (Table 45)	37	65.0	+65
From Norwegian waters	— to FR Germany	—	3.3	3.3
	— to Netherlands	—	1.0	1.0
	— to France	—	1.2	1.2
	— to Belgium	—	1.2	1.2
From Dutch waters	— to Holland	6.2	20.8	?
	— to FR Germany	6.2	6.2	?

* This project has still a number of unresolved features.
Note: Both signed, existing schemes and the schemes under discussion are subject to change, but this table is indicative of the order of imports into the countries named, as published up to May 1976.

Source: *Petroleum Economist*, various issues to 1976.

for their industrial category of reserves.[315] The Soviet Plan, 1976–1980,[286] suggests the following increases, measured in cubic kilometres

— production from 290 in 1975 to 400–435 in 1980,
— domestic consumption in 1974 at 91 to 126 in 1980,
— imports from Afghanistan and Iran, from 11.9 in 1974 to 14 in 1980,
— exports to Eastern Europe from 8.6 in 1974 to 28.4 in 1980,
— exports to Western Europe from 5.4 in 1974 to 23.3 in 1980.

A base figure for net exports of natural gas to Western Europe of 20–25 km^3 can therefore be taken for the 1980s and into the 1990s, but beyond that amount it is difficult to estimate because natural gas is an increasingly effective[315] energy source within the USSR, with currently 150 million residential gas consumers, compared with the 135 million in the USA, and an Interconnected Closed Gas Supply System, (called ESG in the USSR), has been established as a centrally controlled grid system. Furthermore the logistics get more difficult as local supplies decline in the industrialized west and the centres of production move eastwards. Hence whilst exports eastwards to Japan, which have been talked about for a long time, may materialise, it is more difficult to see very large increases in exports westward beyond Eastern Europe.

Iran has the reserves to satisfy its own domestic needs for residential and industrial use, and still allow large exports. The first exports through the IGAT I pipeline to the Soviet Union from the main oilfields started in 1970, and were running at 10 km^3 per annum. A projected second line, IGAT II, would deliver from the onshore Kangan area on the coast of Southwest Iran, some 13.4 km^3 to the Soviet border, having also supplied 10 km^3 for domestic consumption; Czechoslovakia are said to be wanting supplies from IGAT II which would have supplied Western Europe, via the USSR, so that the total system would be carrying 25 km^3 per annum. The Islamic regime halted this scheme in 1979.

Two export schemes of liquefied natural gas, (LNG), were under discussion. The first, the Kalingas project, for which the Pars field offshore Kangan has been designated as the source but is not yet fully delineated, would export a total of 8.4–12.4 km^3 per annum as LNG to the USA and Japan. A second project would divide between the USA and Europe a total of 20–30 km^3 per annum as LNG from a plant near Kangan, using the same Pars field, or alternatively be piped to Iskanderun in Turkey, where it would be liquefied and then shipped. This is the scheme which was earlier reported as being for 30 km^3 to Europe and is noted in table 46 as speculative. The preliminary costs estimates are of the order of $6 billion; if the implementing contracts had been signed during 1976, and work progressed on schedule, the project could have been operational in the 1980s. The total export from Iran, if all the schemes[316] under consideration had come to fruition, would have been some 67 km^3. However, all work halted in 1979.

Exports from the other states around the Persian Gulf are most probable as LNG schemes, even though there has been a recent emphasis on discussions about the local use of their surplus gas, particularly associated gas, for the development of petrochemical and other industries. The LNG export from Abu Dhabi to Japan, in a deal with the Tokyo Electric Power Company, was to start operating in 1977 and had one most important feature, in that the gas prices have been revised so that the price of the LNG will be related to the landed price of crude oil imported into Japan, and be in full equivalence in 1980. The OPEC organization is said to be considering how best its members can regulate the price of their natural gas, and this move by Abu Dhabi would indicate one likely way in which it may be done. Thus with natural gas tied to crude oil equivalence, itself tied in principle, though loosely in practice, to the cost of manufactured imports into OPEC, there would seem every possibility that OPEC controlled natural gas, which is roughly 40% of the world's proved reserves, will not be a cheap source of energy, and will be subject to the same security of supply as OPEC crude oil.

Algeria is a member of OPEC, but gas is its main energy resource, having proved reserves of some 3,000 km³, (some estimates[75,313] vary from 2,614 km³ to 3,569 km³) which are over twice the crude oil reserves equivalent. Algeria is nearer the European market, and is reported as now favouring pipeline disposal, because of the difficulties of financing, building, and maintaining increasingly complex liquefaction plants. Italy signed a 25-year agreement in 1973 for the supply of 11.75 km³ per annum gas from Hassi R'Mel by a 2,400 km pipeline to La Spezia in Northern Italy, including 175 km of deep sea line crossing to and from Sicily. Full-scale laying of both sea and land lines is planned for 1977, completion in 1979 and full capacity transmission by 1982. This scheme was held up because Tunisia would not agree transit terms. Construction contracts were signed in October 1977, for the 2,500 km line which will supply 300 km³ of gas over 25 years,[384] starting in 1981. The technological advance implied in the presumed successful overcoming of the problems of laying pipe to carry such volumes across the sea channels in depths up to 600 metres, may well have an effect on other schemes elsewhere for deep water crossings, as across the Norwegian Trench, or even from Algeria to Spain.

It was reported[392] as of August 1977, that Algeria had increased their efforts to make firm European contracts for natural gas supplies. Contracts approved by the respective Governments include:

- 5.0 km³/year to Zeebrugge, Belgium   from 1980/81   for 20–25 years
- 5.15 km³/year to Montoir, France   from 1980/81   for 20 years
- 4.0 km³/year to Maasnlakte, Netherlands   from 1982   for 20 years
- 4.0 km³/year to Wilhelmshaven, FR Germany   from 1982   for 20 years

The Chase Manhattan Bank forecast[392] that in 1980 Algeria would be exporting 30 km³ annually, which would be 60% of the world's exports. They estimated that 17 km³ would be to the US East Coast. They expected Algeria's annual exports would be 67 km³ in 1985, half the world's exports, and 50 km³ going to the US East Coast. Another forecast put Algeria's exports in 1980 at only 20 km³ with about 13 km³ to the US East Coast. It is difficult to judge, even in late 1979, which of these forecasts will be nearest the actual figures, because there are so many technical problems which have plagued plant construction at both the exporting and importing ends and, particularly in the USA, so many regulatory bodies are involved that delays are unpredictable and can be serious.

Nigeria has ample supplies of associated gas to support a strong export trade, but collection from the many comparatively small fields and the unsettled political situation have held back schemes which have been under consideration for a number of years for export to both Europe and the USA.

Transportation of natural gas for export often has to be either as LNG in costly special tankers, up to $150 million each, or as methanol, which can be carried in normal tankers but which has not as yet become popular. In the immediate future, with the slackening in demand since early 1974, there is a surplus of carriers. This may continue to 1980, with a total fleet estimated

at 79 ships, with a capacity averaging 83,800 m³, but one forecast[314] is of a
need in 1981-1986 of 90-95 ships of 125,000 m³ each, and 19 ships of this
size are said to be on order. However there is a natural hesitation to build
during the present surplus, with the expectation of slow growth in energy
demand at least until 1980, in most gas consuming countries, and until the
USA policy becomes clear; if there are no restrictions on imports into the USA
and USA domestic energy sources do not expand, there will be a big increase
in ship-borne ocean LNG trade. The trade to Japan from the Middle East and
South East Asia will probably be the main LNG trade in competition with that
to the USA in such circumstances.

Summary

The availability of the indigenous resources emerge as being located in eight
distinct areas, each with different circumstances controlling their development:

- the mature onshore giant of Groningen, declining after 1984 and with possibly restricted future support from new onshore or offshore fields,
- the maturing southern basin of the North Sea, predominantly on the British
  sector and serving Britain,
- the young Ekofisk Group of oilfields in the south of the northern basin of
  the North Sea, in the Norwegian sector, with gas supplies already predicated
  to FR Germany and westward, because of the geographical factor of the
  Norwegian Trench,
- the younger non-associated gas of the giant Frigg field, straddling the median
  line but going west to Scotland because of the Trench and the ready market,
- the associated gas of the large oilfields, on either side of the boundary, for
  which lines west or east are economic and are being built or planned,
- the associated gas from the smaller oilfields or non-associated gas from minor
  deposits which, in aggregate, could make economic and resource conservation
  sense, by either two national collection systems, now under study, or one
  UK/Norway scheme,
- the as yet unknown significance to UK and Eire of the finds in the Irish and
  Celtic seas,
- the outer fringes of the continental shelf northwards and westward, even
  to Rockall.

This subdivision of the indigenous natural resources, considered in the background of the international legal aspects and the varied expressions of the involvement of the different governments and private interests, emphasizes that
there is danger in oversimplifying the problem of the size and possible effective
life of these resources.

The availability of the external supplies which may augment these indigenous
supplies of West Europe depends mainly on:

- the success of the Soviet Union in finding supplies in sufficient quantity
  to be surplus to the growing requirements of itself and Eastern Europe
  beyond the 20-30 km³ per annum base of the probable 1980s exports,
- the willingness of Western Europe to accept supplies from OPEC by pipeline
  and tanker at prices based on equivalence with imported crude oil and with
  similar security of supply in a market in which there will be competition
  from Japan and could be from the USA,
- the enigma of future discoveries in new areas,

- the possibility of a policy, (whether national, e.g. UK; regional, e.g. EEC; or wider, e.g. OECD/IEA), to develop substitute natural gas, SNG, from coal. This could be an integral part of a long-term policy to ensure that the great coal resources of the world are available in fact, and not just in theory, to cover the potential energy gap straddling the end of the century, and so comes full circle back to the coal resources.

THE UNCONVENTIONAL NON-RENEWABLE ENERGY RESOURCES

The petroleum locked in the tar sands or shales and in the very heavy oils is not inexhaustible but the quantities are so great that it is most unlikely that they will ever be worked-out. There is much talk of their exploitation and also of the natural gas in gas hydrates and in geo-pressured deep strata. Research is continuing, particularly in Britain, FR Germany and the USA, into coal-based liquid and gaseous fuels. The USSR is also reported to be dusting off the projects in these fields which were abandoned, as elsewhere, when petroleum was in ample supply. However, the exploitation of oil sands is still restricted to that of the Athabasca area in Canada, and of oil liquefaction to the SASOL plant in South Africa. There are continuing news items of new processes to utilize the very large resources of the oil shales, particularly in the USA. However, the economic, physical and environmental constraints have as yet inhibited the application of the needed massive investments. Thus in the OECD 1977 study[329] no production is forecast from very heavy oil, oil sands or oil shales in the USA in 1985; nor is there any allowance for synthetic natural gas (SNG), from coal, or gas from tight formations. These sources are considered not to be currently competitive with imported oil. The high investment needed is also considered as uneconomic in diverting capital from indigenous natural gas, once prices are allowed to rise, even if not fully decontrolled. In the Accelerated Policy Case of the OECD Report, the price of gas in the USA in 1985 was $2 per thousand cubic feet, which equals £8.22/barrel of oil equivalent — some would suggest a much higher price being common in the USA in 1985.

Despite the talk and the work being done on both the inexhaustible energy sources and the unconventional fuel resources, and despite the point that habits can change fast, it is still difficult to see how these sources could make a significant global contribution before the end of this century. In a local sense they may well be important. In the world-wide discussions which have been common in both the USA and Europe, but particularly in Britain, FR Germany and France in 1977, about nuclear power, one popular argument by its antagonists is that the diversion of vast sums of money from nuclear RD&D could accomplish the technological break-throughs which currently inhibit that significant global contribution to energy supplies. It is not possible to deny this thesis categorically, but the counter argument is that it would be foolish to stop the development of a known and safe source of power and gamble on the unknown when there is no time to lose and the dangers of conflict attendant on an energy shortage are obvious. This argument has been displayed in a 1977 book by Professor Sir Fred Hoyle[404] with his customary vigour. It is a very understandable feeling which the layman often has that there is so much natural energy 'going to waste' which 'they' should harness for the benefit of mankind. The hard facts are that, as yet, no viable commercial system of wide application has been developed. It might be opportune here merely to note the European research on the Joint European Torus for the harnessing of nuclear fusion mentioned ealier, which it was agreed in October 1977 would go ahead at Culham, England. Recent work on magneto-hydro dynamics (MHD) is reported[405] as showing promise. Thus whilst believing that there is little chance for the 'new' sources to contribute much before the next century yet it is very necessary to keep development in all forms of energy under constant review in order to detect trends which might alter that belief.

FISSILE FUELS

The pattern of production, shown historically and graphically in figure 23 is an important feature of the uranium mining industry, for the fall from the peak production of 1959 was a traumatic experience for the producers.

The rapid rise from 1954 to the peak was due to military procurement and 90% by the USA. In 1959 military needs were satisfied and domestic discoveries in the USA had been so successful that a protective embrago on imports was maintained. Production had been nearly all from mines in the USA, Canada and South Africa. Some small sales had been made from 1958 onwards and these continued, together with uncompleted military contracts.[287e] South African production paralleled the world production but the uranium producers were cushioned by the uranium being a by-product of their gold production. The USA producers were bolstered by continued procurement, by the Government, for use and reserve. The Canadian uranium miners were the worst hit; nine

ANNUAL WORLD PRODUCTION OF URANIUM OXIDE   Fig. 23
(excluding the USSR, China and Eastern Europe)

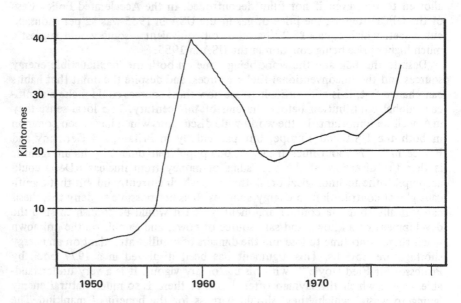

Source: 'The Pattern of Uranium Production in South Africa', R. E. Worrell & S. A. Young, International Symposium on Uranium Supply & Demand, Uranium Institute, 16 June 1976, London

out of eleven mines were shut down; prices fell to $5/lb of oxide; only high grade ores were mined, and often wastefully; the Canadian Government brought in support schemes, which resulted in a 7 kt stockpile by 1970; investment in exploration and new mining virtually ceased, because, simultaneously, the early hopes faded of nuclear power being the panacea for all energy ills, as reactor problems increased. The rise in production was resumed from 1965, but was hesitant, and in the later 1960s was inhibited, particularly in the USA, by the holding up of power plant plans by environmentalist opposition. Exploration was increased markedly by the 1973 oil shortage, and production moved slightly upwards. With this pattern of past production, there should be little surprise that uranium producing companies view the future with caution, and that scepticism is shown at the Lieberman exercise. Government involvement has always been greater in the uranium mining industry than in any other energy resource supply industry, at least until recently, and the desire of the Australian Government for orderly marketing, mentioned when discussing reserves earlier, is echoed by every producer and consumer and Government.

Potential production
Table 47 sets down the past productions in 1972 and 1974 and then the estimated and planned productions in 1975, 1978, 1980 and 1985 with a few figures of 1985 potential from sources other than the basic source of the 1976 OECD Report. The last three columns of that table are under the proviso 'market conditions permitting', and this is a very important caveat. In 1985, the four major producers, excluding the USSR, China and Eastern Europe, will be Australia, Canada, South Africa and the USA, with Niger, France and the Gabon following. If Sweden and Mexico are to attain their 1985 targets they will have to move fast, because of the long lead times of ten years or more. This same point is also applicable to the major producers. Australia has already been noted as having a current policy to allow private enterprise to mine and trade, subject to control to maintain a reasonable price structure, majority Australian shareholding, and, also, adequate international safety standards. This policy could change if a less liberal and more nationalistic party regained power. Canada, on the other hand, in 1974 promulgated a policy, similar to that which governs their export of oil and natural gas, which demands sufficient uranium to be allocated to meet the 30-year fuelling requirements of the nuclear plants expected to be operating in Canada in a 10-year forward period. Recent nuclear capacity projections suggest that by 2000 there may be some 75,000 MWe installed but the range could be from 50,000 to 115,000 MWe, which would entail the allocation of from 200 to 500 kt uranium for domestic use by the year 1990. Canada's measured and indicated resources, reasonably assured, are only 173 kt and will support only the programme of 11.5 kt as the production attainable in 1985. Hence there is every incentive for Canada to develop additional production facilities to exploit the new finds at Key Lake, Saskatchewan and explore for more. Fortunately the Key Lake deposit opens up possibilities for further ore bodies in similar geological circumstances over a wide area. Canada therefore could be in a position to export but only if new discoveries do adequately cover its domestic needs. For the United States, the study for the Uranium Institute[287a] suggested that the maximum feasible growth rate of production capacity would be between 8% and 12% per annum, depending on the severity of the competition from other industries for capital, labour and material, and the timeliness of decision. On such estimates the 40 kt capacity in 1985 in table 48

Table 47
Annual uranium productions and production capacities
(excluding the USSR, China and Eastern Europe)
(tonnes of uranium)

	Productions			Production Capacities			
				Planned		Projected*	Attainable*
	1972	1974	1975 est	1975	1978	1980	1985
Argentina	26	50	60	60	120	600	720
Australia[1]	–	–	–	–	760	3,260	5,000
Canada[1]	4,000	3,420	4,700	6,500	8,500	10,000	11,500
Denmark	–	–	–	–	–	–	1,000–1,500
France	1,380	1,610	1,700	1,800	2,200	3,000	3,000–3,500
Gabon	210	436	800	800	1,200	1,200	1,200
Germany	–	26	?	250	250	250	250
Italy	–	–	–	–	–	120	120
Japan	16	9	4	30	30	30	30
Mexico	–	–	–	–	210	320	1,000
Niger	870	1,250	1,200	1,200	2,200	4,000	6,000
Portugal	81	89	115	115	130	130	130
South Africa	3,080	2,711	2,600	2,700	9,200	11,250	13,800
Spain	60	60	144	144	340	680	680
Sweden	–	–	–	–	–	–	1,300
USA	9,900	8,800	9,000	12,000	19,000	25,000	40,000
Yugoslavia	–	–	–	–	–	120	180
Totals (rounded)	19,623	18,461	20,323	25,600	44,100	60,000	87,000

* Market conditions permitting

Sources: 'Uranium: Resources, Production & Demand, etc', Joint Report NEA/IAEA, Tables 3 & 4, December 1975, OECD, Paris, 1976.
[1] Papers in the International Uranium Supply & Demand Symposium, Uranium Institute, 16 June 1976, London, gave 1985 potential production for Australia of between 10 and 20,000 t and Canada of 11,500 t.

would not be attained until 1988 on the high growth rate and until 1994 on the low rate. The major uncertainty is USA domestic requirement. In South Africa, on the other hand, the domestic demand will be low, for the contract for the supply of the first nuclear plant was signed in May 1976, and there is only discussion about an enrichment plant. The production uncertainties, as has been said, lie in the position of the Rossing Mine, the production rates of gold and copper, the progress in installation of new machines to expand existing extraction plant and start new processing at other mines and tailings dumps in the goldfields. South Africa should be in a position to export. The French-linked reserves and production facilities in France, Niger and Gabon are, probably, dependent for their expansion on the drive from France, but it should be noted that the total attainable, shown in table 47 as from the 1976 OECD Report, which means as reported to the NEA/IAEA by the member countries, is 10,200–10,700 t, which is approximately 12% of the 'world' total of 87,000 t.

There are three factors which may affect the attaining of the 'world' total. A possible manpower scarcity in mine operating labour is a present concern of the industry.[287e] Many mines are isolated and unattractive to modern labour, so that measures to improve living conditions and safety are being introduced, similar to those which the oil industry has had to develop over the past 20 years, but which still seem novel to the uranium mining industry. Secondly, as in most other industries, there is need to improve the whole extraction process; some new methods are being tried, e.g. *in-situ* bacterial leaching will be introduced in the Espanola Mine in Ontario, Canada, in 1977,[287a] and new separation methods are being developed in South Africa.[287d] Thirdly, the possible production from the almost ubiquitous very low grade deposits are continually being quoted and tend to add to the complacency with which many view the future energy scene and which mask the urgency for sustained effort. Production of uranium from granites and seawater was said earlier to be of academic interest only, and this view can be stressed by considering the size of the production facilities necessary, rather than the size of the ultimate resources. It has been noted by von Kienlin[287g] that in order to produce 5 kta, (which is less than 6% of the 'attainable' 1985 production capacity in table 47), from seawater, by the only method at an appreciable stage of development, using titanium hydrate with activated carbon as the adsorption agent, would require one million cubic metres of the agent to be put into the seawater each day on a two to four day cycle, requiring a storage capacity equal to that of a trough 100 km long, 5 m deep and 6 m wide. The processing of granites to extract 5 kta from 5,000 km^3 of rock at 30% of the uranium content, would require a mine ten times the size of the greatest strip mine in the world and require an investment of $1,000 million. Similar operations in phosphates or uraniferous black shales would require equally enormous facilities and the uranium from a tonne of many shales has a heat value, when used in a thermal reactor, of only a tonne of coal. Therefore it must be emphasized that the extraction of significant amounts of uranium from such low grade sources 'during the next few decades is an illusion'.[287g]

In summary, the uncertainties which affect the supply side are conditioned by:
1. the success in finding new reserves, which depends on the quality and the quantity of the effort applied, and on the concepts guiding the exploration as well as the funds involved,
2. the improvements in extraction techniques, and reduction in the lead times from discovery to production,
3. the degree to which the effort can be sustained and fluctuations smoothed out,
4. the action of governments in encouraging or hindering exploration and development, and controlling production and exports,
5. the availability of manpower, money and materials for an industry with an especially unstable history to add to normal high cost mining risks.

The first two points can be considered as internal to the mining side of the uranium industry; the last two points are external to the mining industry; whilst the middle point is inherently a problem for both the mining and consuming sides of the nuclear energy industry and may be alleviated by better communication between them. These points are discussed further in chapter 9.

The International Nuclear Fuel Cycle Evaluation Group (INFCEG) was set up by the Carter Administration of the USA in order to assess the need for expanding the facilities for enriching and reprocessing nuclear fuels worldwide. This action was in support of the idea that such should be restricted because of the dangers of the proliferation of nuclear arms capability. One task was to assess possible uranium production and table 48 is taken from their preliminary papers as reported in *Nuclear Fuel*, 20 August 1979. It is stressed that these are projected production capabilities and are not production

Table 48

Projected uranium production capabilities, 1980–2025, based on known resources of major producing countries outside the communist countries (kilotonnes)

	1980	1985	1990	1990 (a)	1995	2000	2005	2010	2015	2020	2025
Australia			20	13.6/16.5	17	10	2	11.0	9.5	9.0	8.8
Canada	7.6	12.8	13.5	12.0/15.0	13.5	13.1	12.3				
France	3.6	4.0	4.4	3.5/4.3	3.1	1.6					
Namibia	4.1	5.0	5.0	*	4.6	4.6					
Niger	3.7	9.9	9.9	8.0/11.0	10.2	5.5	5.5				
S. Africa	6.5	10.6	10.3	12/16*	7.2	4.3					
USA	19.2	29.2	40.8	24/35	47.7	51.6	46.7	40.7	25.4	13.3	
Others	3.0	6.5	13.5	4.8/9.7	10.0	11.0	11.0	9.5	7.0	1.0	
Total	**47.7**	**78.0**	**117.4**	**78/108**	**126.6**	**101.7**	**77.5**	**61.2**	**41.9**	**23.3**	**8.8**
Phosphates	1.0	3.5	5.0		6.0	8.0	10.0	12.0	13.0	14.0	16.0

Sources: International Nuclear Fuel Cycle Evaluation Group, preliminary papers, as reported in *Nuclear Fuel*, 20 August, 1979.

(a) 'The Balance of Supply and Demand, 1978–1990', February 1979, Uranium Inst. London.

* Namibia and South Africa combined.

forecasts, and are based on the resources as known at the time of compilation, without any allowance for new discoveries. To the table have been added the estimates for 1990 of the Supply and Demand Committee of the Uranium Institute, February, 1979.[515] The ranges given encompass the INFCEH estimates for Canada, and South Africa and Namibia combined. France's production is close to both estimates. The estimates vary greatly for the two major producers, Australia and the USA, in 1990, with a resulting much lower range for the total for the world outside the communist areas in 1990. This would seem to be a very important divergence of opinion, despite the caveats placed on their figures by both authorities.

Thorium production as a metal was given in the 1974 WEC Survey as 77.2 t from Africa, 342 t from Asia, 359 t from Brazil and 228.8 t from Australia, making a total of 1,007 t. The source was given as the Mineral Yearbooks 1968 and 1971, of the US Bureau of Mines and 'based on average thorium oxide content of 6% in monazite that might be recovered'. Monazite is a rare earth thorium phosphate exploited mainly in Brazil, Australia, India, Malaysia and Malagasy, but also in the USA and other countries about which there is no data. Canada was said to have a production of 56 t. World production even outside the USSR, Eastern Europe and China cannot therefore be taken as 1,007 t for the data is incomplete. The data was little better in the 1978 WEC Survey, when productions were given as 200.6 g (1973) from Zaire, 150 t (1976) from Brazil and 289 t from Australia, together with some notes on these and other countries. Demand for thorium at this time is so small and production only a byproduct so that information is poor. Hence these figures are in no sense any guide to what production could be, when and if the demand for thorium as a fuel for nuclear reactors materializes.

6
Conversion

To this point progress in pursuit of the understanding of the availability of the world's energy resources has been from resource base to resources to reserves to proved reserves and so to production of the primary energy sources. But these must be converted to useable energy either through some intermediate form (coke, manufactured gas, petroleum products, electricity, steam, or hot water) or directly to heat. The value of an energy resource is dependent on the efficiency of the conversion method, whether in one or two stages.

Since October 1973 it has become obvious that the era of the supply-oriented economy is over. Energy conservation in all forms will be increasingly important for decades. A normal energy conversion route is from fuel to heat to working fluid to mechanical energy to electricity. The average improvements in efficiency using this route are said to be from 15% in 1922, to 20% in 1937, to 30% in 1956, to about 33% in 1975, with about 42% being the maximum achievable efficiency using current techniques. The use of steam as the working fluid has probably reached its asymptotic limit of efficiency. New technologies must be developed.[143] These will be made through use of different working fluids, through combining, staging or, in some cases, by eliminating one or more devices in multi-device systems. Examples of new or improved approaches include fuel cells, gas turbine topping cycles, and binary steam-ammonia cycles. There will have to be further research into such systems as magnetohydrodynamics, thermionics, thermo-electric conversions, thermogas dynamics, photovoltaic systems etc. Some systems need a single technical breakthrough to become viable; others need more.

COAL

The prime role of coal has been in recent years and will be for a long time, to provide fuel for the generation of electricity. Cheap oil began to replace coal from the 1950s onwards, but coal has been fighting back since 1973, even though the original price advantage, after the increase in oil prices, has been eroded by coal prices moving up, particularly with increased labour costs. However, in Britain, for example, the National Coal Board (NCB) still considered in March 1976 that it had a 1½ pence/therm advantage over oil, whilst in the USA in 1974 the coal price was under half that of oil and the Bankers Trust Company study, mentioned earlier, estimated that some 707 Mt would be needed for power stations in 1985.[263]

If coal is indeed to act as the major energy source from 1990/2000 to 2000/2010, there must be improved combustion efficiency, world-wide. Despite the claim of some engineers in the industrialized countries, such as Britain, that the combustion efficiency in all except old power stations is near to the theoretical maximum, this is not a universal situation; operators also must not be complacent. There are a number of methods under development aimed at cutting costs. One of these is the fluidized bed combustion technique[149] in which powerful air jets, in a specially constructed boiler, turn burning coal into a bubbling bed of red-hot ash which looks like and acts like a boiling liquid. This technique was developed in Britain and is now the subject of a project of the coal technology group of the International Energy Agency (IEA). Some £10 million will be invested by the USA, FR Germany and the UK, and a plant built at Grimethorpe, Yorkshire, England, to investigate the special features of the method. Among the advantages in raising steam for industry and generating electricity are:

— compactness, with lower capital costs than conventional plants,
— high thermal efficiency, even when burning low-grade coal,
— reduction in air pollution, as sulphur is retained in the ash bed.

The general technique of fluidized bed combustion,[150] for there are a number of versions, will undoubtedly improve the economics and overall efficiency of coal utilization.[448]

Gasification

Much work is being done on conversion of coal into gaseous hydrocarbons. Some coalmen consider that the conversion and distribution of synthetic or substitute natural gas (SNG), is more economic in both money and energy than the conversion of coal to, and distribution of, electricity.[266] Before the advent of natural gas from the North Sea, the British gas industry, particularly through the efforts of the late Dr. Frederick Dent, was a leader in gas-making technology, which had started in 1936 with experiments in the Lurgi gasification process developed in Germany. Therefore it was logical for American interest in British experience to increase as their own domestic natural gas supplies declined and energy began to cost more. On the one side, some 14 plants, based on the British catalytic process (CRG), of producing SNG from naphtha, have been installed in the USA under licence, and are producing 33 million m^3 per day, (12 km^3 pa). On the other side, there has been US sponsorship in the development of British ideas in the gasification of coal. The original Lurgi plant at Westfield, Scotland, which had been acquired from Germany, produced town gas from 1960, using coal from an adjacent open cast mine, until due to be closed down when natural gas supplies were scheduled to arrive in June 1974. The US-based Continental Oil Company (Conoco) with its strong coal interests, made a deal with the British Gas Corporation (BGC) to extend the plant with a purification element and a methane synthesis element, based on a process invented by Dr. Dent in 1937, and so produce SNG commercially. Modification of the plant started in late 1972, and ended with the production of SNG for two months to consumers in Fife, Scotland, over the public distribution system and, incidentally, also proving the interchangeability of various gas appliances. The project ended in October 1974, having cost $6 million, but the extra plant was purchased back by BGC for further work.

From August 1973 to April 1974, another series of trials were conducted on specially imported US coals at Westfield in one of the other Lurgi gasifiers, after modifications agreed between BGC, the American Gas Association (AGA) and the Lurgi company. The objective was to widen the scope of the Lurgi process in order to handle not only the preferred coals, which are weakly caking, have a high fusion point, and are highly reactive similar to those of the Western USA, but also the dissimilar coals of the mid-West, as in Illinois, and of the East, like the Pittsburgh coals. Such coals had been considered unsuitable for the Lurgi process, but their fields are nearer to the main US market, and, furthermore, if gasification of such coals was to be on a big scale, then run-of-the-mine and not graded coal must be the feed. The trials were successful and 'represented important advances in fixed bed, high pressure gasification technology and greatly extended the range of coals available to this type of gasification'.[299]

A third, important and continuing experiment was started, after the release of the gasifier from the AGA trials, by converting it to a commercial version of the slagging gasifier. The principle of the process had been seen by Dr. Dent in the 1930s; another Lurgi plant was obtained from Germany in 1955 and installed at the UK Midlands Gas Research Station; pilot plant trials were successful, and by 1964 development had reached the point of being ready to scale up to commercial size, but work was suspended as natural gas became available to the Midlands from the North Sea. Crushed coal, steam and oxygen are combined in the slagging gasifier, at a higher temperature than in the conventional Lurgi process and a liquid slag is formed, which is easier to handle than dust. The advantages are high input, low steam consumption, low aqueous liquor production and so less disposal problem, high gasification efficiency – all leading to lower costs – and the capability to widen the range of suitable coals. The current project is a three year, $10 million, joint effort of some 14 American companies, (from the oil, coal, gas and pipeline industries), and the first trial was run in April 1975. The Energy Research and Development Administration in the USA has awarded a contract to a consortium, led by Conoco, to build a large demonstration plant in the USA, based on the Westfield slagging gasifier.

There are a number of other gas-making processes being researched, developed and tested, particularly in the USA, many of which are, however, at a less commercially advanced stage than the British types. There are, for example, processes for making gas of low calorific value for power stations, and ERDA has called for bids to build a demonstration plant in this field also. There is every expectation that improvements will materialize for the second generation gasifiers. Coal gasification is most important for those countries which have an existing gas distribution system and the coal reserves, but is of less value to other countries, like India, which only have the coal, and neither distribution system nor the appliances, perhaps millions, at the consumer end. Gasification therefore is of immediate interest to the USA and may well be for the UK, Germany, Holland, Belgium, France, Austria and Italy when, in fifteen or twenty-five years time the supplies from the North Sea may be declining and the only alternative might be gas or oil from the Middle East.

Work continues on the gasification of coals in-situ, without mining, in order to utilize coals of poor quality or in thin seams. Most methods, however, require many holes or shafts, and the high cost of drilling is a major problem.

Quite large sums of money are now being devoted to research, development and demonstration (R,D&D) in coal conversion. The ERDA budget for 1975–1976 was $210 million; the FR German Government allocated DM 600 million over the 1974–1977 period. Hence, with the description of the state of progress which has just been given, the point should be clear that the gasification of coal at the surface is a technical and economic reality. Nevertheless, it will require determination to maintain, over the next few years, the momentum necessary to ensure the early commitment of funds to commercial projects which may need 5–10 years lead time, if conceived as new mine and plant complexes. An illustration of the problem is that the USA administration have recently lowered their targets, in that in January 1976, they spoke of the need to build up synthetic fuels production from 1 mbdoe, from all sources, in the mid-1980s, to 5.5 bodoe in 1995, whereas the 1974 study of the US National Academy of Engineering[54] was aiming at a production of 1.7 mbdoe in 1985 from coal alone.

Liquefaction
Progress is being made in the field of liquefaction for the production of synthetic liquid petroleum fuels and is most marked in South Africa. The South African Coal, Oil and Gas Corporation (SASOL), is the leading world producer of liquids from coal, and merits attention because this source of energy may be very important in the future and therefore it is valuable to gain an impression of the scale of current operations and plans. SASOL was formed as a public company after the passing of the Liquid Fuel and Gas Act in 1947.[297] The Sigma Colliery was designed to supply coal, for the liquefaction plant set up near-by, originally at a rate of 34.5 kt/week, raised to 88 kt/week by 1976 and currently being raised to 100 kt/week. The first oil was produced from SASOL I in 1955. In the Fischer-Tropsch process as used in this plant the coal is gasified with steam and oxygen, purified and then converted into gaseous, liquid and solid hydrocarbons by two different processes – the fixed bed iron process of Lurgi-Ruhrchemie, and the iron entrained fluidized bed process called 'Synthol' by Kellog-SASOL, which uses somewhat higher temperatures. The fixed bed process produces paraffin wax, diesel oil, gasolines and gas, whilst the other process produces increased quantities of gasolines of higher motor fuel quality.[142] SASOL I has a current production of about 250,000 ta. Following the 1973 threatened oil shortage, a second plant was proposed, approved in December 1974, and SASOL II is now being constructed. The plant will use the 'Synthol' process and the building in of the lessons learned over the past 20 years, will result in a more efficient plant which will be in effect a 'coal refinery', sitting on top of a highly automated mine specifically designed to supply its raw material. It is estimated that the mine will be marginally profitable with coal at $3–4/t. More recent plans by private interests have been floated for the building of complexes of mine and treatment plants, of which one such would treat 15 Mta of coal to produce 1.25 Mta of synthetic crude oil, 2.2 Mta of formed coke, 8 Mta of high ash char, and '27 million heat units of fuel gas'.[297] South Africa is still without any other energy source and so security of supply plays an important role in Government thinking. In the USA, ERDA has awarded a $237 million contract to an American Group to build a demonstration sized 'coal refinery', producing a range of liquid products from fuel oil upwards, as well as SNG; an ERDA official[300] foretold that fuel oils would be produced at $15/barrel and gasolines at 33¢/gallon.

Chemical feedstocks
There are many examples of chemicals from coal but, in general, it would seem that coal will have to compete for a long time with petroleum in the organic chemical field, for petroleum has the advantage of having hydrogen already, and it is to this field that petroleum will be increasingly turned as it becomes scarce and too valuable to be wasted as a primary fuel.

South Africa can illustrate both one type of chemical use for coal and also substitution of raw materials as circumstances changed. Calcium carbide was first made from coal in South Africa in 1918 and the acetylene route for the manufacture of polyvinyl-chloride (PVC), was used in the 1950s. In 1967 a PVC plant, based on imported naphtha, was built at Sasolburg, but the 1973 threat to oil imports revived interest in the acetylene route and two new 33 MW furnaces are now planned at Sasolburg, to produce 147 kta of carbide, which will be purified, reacted with hydrogen chloride and polymerised to PVC.

International collaboration is becoming a marked feature of the coal industry.

Mention has been made of several instances of collaboration between the British Gas Corporation and the USA. The UK National Coal Board (NCB), has collaborative agreements with the USA, Canada, Poland, and the USSR, as well as with FR Germany, France and Belgium in the European Coal and Steel Community (ECSC). The NCB's Mining Research and Development Establishment in February 1976 had 28 contracts for projects receiving support from the ECSC.[301] The programme of the coal technology group of the IEA, in addition to the fluidized bed combustion project, mentioned already, has four gas research projects, costing about $1 million/year and financed by ten countries.[302] These projects illustrate an attempt at collaboration which, if successful, will be on a par with that which the uranium industry has achieved. These projects entail the establishment of four Services, for the benefit primarily of the members of the IEA:

− **Resource assessment** − to coordinate the different techniques by which countries assess their coal resources,
− **Technical information** − to provide a rapid international dissemination of the increasing amount of information on all aspects, from exploration to end-use,
− **Economic assessment** − to deal with costs involved in mining, conversion and utilization, by conventional and new methods e.g. coal refineries or coalplexes,
− **Mining technology** − to study the range of equipment available and being developed, world-wide, and the development of total systems from coal face to surface.

A measure of the interest in coal conversion, world-wide, is the issue in 1975 of a book describing some 170 processes in the US Patent Office for obtaining oil from coal[509] and one book in 1976 which described[510] 100 processes on coal conversion technology dealing with:

− desulphurizing high sulphur coals,
− demineralizing high ash coals,
− carbonizing or volatilizing coal in the absence of air,
− gasification with partial or controlled oxidation.

Another example is the suggestion[511] that in the next century the British gas industry may need 50 to 150 tonnes of coal annually to provide gas to supplement the then declining North Sea production. This amount does not include any possible gas export supplies.

OIL SANDS

Mention has been made earlier of current commercial production by known methods of conversion of the heavy oils of the Albertan oil sands to 'synthetic' crude. Another conversion method may be mentioned here which is only in the laboratory test stage which uses a solvent and infra-red radiation method to process oil sands.[151] This Javelin Environmental Protection Oil Sands System (JEPOSS) is said to have the advantages of on-site processing at the open pit, uses no water in separating the oil from the sand, has a high oil/bitumen recovery (90% is claimed) and uses less energy than any other process. Such schemes may be a long way from commercial production but they should not be ignored. Non-conventional sources of energy may need non-conventional conversion/recovery technology.

OIL

Crude oils are the most versatile of the energy resources. They vary greatly in characteristics, and so in the products into which they can be made by simple or complicated, cheap or expensive methods by many varieties of conversion routes. All are well-documented except when there may be proprietary interests. Here, however, discussion will be confined to illustrating one or two effects of broad differences in crude oils on the market and refining patterns and to noting economies of scale on the conversion field.

There is first the classic comparison of the refining pattern in the USA, aimed at maximizing gasolines yield, and the European pattern aimed at maximizing fuel oils and middle distillates (the gasoil/diesel oil range). The USA domestic crudes were converted to the high-priced gasolines and demand for fuel oils, particularly on the cold and industrialized north-east coast, has for long been satisfied by imports from Venezuela. Kuwait crude is a very heavy crude, like many of the Venezuelan crudes, and found a ready market in Europe because it could yield a high percentage of the heavy ends on distillation. North Sea oil is a light, low sulphur oil and this fact has a triple effect on the European and particularly British refinery pattern. In 1973 the UK produced 0.252 Mt of natural gasoline (from the North Sea gasfields) and 0.088 Mt of crude oil and imported 116.4 Mt of crude oil and 19.7 Mt of refined products. The crude oil came mainly as almost 26 Mt from Saudi Arabia, 21 Mt from Kuwait, 21 Mt from Iran, almost 12 Mt from Libya and 9 Mt from Nigeria. Some 16 Mt of refined products were re-exported and 3.8 Mt of crude oil was re-exported. The 16 Mt of refined products went mainly to north-west Europe.[152] This type of flexibility is typical of the European refining pattern. Because of the influence of the integrated major oil companies, the western European market has been supplied by a network of refineries built to complement each other and to match the total market. Each refinery was not built to match its own immediate market.

This type of flexibility has been of great benefit to the whole European market. However, Norwegian oil well tend to supply Norway and, possibly preferentially, Scandinavia rather than other outside markets. The British Government has already informed British refiners that it would expect Britain to become more self-sufficient, to refine at least two-thirds of indigenous production and to re-vamp the refineries to handle the light crudes and make more light-ends. In this way imports could be reduced of the high-cost gasolines and petrochemical feed stocks and so improve the balance of payments. One refiner installed new cracking plants at two plants at a cost of £45 million. New refineries or additions will obviously have to be geared to producing more light ends from the lighter crudes. This will reduce the import of the lighter crudes but some heavy crudes will be needed to provide the fuel oils, unless they are imported as products. There is obviously going to be a change in markets, and a change in refining pattern and, to some degree, an imposed loss of flexibility as a result of North Sea oil.

The 1970 and 1978 patterns of demand, in percentages by weight, of the USA, Europe and Japan, confirms the differences in patterns with the exaggerated Japanese call for fuel oil. In Japan 80% of its oil is used for electricity

Table 49
Main oil product consumption 1970 & 1978
(percentages by weight)

	USA		W Europe		Japan	
	1970	1978	1970	1978	1970	1978
Gasolines	48	46	19	23	17	20
Middle distillates	31	32	37	41	18	26
Fuel oil	21	22	44	36	64	54

Source: *BP Statistical Review of the World Oil Industry*, p. 22, 1978.

generation compared to 30% in the UK and 20% in the USA — a timely reminder that conversion is required to fulfil end-use demand.

In the USA in 1972[121] the total production of products was 472.5 Mt. This total was made up by 62.5% gasolines (including 22.8 Mt natural gasolines) and 37.5% middle distillates and residuals with the balancing to the inland demand pattern by imports of some 106 Mt of the heavy-ends. The contrasts between these patterns and crudes illustrates the variations possible, but of course, the economies vary according to the route taken to satisfy demand and the size of the market for there are marked economies of scale in petroleum refining.[153]

The size of any refinery is set by a combination of factors peculiar to the owners, the size of market, the risk/reward ratio, long-range plans and judgement. Other more general factors indicate that there are clearcut economies of scale up to at least a level of 15 Mta. In 1970 the average of all new refineries announced was 4 Mta, in 1974 the average was 6.4 Mta with 20% of the projects over 10 Mta and the average before 1980 may be 10 Mta. This trend to greater size and, indeed to greater complexity, can be expected to continue into the mid-1980s. Not only are greater volumes foreseen but also greater efficiency in end-use and greater environmental protection will necessitate tighter product specifications and hence more complex processes. New refineries up to 25 Mta capacity may be expected: a number of refineries of this size do exist, some have done for many years, like Abadan, Iran, Ras Tanura in Saudi Arabia, etc. But they have grown gradually to that size. New large refineries will cost very large sums. A 25 Mta refinery committed at the beginning of 1975 is estimated to require a capital outlay of between US $1,260 million and $1,340 million on completion in 54/60 months, compared to a total outlay of US $238 million for a 2.5 Mta refinery. The high costs may cause cost-sharing with partners or, in the Middle East, partnerships, with the oil companies contributing expertise.

Frankel and Newton wrote a very important paper[533] in 1979. They had discussed in earlier papers to the World Petroleum Congresses, in 1959 and 1963, the shifts in location of oil refineries from the producer end to the consumer end, with some intermediate refineries in the Mediterranean, Aden, the Rotterdam/Antwerp area and Singapore. Processing for re-export to adjust imbalances in individual countries declined in the 1970s in the Mediterranean and Aden; the Rotterdam/Antwerp complexes became integrated into North-West Europe; but Singapore and some of the Caribbean refineries increased in importance. The shift of refining capacity to the consumers had halted by 1979. However, producers had not gone as far into refining as some had expected. This was due to existing over-capacity, through low demand, the scale of finance required and, most important, the appreciation of a lack of adequate management manpower and skills. Frankel and Newton suggested that there was a case for the producers to convert heavy crudes, say of 27° API, into lighter, say 38° API crudes of lower sulphur content, say down to 0.5%, and ship these to the consumers' refineries, where there would be growing need for the lighter end of the barrel as coal and nuclear power substituted for the heavy end primary fuel uses. The producers could use associated gas for the energy intensive processes required, but still benefit from the economies of using very large tankers.

Such then are the factors in the petroleum conversion side which must be borne in mind when considering petroleum as a resource.

HYDROGEN

World consumption of hydrogen in 1970 was estimated to be about 17 Gt, perhaps 20 Gt, as a reducing agent in the chemical industry, in the production of ammonia and methanol and in hydrogen-desulphurization and upgrading of petroleum products. This market will grow but it is difficult to quantify for much is produced inside the consuming plants, and derived from steam, using fossil fuels, or from fossil fuels themselves.[154] Indeed, accurate costings as well as firm statistics may be difficult to obtain. Table 50, however, gives some figures which may be used as indicators.

In the methanol column it has been assumed that carbon dioxide is available and indeed the report favoured a methanol rather than direct hydrogen route. These figures indicate that even at high fossil fuel prices hydrogen has a greater value as a chemical feedstock than as a fuel.

Table 50
Value of hydrogen as alternative to fossil fuels

Fossil Fuel p/therm	Price $/boe *	Value of H_2, £/tonne, used for:-			
		Fuel	NH3	Methanol	Petrochemicals
2	2.66	23	90	80	115
4	5.32	47	130	120	165
10	13.3	117	250	230	280
20	26.6	243	440	420	470

* Approx. $/barrel oil equivalent @ 1 barrel = 54 therms and 1p = $0.024

Source: Dr. Tatchell at Symposium on Hydrogen at Queen Mary College, London reported in *Energy policy*, vol. 2, Sep. 74, p. 247, with column of $/boe added to give alternative fuel cost comparison.

Hydrogen has been proposed for many years as an alternative to electricity for transporting large quantities of energy over great distances. Hydrogen is not a primary energy source but could be produced from water by a number of methods, both thermochemical and electrolytic. Many ideas are being conceived, designed and worked on in the laboratory, as are transportation and storage systems. On the utilization side there are many ways in which hydrogen might be used in aircraft, automobiles, appliances and in electrical utility systems. There has been much recent discussion about hydrogen as a 'by-product' of electricity generation by a nuclear reactor because this route offers a cheap, 'by-product' source of energy. However, there are crucial problems to wide-scale use of hydrogen, affecting transportation and storage. Hydrogen tends to permeate metals and induce embrittlement. Two suggested solutions to these problems are the use of additives to inhibit attack on metals (a very small quantity of oxygen is one way) and the use of barriers or coatings. Some members of the natural gas industry in the USA (discussions during the Ninth World Energy Conference, Detroit, 1974) appear to look on hydrogen as the fluid of the future to fill their pipelines as natural gas supplies grow scarcer. This is a 'pipe-dream'. A most efficient inhibitor would have to be developed to overcome leakage in old pipelines systems. The heating value of hydrogen is low, being only 270 Btu/cu.ft. compared to 900 Btu/cu.ft. for methane. Hence, with its lower density, larger pipelines or higher pressures would be required to provide the same energy as natural gas or substitute natural gas, involving considerable capital investment. The storage problems arise from the same basic difficulties. Hydrogen can be stored in expensive Dewar Flasks, or under pressure but this is costly. Some end-of-century fuel production costs have been guessed (*Nature*, Vol. 249, No. 5459 p. 275, 21 June 1974) as 135–175 pence/GJ for hydrogen by advanced technology electrolysis (UK) 100–140 p/GJ for hydrogen (USA), 110–130 p/GJ for motor gasoline from coal and 140–220 p/GJ for motor gasoline from crude oil at two to three times the prices of early 1973. However, despite these very grave difficulties there is a lot of work, but perhaps more talk, about various sophisticated ways in which hydrogen could be utilized. Perhaps the early twenty-first century will see some of them in action but the so-called 'hydrogen economy' seems a long way into the future.[155]

At a symposium in February 1979[446-449] discussion of the hydrogen economy concept brought out some points of interest. It is usually assumed that the primary sources of energy – fission, fusion, solar, geothermal etc. – would be available to make hydrogen at prices competitive with fossil fuels by any one of three methods, direct thermal, thermo-chemical or by electrolysis, which would separate water into hydrogen and oxygen. One opinion[446] was that improvements in electrolizers could be effected by raising the working temperature to 300°C or even 1,000°C and adapting some of the developments from fuel cells. It should then be possible to produce pure hydrogen at only 40% above the refinery price of US gasoline in 1974. Another opinion[447] estimated costs as follows:

Fuel	Costs in 1979 US $ per GJ	
	2000	2020
Gasolines from coal	9.37	10.68
Gasolines from shale	10.05	11.87
Gaseous hydrogen	10.35	10.35
Liquid hydrogen	12.76	12.76

The advocates for the widespread development of hydrogen usage stress that hydrogen is not a primary energy source, but as an energy carrier, between an abundant source, water, to the consumption point, is storable, economically produced by a number of methods from any primary energy source, is pollution free and renewable, as both the source and end-combustion product are non-toxic water, is a safer fuel than gasoline, (ignition temperature 574°C compared with gasoline 257°C, methane 577°C), and can combine in a fuel cell with oxygen and produce electricity at a conversion efficiency of 60-70% or perhaps more.

In the transportation field there are a number of projects in the USA testing the use of hydrogen in aircraft, cars, buses, jeeps, and railway trains. In the industrial field hydrogen is used and its use could be expanded in the petroleum, chemical, fertilizer and food industries, and be used for direct reduction of iron ore and for steam raising. In the residential and commercial fields, hydrogen could replace natural gas in many uses, including very efficient air-conditioning devices using hydrides. In the electricity utility sector hydrogen can generate electricity in fuel cells or in gas turbines and can be a storage vector. In gas utilities hydrogen could be a supplement to 20%, without any change necessary to burners or distribution system.

Methanol, or methyl spirit is the hydrocarbon which, being mid-way in many respects between 'carbon economy' and a 'hydrogen economy', may have an important future role. Methanol suffers from none of the defects of hydrogen as to calorific value or transportation and storage problems, being a liquid at normal temperature and pressure. The cost of manufacturing methanol from natural gas is more expensive than simple liquefaction of natural gas. However, shipment by normal tankers is possible rather than requiring special, insulated tankers, and so methanol had a distinct economic advantage. To date methanol has not gained many supporters and very often is dismissed lightly as just 'another possibility'. An interesting recent paper[156] suggests that methyl spirit could be a substitute for crude oil in the medium-term and act as a price setter for crude oils. Perhaps methanol should not be lightly dismissed.

ENVIRONMENTAL CONSIDERATIONS

The conversion phase in the energy resource chain to useful energy is one which has a greater impact on the environment than the other phases.

In the nuclear energy industry there are the two major problem areas of safety and waste disposal. Safety is a subject on which there is much ill-informed and often emotional discussion. The International Atomic Energy Agency is designing rules on safety. Waste disposal is also a problem which has plagued the nuclear industry from inception and a number of cases in both areas were discussed at the World Energy Conference in Detroit in 1974.[157] Discussion will undoubtedly continue and fears will undoubtedly inhibit development until they are dispelled. It is not intended to estimate here when that will be.

The coal industry has the waste disposal problem of fly ash and again there are many ways of tackling the problem, again discussed at the 1974 World Energy Conference[158] in papers from many countries, including UK, USSR, Hungary and Poland. One interesting fact is that the ash content of coal has little effect on lowering combustion temperatures when used in the magnetohydrodynamic (MHD) process of energy conversion but it does have an effect on the electrodes and generator walls. Therefore for this process either a low-ash coal must be chosen or the ash reduced.[159]

Pollution by waste heat from power stations is being tackled in many ways. The proponents of MHD electrical power generation[160] calculate that whilst a nuclear power plant at 32% efficiency contributes 2 units of waste heat generated, fossil fuel plants with 40% efficiency reject 1.5 units of waste heat, yet MHD system at 50% and 60% efficiencies reject only 1.0 and 0.7 units of waste heat per unit of electricity generated.

Sulphur emissions plague the conversion and utilization of coals and oils because of environmental pollution. Low sulphur fuels are at a premium and only desulphurization can make high-sulphur fuels useable. Methods of desulphurization of coals, and particularly of oils, have improved very much since the mid-1960s when the problem became of considerable public concern. High stacks solve dispersion difficulties in some areas; direct and indirect desulphurization of heavy fuel oil and the desulphurization of stack gases have been introduced into some refineries; sulphur dioxide removal from stack gases in power stations is possible; many other more sophisticated measures are possible. These, together with the environmental problems caused by the end use of petroleum products, were discussed at the Ninth World Petroleum Congress, Tokyo, May 1975.[161] All beneficiations cost money and the consumer must pay the 'social cost' as a part of the product cost, whether as a litre of gasoline or a unit of electricity. Costs are and will be reduced, for there is much competition and there is a marked reduction in supply flexibility by restrictions on the use of the large quantities of sulphur-bearing crude oils and coals which are available.

These examples are only a few of the practical problems arising from the need to upgrade the primary or intermediate or final energy source because of environmental considerations. Another way to express the point is that a barrel of crude oil with 4% sulphur and one of crude oil with 0.04% sulphur may have the same volume but very different values.

THE INEXHAUSTIBLE ENERGY SOURCES

The inexhaustible or renewable energy sources are the direct and indirect solar, lunar and terrestrial sources which have an enormous or infinite resource base but are predominantly very diffuse and have to be concentrated. The basic problem is therefore one of conversion. There are many types of conversion applicable to the inexhaustible sources, many long established, but many on which research and some on which development intensified after 1973. When considering the published results of such work there are at least five aspects which should be borne in mind, no matter what the subject:

1. **Costs.** Equipment used in the conversion of renewable energy is, and probably will be until the 1990s, manufactured using non-renewable fuel primary sources. There is an intricate feedback, for instance, of rising oil prices on materials, appliances and manpower, so that costs require concomitant reviewing with prices/costs of other energies.

   As renewable energy sources make an increasing contribution there may have to be a change in conventional economic criteria. Energy accounting using energy input and output balances is of significance when the energy resources are finite energy stocks. Renewable resources, however, provide energy flows, not stocks. Conversion plant efficiency may often be very dependent on location. Capital and operating costs, as well as specific use. may vary widely with the local societal/environmental circumstances and on the impact of the new energy source on the existing energy supply system, Perhaps in the 1990s, but certainly from the 2000s, economic criteria will have to take account to a greater extent than in the 1950s the differences and differing rates of change in societies and their attitudes to environmental protection. Renewable source investments can be very large and the risks inherent in new technologies so great that government funds are essential initially, even in free-enterprise countries. Hence renewable source investment policy has other factors to add to the usual ones of judging the future rates of interest, inflation and devaluation, and a plant's conversion efficiency and life expectancy.

2. **Research.** The frontiers of science are always an exciting attraction to the scientist and technologist. Just as in the early 1950s the challenges of nuclear physics and the hammering of weapons into ploughshares attracted the best students, so the challenges of the unconventional energy sources and the ideal of non-polluting, inexhaustible, environmentally benign fuels attracted the best students in the 1970s. The development of nuclear energy was not as fast as many expected, so that in the late 1960s and early 1970s many nuclear physicists turned to other fields. Some, for instance, applied their numeracy, bent for precision and scientific logic to energy demand forecasting. Unfortunately, no matter how dextrous the manipulation of figures in computers the sociological data on which demand rests is very vague. Research into the inexhaustible energies' conversion is probably in 1980 at a stage when disillusionment at slow progress may well divert the best brains into other fields. The challenges in the non-renewable energy fields are just as great as ever and one hopes will attract back some of these former 'best students'.

3. **Development.** In some areas, such as in the harnessing of the winds and waves, the basic science and technology seemed to be ready made. In these areas in 1979 it was noticeable,[448] particularly for wave-power, that sufficient had been learned in the previous five or so years, to show that the problems were far more difficult than at first thought. Materials, for instance, were not available which could withstand the stresses to which they were subjected in some new systems. This appreciation marked progress but also gave confirmation to the opinion that significant global contribution from many of the inexhaustible sources will be slow to develop.

4. **Market penetration.** Any novel energy system has to penetrate a market which is organized, mentally and physically, to an established source in which has been invested untold money. It was 75 years from the earliest oil wells before coal was ousted as the dominant world energy source. Aircraft took roughly 50 years to become a significant mode of world travel. Transistor radios, as a mass-communication appliance, swept across the world in say 20 years. Mini-skirts circled the world in perhaps 5 years. Hence, some might argue that small, cheap units, can effect both more rapid and more significant changes in social habits than big units and that the speed of possible change is increasing.

5. **Appropriate technology.** The doctrine which Dr. Ernst Schumacher crystallized in the phrase 'small is beautiful' made ground in the 1970s.[450] The oil shortages of early 1979, following the revolution in Iran, accelerated the funding of R & D into alternative energy sources by governments. There was also an apparent change of attitude amongst many in the industrialized, non-Communist world against the idols of big business, and high economic growth and against the profligate waste of natural resources, reinforcing the

environmentalist attitudes which grew in the late 1960s. The 1980s may well see greater progress towards adoption of more appropriate technology, to the adoption of smaller energy units, which implies a change towards societies more appropriate to their natural environment and a revival of interest in individual personal fulfilment. Therefore it is very necessary, when looking at the many ways in which the inexhaustible energies may be harnessed to bear in mind the possible effects on trends in social behaviour.

Solar energy

Attempts to harness solar energy have attracted most widespread attention. The Israelis were early in the field and the Australians have done and are doing considerable work for their domestic market but with the possibilities of export markets well appreciated. Some of the oil-producing states are also showing interest in fostering developments. In the United States of America the research has mainly been aimed at massive installations. The critics of this approach suggest[356] that it might lead, like nuclear research, into systems which have an adverse environmental impact and be of little value outside the industrialized countries. Many consider that it is through the world-wide spread of small units that the 'new' energies will have the greatest impact. This view is undoubtedly increasing as the articulate environmentalist enthusiasts take up the theme.[357]

The 'organic route' is increasingly fashionable with the environmentalists, and, perhaps fortunately, also amongst some practical engineers. In the developing countries there are the problems of firewood scarcity, the dangers of deforestation and the difficulty of reafforestation or the establishing of plantations of fast-growing vegetation to be cut periodically for use as fuel or even production of alcohol fuels. Nepal can be instanced as an energy-poor country where use of wastes could be valuable. India is a larger country which is beginning to appreciate that the burning of dung deprives the land of a useful fertilizer, whereas methane production from biomass can provide both fuel and leave a fertilizer. These and other biological systems are getting increasing attention. As mentioned earlier, some argue that the way in which such schemes can best be fostered is not by instruction or aid from central government but be 'seeding' methods, leaving the local initiatives to follow up and improve local conditions.

In Britain the 1974 opinions of the scientists of the Central Electricity Generating Board were reported[44] as being that direct use of solar energy for domestic water heating was just about economic in south-west England at a cost of about £95.00 per unit per house; however, direct conversion to electricity seemed likely to remain uneconomic in the UK at least until the end of the century. A UK Department of Energy review in 1979 essentially confirmed[478] this finding in its description of its research projects.

For electricity generation there are three obvious techniques. First, as used in satellites, the photo-voltaic technique of direct light to electricity at low voltage by expensive and relatively inefficient solar cells at a 1973 cost[45] of about US$20/w. Cruder, less expensive versions are used for re-charging batteries on offshore oil rigs etc. Second, the photo-thermal, solar thermal collection systems.[46] Various types of collector are used, flat, concave, trough, cylindrical or paraboloid, with or without dynamic positioning mechanisms and with or without sensitive receptor coatings. The collected energy heats a transfer medium which, for electricity generation, is usually water to convert to steam to drive a generator. One scheme, however, would use liquid sodium, see later. Thirdly a more speculative scheme would use synchronous-orbiting Satellite Power

Stations (SPSs), to collect, transform and transmit power by microwaves to earth receiving stations.

The concept of collecting solar energy in space and transmitting it to earth as a microwave beam[449] was first proposed, on the basis of mid-twentieth century technology, by Dr. Peter Glaser in 1968. Detailed designs for SPSs were developed in the USA and FR Germany; the USSR declared its interest in April, 1978, and the British Aerospace Dynamics Group presented a commissioned study to the European Space Agency in June 1978. The US Congress in mid-1978 provided $20 million for an accelerated SPS programme, and the US Senate provided $1 million for a specialized feasibility study over two years. In the Carter Energy Plan in mid-July 1979 (see discussion under USA in Chapter 9) solar energy was to provide 20% of US energy consumption in the year 2000, and SPS advocates will undoubtedly try to contribute. A number of problems, however, would seem to require solution, including, as of 1979:

- the need to minimize ionospheric and communications interference and critics contend that these have not been given adequate consideration.[451]
- the size of an SPS; the US prototype was for two rigid rectangular girder structures some 6 km × 5 km × 0.2 km. Another design was for 24 pairs of solar arrays of 2 km × 0.5 km, feeding microwaves into a 72 km circular waveguide. Such structures would be in geosynchronous orbit about 22,300 miles from the earth. Note however:

  a. the return in mid-1979 of the US Skylab from orbit on only a partially controlled path, drew attention to the material waste which was in space, which had been and was continuing to be dumped with the wanton abandon with which the early coal industry despoiled, for example, the valleys of Wales.
  b. the transportation of the mass required for a full system of say 50 SPSs of the US prototype would require propellants which could pose a very serious pollution problem.

- the biological effects of microwave impact were difficult to foresee but it was said that safe exposure limits could be defined by 1980.
- the costs would be several times the $25 billion cost of the Apollo project.
- the unknowns of the possible use of asteroidal or lunar material or lunar bases for transportation launches, in order to lessen the launch problems and as a study towards 'space industrialization', were to be considered as part of the US Senate inspired feasibility study.

This summary review illustrates one of the solar projects which was being actively worked on in 1979 and also the very great problems to be solved or the SPS idea abandoned in the 1980s.

Another major project more within current technology is the concept of 'power towers' or 'heliostats'. A solar central-receiver power plant to produce 100 MW of electricity would consist of a sun-tracking array of 2000 mirrors spread over about 3.5 km^2 which would concentrate heat on a central elevated tank. If this tank were filled, for instance, with liquid sodium rather than with water, the operating pressure need only be some 40 psi, in contrast to a water/steam system at 1,400 psi, yet by heat exchangers provide superheated steam to drive electricity generating turbines. US General Electric argued that electricity produced by a solar thermal electric conversion (STEC) system would be cost-competitive in the 1990s if oil prices continued to rise.[453]

Elsewhere there are other projects. One scheme discussed for many years involves the flooding of the Qattara Depression in Egypt to a depth of 60 m. below the level of the Mediterranean.[48] The level would be maintained, by the high evaporation rate of 1.7 m/year, over an area of 12,000 km². The differential in level, it is said, could operate a hydro-electric plant of about 4 GW capacity. Another project concerns stratified brines: observations of density in stratified brines in Lago Pueblo, Gran Roque, Venezuelan Antilles,[49] suggests that such brines, in natural or artificial reservoirs, should be considered as solar energy collectors, because appreciable amounts can be stored at low cost. Such methods have been used in Israel. However, to be significant, such schemes must be very large and the effort necessary may well keep them in the possible class for many decades.

Solar energy is already used in millions of individual solar water heaters, mainly in Japan, Australia, Israel, USA and USSR. No scientific breakthrough is required. Important technical breakthroughs, particularly as to storage, are required, if large-scale schemes are to materialize. Perhaps, though, large-scale schemes are not really essential; perhaps the best conversion of solar energy to man's benefit is through small units. Some 20% of total US energy is consumed in the heating and cooling of individual residential and commercial buildings. Some 80% of the world's population lives between the 40° parallels of latitude and air cooling in the developing, semi-tropical countries will be a major sector in improving living standards. Hence the development of small-scale solar energy appliances could have as significant a role in energy availability as did the kerosene/paraffin lamp in the last half of the last century. There were reputed[453] to be almost 200 firms in 1979 in the sun-belt of the USA who were prepared to offer solar concentrators and heat storage systems which would provide working temperatures up to about 325°C. for domestic or commercial heating or cooling, and one firm which could provide 485°C. for industrial process heat. The solar furnace at Odeilo, France, is unique and has achieved 3,800°C.

Photovoltaic energy conversion to electricity using silicon solar cells developed in 1955 by Bell Laboratories, USA, has been the main spacecraft power system since then.. Costs at $10,000 to $15,000/peak kW, with proved efficiency of 12–15%, make electricity 50–100 times more expensive than by conventional methods.[424] A technological breakthrough, of the type of the miniaturization of computers, is required before photovoltaic energy could make a major contribution to world energy. Such a breakthrough may materialise from work sponsored by the US Electric Power Research Institute (EPRI) to the tune of $25–30 million, 1978–1983, particularly on thermophotovoltaic (TPV) converters[463] in which a metallic element between sunlight and cell increases the infra-red impact on the cell. By 1977 work at Stanford University was said to have raised the conversion efficiency from the usual 12% to 26% with hopes for 35% efficiency eventually, which was thought to be economic for large power stations. There is a lot of work elsewhere and there have been claims made for solar cells[464] using special amorphous alloys in thin film as being competitive in 1979.

The WAES report[322] suggested that the total solar energy contribution in 2000 in the world outside the communist areas (WOCA) might be of the order of 2 million barrels a day oil equivalent (mbdoe). President Carter's proposal of mid-1979 was that solar energy should contribute 20% of US total energy demand in 2000. If total US Energy demand in 2000 were 55 mbdoe (See USA in chapter 8 on demand), then 11 mbdoe would be required from solar energy.

The report of the Conservation Commission of the World Energy Conference[424] (CoCo) estimated a range for 2020 for the world solar energy contribution between 8 and 33 mbdoe. An estimate which the author gave in early 1979 was for a total contribution from all unconventional energy sources in North America[456] of some 5 mbdoe, and this was considered by many to be an overly optimistic estimate. President Carter's proposal must therefore at least represent a very high target. A Solar Energy Research Institute was established at Golden, Colorado, in 1977, but there was little sign of progress two years later. Whether the new Solar Bank will be more effective was a moot point in late 1979, despite the suggestion that the Department of Energy Programme alone would involve contracts of $200–$300 million by the early 1980s.

The less developed countries (LDCs) were the hardest hit by the rising oil/energy prices after 1973 and, as most have abundant sunshine, though not all continuous, solar power is an obvious alternative energy source. The significant research work is, however, being done in the developed countries and most by very large companies diversifying into this new field and appreciating the big potential markets of the LDCs. Much of the research is increasingly sophisticated and uses materials, manufacturing methods and patented processes well beyond the capabilities of most LDCs.[498] Moreover 'big business' has thrived on economies of scale which preclude adaptations to local conditions. Hence there seemed a real danger that the developing countries in the 1980s might be subjected to high-powered salesmanship and become importers of technology, whose merits might be difficult for them to assess and be almost as expensive to them as oil, without either helping to establish local industry or satisfying special local needs. The problem of the adaptation of solar energy to their needs was both greater than many in the developing countries appreciated and had been given probably too little consideration by the research, development and management staffs of government or industry in the developed countries.

Winds

There are many schemes being discussed to harness the winds to a much greater extent than commonly used in the 1970s. At the one extreme are the schemes for chains of windmills offshore, and, at the other extreme, the incorporation of a windmill in the design of an individual energy-self-sufficient house. The energy output from an aerogenerator depends on three main factors, the cube of the wind speed, the total time over which the wind blows, and the square of the rotor diameter. The first two factors place a considerable premium on the windiest sites, but these are often sites of great natural beauty or far from conurbations. The third factor suggests economies with very large machines,[478] which have an environmental impact, for example they can cause TV signal distortion up to 0.5 km from a 0.1 Mw machine. The late 1970s saw a lot of talk about the need to develop windpower, yet the UK Department of Energy was spending only £500,000 on research and development of wind power in 1978/1979. In January 1977 it was said[479] that Canada by 1990 will have installed between 1,000 and 3,000 windmills, each producing 400–600 MWh/year. Sweden, Denmark, Holland and the US also seem to favour wind power; one estimate was that wind should provide 10% of US energy needs in 2000. In 1977 also[480] it was said that windmills, in mass-produced 'cells' of 200 kW with 25 mph average wind speeds, would have a per kW installed capital equipment cost of just over £250, against oil stations from £150 to £230, and nuclear from £220 to £375. In September 1977 the West German Ministry of Research and Technology commissioned production plans for a windmill with a 113 m rotor with an expected output of 2 to 3 MW.[481] The year 1977 was one when windpower was a strong topic.

The winds are an intermittent source of energy and seasonal but in temperate climates the advocates of their utilization claim this as a cost advantage, providing most energy in the winter when the need is greatest,[499] in contrast to the solar energy regime. The intermittency of the power source creates a major problem which is generally agreed as a serious underlying factor in wind exploitation.[500] (See chapter 7 on transportation and storage for further discussion of storage.) A dozen papers on various aspects of wind power conversion were presented at a conference on future energy concepts[448] early in 1979. They illustrated the wide range of ideas being considered and also the wide range of claims of cost effectiveness of schemes rarely at that time beyond the experimental stage. As with solar energy conversion plant, the 1980s will see much more research, development and demonstration and, probably, commercial exploitation. Also as noted for solar plant, the greatest benefit to those countries with least indigenous available energy, the non-oil producing developing countries, may well be through small simple machines which can be manufactured locally, do not depend on high-cost, high-technology materials, such as special steels or complex glass reinforced plastics for rotor blades, and are suited to local conditions, perhaps as part of hybrid energy systems.

Waves

The inherent power in the oceans and in the 'potted wind energy',[357] both as wave energy and ocean swell, is obvious, especially in the United Kingdom, which in the 1970s was probably leading the world in work towards harnessing this form of energy. The resource base is enormous, the proved reserves and production nil, for there was no exploitation economic under current cost conditions with current technology. Research in the UK has been directed mainly, but not exclusively, to four main systems, three named after their inventors. The Salter 'ducks' and the Cockerell wave contouring, articulated rafts, use the motion of one component relative to another, (an axle or another raft). One tenth scale models were tested in 1978. The Russell rectifier uses a constant head of water between a high reservoir filled by the wave crest and a lower reservoir emptied in the troughs. The Hydraulic Research Station worked on this system. The Oscillating Water Column (OWC) compresses air to form an energy storage used to drive a turbine and was worked on in the National Engineering Laboratory. Several universities were also experimenting with various ideas, such as an airfilled bag system invented by M. French, with the air compressed by wave action forced through a turbine. There are other omnidirectional devices, such as airbuoys and submerged ducts,[501] some floating and some fixed to the seabed, still not having moved far from the drawing board by 1979. Theoretically 'machines can be built which will deliver at least 25% of the wave energy that comes at them as useful electrical energy'.[357] Discussions[448] in early 1979 showed that at that time, and already mentioned earlier as a general point, research into all the systems had reached a critical point with the problems much better appreciated than in the enthusiastic early stages in the early 1970s. The difficulties of conversion in the ocean environment lie with the ranges between calm and tempest, the directionality and irregularity of the waves, cost effectiveness, anchoring, mooring, maintenance and eventual removal of fixed installations, corrosion, fatigue and fouling, environmental disturbance to fishing, birdlife and marine and coastal ecosystems and shipping, and transmission to market of the energy generated offshore sparsely populated areas, like the Western Islands of Scotland. It had become clear that a team was involved in almost all systems, drawn from civil, mechanical, marine and electrical engineering and geology as well as basic science research as into fluid mechanics.[502] Hence the 1980s will see a lot of further work but probably it will be the late 1990s at the earliest before wavepower will be contributing significantly even to Britain's energy needs.

Tidal power
Tidal energy can be harnessed only when adequately controlled. Only two tidal power generating stations operated commercially in the 1970s – one at Rance[57] at the eastern foot of the Cherbourg or Cotentin Peninsula of France and a small experimental unit in the Kislaya Inlet of the USSR. The Rance plant was never expanded to its original size. These are two of the 25 sites world-wide which were considered to be viable in 1978.[424] In the UK the Severn Barrage and Morecambe Bay schemes have been discussed for many decades but may not even be constructed in the 1980s. Perhaps though some of the small installations which were working along the European Atlantic seaboard in the Middle Ages may be resuscitated.

Ocean thermal power
The difference in temperature between the sunheated surface water and cold water at depth can be utilized. In the tropics the difference may be some 25° Celsius between the surface and at 600 m depth. The French have operated an ocean thermal energy converter (OTEC) off the West coast of Africa. Other large OTEC schemes have been discussed. The US Department of Energy funded in 1977 a project in the Pacific off Hawaii. But the structures must be large, are costly and have a low theoretical efficiency of 5%–7%, or lower.[500]

Bioconversion systems
Four aspects of the use of animals and plants as biological raw material sources for conversion merit consideration – domestic, municipal, agricultural and industrial wastes, energy crop plantations, genetic engineering to improve plant yields, and simulation of photosynthesis.

Methane from digested municipal sewage sludge and solid municipal refuse as fuels towards electricity generation are well established practices in many places. A scheme in Jersey, English Channel Islands, has shown that 175 kWh per head per annum can be produced from municipal wastes with a plant pay-off of 3–5 years.[448] More plants will be commissioned with time. Connecticut, USA, had a plan in 1978 for a 10-plant, $250 million system to convert 84% of the state's solid wastes into electricity equal to 10% of the state's demand.[462] New techniques in pyrolysis of refuse for production and storage of gaseous, liquid and solid, pelleted fuels will be developed. Germany, France and Switzerland are in the forefront of development, yet only about 15% of available refuse was estimated[448] as being used in 1979.

Animal dung has been a fuel through the ages and is a very important energy source in some countries, e.g. India. This practice is itself wasteful because the land does not get a natural fertilizer. Machines to produce methane from animal wastes are made in many forms and their installation on farms can be expected to increase. The value of the utilization of human and animal wastes in a country as energy-deficient as Nepal have been demonstrated.[402] In 1977 China claimed[459] to be the world leader in the utilization of bio-gas.

The utilization of most industrial wastes as fuels might be better discussed as a form of energy conservation. But the use of wood wastes in the timber industry and cane waste, or bagasse, in the sugar industry are well-established and sometimes are included in non-commercial energy production/consumption data, e.g. one tonne of bagasse is estimated as equal to one barrel of oil.[460]

Wood is the third most important world fuel after oil and coal but is still the premier fuel in the number of consumers. World production of wood as fuelwood was estimated by the UN[421] in 1976 as 1,200 million^3 and charcoal production as 3.5 Mt (both figures rounded). The FAO figure for consumption in 1976 of wood as fuel was some 1,184 million m^3 of roundwood.[465a] Taking the FAO figures[466] for general fuelwood density as 725 kg/m^3 and 1.77 t of wood equal to 1 tce,[465b] the energy value of 1,200 million m^3 of fuelwood would be almost 500 Mtce. However actual energy consumption is said to be possibly three to four times recorded consumption because so much domestic consumption, particularly in forested countries, e.g. in the Soviet Union, is not recorded. Recorded production of fuelwood was estimated to double from 1978 to 2000. This is a growth rate faster than the normal natural increment. Eckholm[457] in 1976 reviewed the need for firewood and the degradation of woodland throughout Asia, Africa and Latin America. Minister Djojohadikusumu[457] in 1977 stressed that many developing countries, including his own Indonesia, will suffer serious environmental and economic degradation unless drastic remedial action is taken in the near future. Furthermore, although blessed with the necessary soil and climate, the immediate needs of the population could make difficult the protection of cultivation of even fast-growing crops for use as raw fuel or for conversion to power alcohol. The last two points are further illustration that the economic assessment of investment policies for renewable energy sources can be very complex.

Energy crops became of increasing interest as energy prices rose with oil prices after 1973. Brazil, for instance, has extended its sugar and molasses industry to make ethyl alcohol (ethanol) by fermentation and is encouraging the use in cars of a 20% alcohol/80% gasoline blend. In 1979 some 15 % of its cars were said[460] to be using this 'gasohol' and alcohol production at about 1 Mt was aimed to increase to 3 Mt by 1982. Brazil has stopped exporting molasses, which may have an effect on world prices. Big quantities are required for significant production. Australia's raw sugar crop in 1974 was reported[467] as worth £1,755 million (1974 money), but it was said[461] in 1979 that if the entire Australian sugar crop was converted to alcohol it would only produce 1.5 Mt, which was only 5% of Australia's oil consumption in 1978. In the USA some 1,000 gasoline stations were selling[468] a 10% maize alcohol blend in 1979. Finland was experimenting with turpentine from pine trees as a fuel.[458] A study by the Battelle Memorial Institute is reported to have argued[458] that cane sugar was the best energy crop and competitive with other sources for alcohol production. The Sudan is interested because the long haul to seaboard of its expanding cane sugar industry makes molasses a waste product unless it could be used to produce power alcohol. But almost any crop can be used. Thailand is interested in plant which could convert to power alcohol any crop which might be surplus and so give stability to its agriculture and have a powerful economic regulator.[458]

David Fishlock, Science Editor to the *Financial Times*, London, reported

that[458] Sir John Corncroft, a Royal Society Research Professor at Sussex University, was offering four reasons why a multinational company with wide-spread interests might find energy cropping of interest:
- a way of living off energy income instead of energy capital,
- a simple technology adaptable to developing countries and agricultural communities,
- a non-polluting process and product,
- the energy conversion machines already exist.

This last point is emphasised by the US Government having budgeted[468] $11 million in loan guarantees for stills and the Senate approving a bill to increase that funding to $500 million, with no need for funds for R & D. The existing large areas under, for instance, maize in the USA may be a start, but elsewhere, particularly in densely populated areas energy crops might have to compete with food crops for adequate space for economic production. A number of large British companies, like Tate & Lyle, (the big sugar refiners), Unilever, Imperial Chemical Industries and the Royal Dutch Shell Group are interested, as well as a number of governments and their research establishments.

It is from research establishments that will come results from attempts at genetic engineering, 'creative botany', to adapt energy crop plants to produce the fuel or chemical required. However, there is usually a long lead time from research to wide-scale adoption and, perhaps, bad side effects to be overcome.

Similar caveats also apply to research into developments of artificial materials which might effect photosynthesis, as suggested by Professor Melvin Calvin, the Nobel Prizewinner,[458] who claimed that efficiencies up to 75% might be attained.

In brief, the 1980s should see a proliferation of practical schemes for the bioconversion of various materials, but also probably more paper schemes of less practical merit which will require much greater early and scientific investigation than, for instance, the ill-fated ground nut scheme in East Africa after the Second World War. Moreover, the five aspects regarding renewable energy projects, mentioned earlier are applicable. It must also be stressed that whilst the primary solar energy is inexhaustible yet energy through bioconversion is renewable only by the application of effort in an essentially second conversion process.

Table 51
Some estimted energy production costs
(dollars per barrel oil equivalent)

US Coal	3—5
Imported coal	8—14
Indigenous coal, NW Europe	10—15
Middle East oil	0.25—1
North Sea oil	7—12
Low btu gas from US coal	30—37
Liquid oil from US coal	30—45
Liquid oil from imported coal	30—45
Liquid oil from oil sands	15—30
Liquid oil from oil shale	15—35
Biomass as a liquid fuel	30—60
Solar	50—1000+
Sugar distillate	60
Methanol from wood	60

Source: quoted as 'oil industry estimates' by Jonathan
Wood, *Sunday Telegraph*, 22 July 1979, London.
See also Figure 35.

Geothermal energy

The earliest commercial use of natural thermal energy, on any significant scale was at Lardarello, 100 km south of Florence, Italy. The first use from 1827 was for boric acid, then from 1904 for both chemical and electrical production and since 1964 for electricity only.[17] Operations in this area can illustrate a number of important points.

i. The local concentration of heat can be very large. The heat extracted from the wells in the 10 km² of Lardarello amounts to 1.06 GW.

ii. The specific density of electrical energy which can be produced can vary considerably even in the same area. In the 10 km² of the main Lardarello production area, a peak value of 12.5 MW/km² is attained but the average for the larger area of about 170 km² is 1.6 MW/km². At Cerro Prieto, Mexico, the average specific density is calculated as 18.7 MW/km².

iii. The exploitable amounts and characteristics of the fluids differ widely between areas. In Lardarello, the Geysers of USA and Matsukawa in Japan, the fluid is superheated steam but with different amounts of incondensable gas — Lardarello 6%, the Geysers 0.6% by mass. In other fields the fluids are dry, saturated steam or a mixture of steam and water, e.g. Wairakei, New Zealand. Some fields deliver water at about boiling temperature, e.g. Reykjavik, Iceland. Some have hot brines with much dissolved salts, e.g. Salton, USA.

iv. Steam or hot water cannot be transported very far without losing their heat and so heat must either be converted to electricity or used directly and locally, though some dispute this point.

v. Dissolved mineral salts may provide a secondary revenue but also may make utilization difficult from corrosion, precipitation, etc. Large amounts of incondensable gas may have to be extracted before the steam can be used.

vi. Experience, improvements in adopting and adapting new prospecting methods, and application of systematic thinking can lead to discovery of hidden sources, away from surface seepages. The discovery of the Mount Amiata field in Italy was an acknowledged breakthrough in geothermal exploration.

Electricity is also generated from geothermal hot springs areas in USA, Mexico, Japan, USSR (as noted in the resources section) and Iceland, whilst design work for a station in El Salvador was under way in 1975. In New Zealand, in the 36 km² around Wairakei, steam can be obtained at shallow depths, but only by drilling to about 610 m close to a major fracture can high-pressure steam be obtained. About 80% by weight of the discharge from the bores is hot water which has to be separated from the steam before use in the steam turbines. The average annual energy input into the national system is 1,100 GW. The installed capacity is 192,600 kW but the high-pressure steam source is declining.[98] At Kawerau there is a 10,000 kW plant. In the USA the Geysers area, California, has 290,000 kW. The grand total of installed capacity in the world was 1.01 GW.[17] The capital and operating costs vary with local conditions from 1.4 to 2.5 mills/MJ. A comparison with other methods, updating some 1970 figures, is given by Leardini as at one place, hydro 1.6 mills/MJ, steam 3.8 mills/MJ and nuclear 4.04 mills/MJ.

The uses of hot springs for other than the generation of electricity are both ancient and modern. The ancient ones are for domestic heating and for hot baths, often with reputedly medicinal benefits from the dissolved salts. Amongst the modern uses can be cited production of potable water in a desalination plant, e.g. El Tatio in Chile; use in the manufacturing process in a pulp and paper mill, e.g. Kawerau, New Zealand; use in conjunction with a lithium-bromide absorption process in refrigeration, e.g. in the USSR and at Rotorua, New Zealand; drying of diatomaceous earth in Iceland; use for domestic heating, district heating and in hothouses and hot beds in horticulture, as in Japan, Romania (using 400 m^3 water per hour at 85°C in pilot greenhouse schemes), Hungary (hothouse area reported in 1970 to be 400,000 m^2 to be doubled by end 1970), USSR (at Makhach-Kola an area of 25 km^2, containing hot-houses and hot beds, allows two crops of flowers and vegetables each year); commercial fish farming, as in Japan in both Hokkaido and Kyushu. In the USSR it is reported[99] that the possibilities are being studied of using geothermal hot water to facilitate mining in permafrost areas. Those waters with high percentages of dissolved salts may form the bases for chemical works as at Larda-rello, and mentioned earlier.

These examples illustrate the many energy uses to which geothermal waters can be put, depending on local circumstances but, because the heat in steam or water cannot be carried long distances the resource is indeed localized in value. Nevertheless these uses must be added to the value of the future electricity generating plant which T. Leardini in 1974 estimated as 923 MW firm, plus 715 MW planned to start in near future.[17]

Heat pumps

Heat pumps are machines driven to move thermal energy from one place of low temperature to another of higher temperature, as in refrigerators and air-conditioners. Heat in a river or lake or sea or at depth on land can be used to improve performance. Hence one is dealing with geothermal energy in the broad sense. The cost is relatively little in energy terms, yet there has been surprisingly little use of heat pumps.[500] However, with the increased costs of primary fuels, the growing emphasis on energy conservation and the utilization of waste heat, one might expect developments in the 1980s. Thus heat pumps are a means of increasing primary energy effectiveness which benefits from improved insulation reducing the necessary input, say for space heating. A total energy system can be arranged with a heat engine which would increase the efficiency still further. Computer models were developed in Stockholm[446] in 1978 to measure the complex impact of advanced heating systems on single family dwellings, which is an indication of the renewed interest being taken, though it may well be in industry in the developed countries that the greatest energy gains might be made.[424a]

NUCLEAR ENERGY

The conversion of atomic energy into electricity may only have become commercial in the 1950s but by the 1980s there have been many reactor years of operation with a high record for safety. In chapter 3 the problems of safety were noted as liable to inhibit the growth of nuclear power and this point is mentioned elsewhere and is a conversion process and plant problem. A spectacular and widely publicised malfunctioning of a pressurized water reactor (PWR) occurred on 30 March 1979, at Three Mile Island, Harrisburg, Pennsylvania. The nuclear fuel, from which heat is removed by water circulating under high pressure, was inadvertently deprived of coolant during the critical early minutes of the crisis. By the time this was discovered the reactor had reached such high temperatures that parts were damaged and conditions arose which were unexpected to the operators. There was confusion which caused considerable alarm and temporary evacuation of people from the neighbourhood. The alarm caused by what turned out to be a 'nine day wonder' was increased by the difficulty of explaining in simple terms the complex course of events which could only be monitored unseen. Nine similar plants were shut down, millions of dollars and possibly years will be required to restore the plant. This accident caused a coalition of many opponents of nuclear power and may have set back the US nuclear industry for several years. Yet the people of Austin, Texas, voted for more nuclear local power during all the publicity. The proponents for nuclear power can use the specific history of the Three Mile Island accident as confirmation of their belief in its safety which was summarized in August 1979 by one writer[470] as:

'A nuclear power plant cannot undergo a nuclear explosion; the only danger is the significant release of radioactivity, and that danger is localised in a few cubic metres of space, where it can be surrounded by a multi-layered defence in depth. Moreover, the time scale of a possible accident is so slow (melting of the fuel, melting through the pressure vessel, possible failure of the containment building) that there is time to bolster the defences wherever they are in danger of growing weak. And even if this slowly progressing battle threatens to be lost there is time to evacuate the endangered area. Nuclear safety, then, is not based on the infallibility of the operators or on the perfect functioning of gadgets but on defence in depth and slow time scale. No other energy facility has even one of these two protections. . . .'

However the President's Commission of Enquiry, (the Kemeny Commission), into the incident recommended that all work on new nuclear plant in the USA should be halted until new safety regulations were agreed and implemented; the suggested regulations were said in Britain to be similar to those already in force in the UK. The Commission also recommended the establishment of an overall body to ensure coordination of all facets. The Nuclear Regulatory Commission accepted the Kemeny Report suggestions as to a case-by-case examination of all projects under construction and planned.

One result of the Three Mile Island incident was thought[508] in October 1979 to confirm that the USA had lost its lead in nuclear policies. In 1977 the USA had tried to force the world to adopt ill-prepared theories for dealing with weapons proliferation by methods plainly of greatest advantage to the USA, whilst inhibiting the growth of nuclear power elsewhere. Other countries need nuclear power more than the USA and many intended to continue in the 1980s to develop capacity whilst taking greater precautions to avoid incidents like that of Three Mile Island.

The problems arising from the debates about the safety of nuclear reactors are discussed later in the context of demand, and supply and demand. It is not intended to discuss the merits of the different types of reactor and their efficiency in converting nuclear energy to useful energy. The relative merits are often obscured by commercial and/or political pride and prejudice, as well as by incomplete technical and economic data. It is however, considered necessary to look at the nuclear fuel cycle. This is the conversion aspect and has an influence on the required input of raw material which is a factor somewhat different to the conversion processes of other industries. First, though, a word on nuclear fusion.

Nuclear energy by fusion

Fusion is the combining of light nuclei to make heavy nuclei at extremely high temperatures. It is often thought to be the panacea of all energy problems once a technique is developed. The potential advantages seem important. Deuterium, the basic fuel, is found in virtually inexhaustible quantities and available at negligible cost. The fusion combustion product is helium which is non-toxic and non-radioactive. There is no danger of runaway chain reaction. There is relatively low associated radioactivity. Some say that there is no possibility of diversion of weapons materials though others note that tritium, the heavy isotope of hydrogen with mass about three times that of ordinary hydrogen, features in the process with deuterium, the isotope with about twice the mass of ordinary hydrogen. Tritium is used in H-bombs.

The problem with fusion is to develop a technique to contain a gas of charged particles, plasma, at stellar temperatures long enough to release a favourable amount of energy whilst the plasma is isolated from the structure of the reactor. There are two known methods, by magnetic fields or by confinement of the atoms themselves with high powered lasers. The first method is better known and has half a dozen or so variations of which the best known is the tokamak. (Tokamak is a Russian acronym, to-toroidal, ka-chamber, mak-magnetic).[144] In this system the fusion reaction is the easiest to achieve and takes place between deuterium and tritium inside a magnetically confined plasma at more than 100 million degrees Celsius. The reaction products are helium ions (He^{-4}) and neutrons. About 80% of the released fusion energy is imparted to the neutrons. Their high kinetic energy must then be converted to heat and breed tritium fuel by absorption in a lithium blanket. Heat transport and conversion systems are then required which are similar to those for fission reactors. In the laser method a solid pellet of deuterium-tritium atoms is hit by laser beams from several directions and so heated to ignition. A lithium blanket and heat transfer and conversion systems are then required as in the tokamak system. The problems of both methods lie in the degrading effect of neutron bombardment on the reactor structure, in the use of lithium, in liquid metal or inorganic salt form, as the reactor coolant and the need to breed and contain tritium for use in the fuel cycle.[145]

Fusion, being a breeder cycle based on tritium, has an advantage in comparison with the fission breeder cycle based on plutonium, in that the reactor does not require continuous cooling after shutdown as does the fission core. The two systems are comparable in respect of volatile radio-active inventories and non-volatiles when niobium alloys are used in fusion reactors. The main problem, however, is that fusion has not reached the scientific feasibility stage. The Fermi team proved that stage for fission in 1942 and fast breeders may possibly be commercial in the later 1980s, then commercial feasibility of fusion cannot be expected until after 2000. Even the optimists now talk of the 1990s[143] and it is too early to make comparisons with the fast fission breeders. However, the 10% saving in the total cost through the fuel cost being negligible may be offset by more structural materials being required in the fusion plant.

Nuclear energy by fission

Comparisons in conversion costs are possible between nuclear and fossil fuel power plants and the 1973/74 increases in oil prices have obviously stimulated assessment.[146] In energy equivalence it is estimated that 100,000 MW of nuclear power, operating at 70% average load factor, would require the annual consumption of:

187

- 350 Mt of coal at 8,500 Btu/lb (e.g. western US coal), or
- 143 Mt of fuel oil at 6.3 M Btu/barrel, or
- 170 Km^3 of natural gas at 1,030 Btu/cu.ft.

Table 52
Comparative capital & generating costs of US —
nuclear, oil-fired and coal-fired power stations

	Nuclear 2 x 1100 MW	Oil-Fired 3 x 700 MW	Coal-Fired 3 x 700 MW
Basic capital costs 1975 (US $ per kW)			
Design & installation	187	103	
Materials	103	37	
Equipment	130	90	
Other client costs	42	23	
Totals	**462**	**253**	
Capital costs on completion (US $ per kW) **1985 — USA costing system** Projects start: Nuclear 1.1.1975 oil, 1.1.1978			
Allowance for funds during constructing at 8% p.a.	230	100	
Escalation at 7.5% p.a.	258	267	
Totals on completion	**950**	**520**	**805**
Generating costs (US mills/kWh)			
Plant capital from above	23	12.5	20
Operating & maintenance	2	2.5	5
Fuel (oil at $12/bl, 1974 coal at $1/M Btu 1974)	9	41.0	19
Totals	**34**	**55**	**44**

Notes: Generating costs averaged over first 10 years, 15% fixed charge rate, 70% plant factor, 5% escalation on fuel and operating & maintenance, 7.5% on capital. Coal fuel costs include 6 mills/kW for SO$_2$ clean up.

Sources: Based on 'Economics of Nuclear Power', W. Kenneth Davis, Figs. 1 & 2, Int. Sym. Nucl. Power & Techn. Taipei 13 Jan. 1975.
and 'Converter Reactor Alternatives', W. Kenneth Davis, Atomic Industrial Forum Conference, Washington D.C. 19 February 1975.

Table 52 illustrates the difference in make-up of the capital costs and, also the low costs in nuclear plants of the fuel, operating and maintenance elements. There have been long, and at times bitter, arguments about such costings. This has been particularly so when different types of reactors are being compared. There is rarely a clear cut case because the bases for estimates are continually shifting. In the USA, for instance, the months spent on design and construction, before loading the fuel, have increased because of stricter specifications and environmental protection constraints, as shown in the table below:

	increase in size of generator	design period (months)	construction period (months)
early 1960s	200MW–450MW	13	38
late 1960s	400MW–700MW	20	54
early 1970s	800MW–1300MW	37	63

Table 53
Fuel cycle reference economic parameters and sensitivity indices

		Sensitivity	
	Parameter	Change in parameter	Change in mills/kWh
Inventory basis			
Working capital charge rate	1.25%/month	+ 10%	− 0.08
Plant capacity factor	0.70	− 10%	+ 0.11
Unit prices			
Yellowcake* ($/lb U_3O_8)	$15	$10	0.874
Conversion to UF_6 ($/lb U_3O_8)	$ 2		
Separative work ($/SWU) **	$75	$10	0.20
Fabrication ($/kg U)	$80	$10	0.05
Recovery ($/kgU)	$100	$10	0.023
Fissile Pu value ($/gm)***	$13	$ 3	0.07
Effective lead/lag times (months)			
U_3O_8 and conversion	15		
Enrichment	8		
Fabrication	4		
Recovery	12		
Pu and U credit	15		
Processing losses	1.5%		

*   Mined and milled uranium ore
**   SWU is a separative work unit, not a physical quantity. It defines the size of an
      enrichment plant.
*** (50% of U_{235})
Note: 1 mill/kWH is roughly equivalent to $0.60/bl of oil or $4/t of oil.

Source: 'Economics of Nuclear Power', W. Kenneth Davis, Tables IX and XI,
    presentation at Int. Symp. on Nucl. Power Techn. & Econ. Taipei,
    13 Jan. 1975.

Fossil fuel power plant sizes have also roughly doubled from about 350 to 800 MW capacity in the same period. However, their combined design and construction times have only risen from 35 to 57 months because the technology is more standard.[147] Furthermore, cost escalations through inflation vary according to the time scale and the relative proportions of the materials, labour and fuel elements. A measure of possible variations can be seen in table 53. The fuel cycle economic parameters are unescalated but represent the source's anticipation of US market conditions in the 1980s and his cost sensitivity estimates.

These indicated variations preclude discussions here of the relative merits even of those reactor types which are the best-known, nor indeed of the relative merits of the different uranium enrichment methods. Furthermore, the choice of a particular reactor type in any particular country may be influenced by unique political reasons which will outweigh economic reasons. However, the different types of reactor are attempts to improve the fuel cycle of the thermal nuclear reaction and so improve the energy conversion stage.

The fuel cycle

The different steps in the fuel cycle are integral, costly and controversial. They are shown in figure 34 and are the conversion of uranium oxide (yellow cake) to uranium hexafluoride UF_6, enrichment to a higher concentration of fissile material by one or other process, fabrication of the fuel 'rods', reprocessing of the spent fuel, and disposal of waste. One major example of a technical process which can have a marked effect on uranium requirement is the enrichment step. Most of the world's current reactors use uranium enriched by a gaseous diffusion process and the plants reject some of the fissile U_{235} as 'tails'. The proportion of 'tails' can be varied, depending on the relative price of the uranium ore and the cost of enrichment, customarily measured in separative work units (SWU). Similarly the output of the plant can be varied, within substantial limits, by varying the reject content. Therefore, if there is a shortage of uranium in enriched form and of separation capacity, greater output can be achieved by rejecting tails with a higher concentration of U_{235}, but this requires more UF_6 at natural concentration — the need to cover this contingency was mentioned as one reason for stockpiling. Kostiuk noted[287e] that,

> 'The American ERDA has made and will make substantial changes in the "transaction tails assay" for its enrichment plants. Until the autumn of 1977, the figure is at 0.2 wt percent of U_{235}. In the early eighties it may rise to 0.3 wt percent. And if a shortage of enrichment plant capacity were to develop, it could rise as high as 0.375 wt percent. In terms of uranium ore required for a given enrichment output, this range of tails assays implies an increase of almost 40% from the figure corresponding to the 0.2% base case'.

The US Government, through ERDA, has monopolised enrichment in the USA, but President Ford's Nuclear Fuel Assurance Bill aimed to help private industry to enter this field. The USA restriction on the enrichment of imported uranium for domestic use will gradually be lifted from 1978 onwards until there is no restriction by 1984; presumably the Government consider that their domestic producers will not need protection after 1984 or perhaps even that imports to the USA will be necessary for domestic requirements. The USA will continue to satisfy the bulk of the non-communist world demand for enriched fuel until the early 1980s, when multi-national Western European plants should be in operation, and both Australia and South Africa may have their own plants. The USA has favoured the gas diffusion process, but there

THE NUCLEAR FUEL CYCLE

Fig. 24

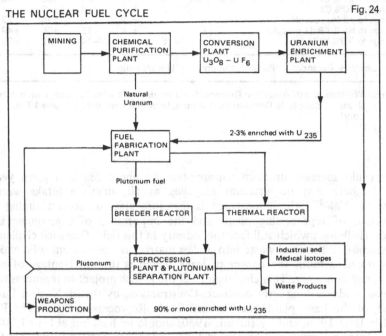

Source: Stockholm Institute of Peace Research, 1975 (The Economist, 6 December 1975, London)

are other processes, e.g. the centrifuge and nozzle processes, being developed in Europe, and the untried laser methods. The principle of the laser method is that monochromatic laser beams separate different isotopes of uranium on the basis of miniscule chemical differences. The promise is that laser methods will halve the cost and save 90% of the energy used in current methods, which is very considerable. Laser techniques are not simple, and an application for a test facility, which might expand into a pilot plant, was costed at $15 million when put to the US Nuclear Regulatory Commission in March 1976. This type of facility would be based on atomic uranium vapour, which is also under research at the USA Livermore Laboratories, whereas molecular methods are being studied at the Los Alamos Laboratories, USA. Research is also being carried out elsewhere than in the USA, but it is still at an early stage and, as just seen, is expensive, but it is said that there are no obvious barriers to success.[280]

The reprocessing of spent fuel, uranium and plutonium, can also have an effect on the uranium intake. An average delay of two years in reprocessing

Table 54
Uranium Demand Sensitivities — USA
(thousand short tons of U_3O_8)

	1980	1985	1990	1995	2000
Base Case; 65% CF, Pu & U recycle 0.3% tails	30	53	81	109	148
Change to 70% CF	+ 2	+ 4	+ 6	+ 8	+11
Change to no Pu recycle	–	+ 2	+ 5	+16	+22
Change to no Pu & U recycle	–	+ 4	+12	+36	+49
Change to 0.2% tails	– 5	– 9	–14	–20	–25

CF = Capacity Factor,   Pu = Plutonium,   U = Uranium

Source: 'Whither North American Demand for Uranium?', Charles F. Luce, International Uranium Supply & Demand Symposium, Uranium Institute, 16 June 1976, London.

services could increase uranium consumption by about 5% in a given year, whilst, if there was no uranium recycling at all, uranium intake would increase by 20%[275] The importance is therefore clear of correct timing of construction of reprocessing plant, and the importance of overcoming the technical challenges which still face the industry in this field. One such challenge is to incorporate the final waste into a glass matrix, as a permanent solid which can be stored without the danger of leaks, which is the disadvantage of any liquid in any container. This challenge faces the British project to reprocess fuel for Japan, under a long-term contract. Construction, by British Nuclear Fuels Limited, of the basic plant, the Thermal Oxide Reprocessing Plant (THORP) started in June 1976, and by the mid-1980s should be handling at least 1,000 t of oxide fuel each year, and be one of the half dozen integrated facilities in the world. The 1976 estimate of the plant cost was £350 million. To this plant will be integrated, when/if successful, the solidifying plant for the waste, using a process called Harvest. The pilot plant for Harvest had been started for completion in 1978, at a cost of £4 million, with the hope that in 1980 a demonstration plant could be ordered at a cost in 1976 estimated at £19 million. The full research and development programme for Harvest over ten years could cost £30 million. British Nuclear Fuels have devised a form of contract for their customers, of the pre-payment, cost plus type, in order to elicit a substantial down-payment towards the cost of the plant and to insure the company against the large uncertainties which still remain over the final capital cost. This example of the uncertainties and the cost are the typical problems of the reprocessing field. The Germans are talking about a bigger plant, reprocessing 1,500 t at maximum, costing £750 million, whilst the French obviously have cost estimates similar to the British, for their bid for half of the £400 million Japan contract quoted only a slightly higher price than the British.[289 b] The fuel cycle capital requirements for the 'world', cumulative to year 2000, were calculated to be US $(1975) 135-104 billion. This calculation was based on the high and low estimates of the 1976 OECD Report, as satisfied by the reactor pattern of table 66, on the characteristics assumed for those reactors, with a ten-year forward reserve fuel allowance, and the necessary mining, milling, conversion, enrichment, fuel fabrication, and reprocessing. The figures again measure the range in demand estimates, but table 54 illustrates again, the variations which can be played technically with the uranium requirements, using, in this case, the USA as the example.

The USA Nuclear Regulatory Commission was to announce its plutonium recycling policy in mid-1977, but the base case of table 54 assumes approval. If the decision is negative, then not only may plutonium recycling be impossible, but the reprocessing of uranium alone may be uneconomic, and so the US demand for uranium in 2000 could be increased by 49,000 short tons of oxide, a 33% increase.

Fast or breeder reactors

The characteristics, the development progress and, most importantly, the timing of the introduction of fast reactors into the nuclear reactor pattern provide probably the most crucial problems in the nuclear energy situation for the next 25 years, even though their actual contribution in 1990 may be small, see table 67.

In the year 2000, however, the OECD allocated 237 GWe of a total 2480 GWe, or 9%, whereas Vaughan argued in November 1975,[270] for a 33% contribution, or 400 GWe from the total of 1220 GWe to which he considered the nuclear share of total electricity would be because of the scarcity of uranium. Vaughan argued that the normal estimate of uranium consumption was beyond the practical feasibility of production. He suggested that a current extraction capacity of 30 kt could double by 1980, treble by 1985 and would be 150 kta at the end of the century, if, but only if, reserves are found and delineated fast enough; this he considered would mean a rate of increase of discovery greater than at any time with more abundant minerals, like copper, nickel, iron or lead, and therefore might not be attainable. Hence if nuclear energy was to satisfy the estimated demand for electricity, the consumption of uranium per kWh must be reduced. Improved enrichment processes, recycling of plutonium, and more efficient use of thermal reactors would help, but, in his view, the only real saving would be through the early introduction of fast breeder reactors and, moreover, the doubling time, or time taken by a reactor to breed more fuel than it has used, would have to be reduced from the current 50 years to about half the doubling times of the breeders now being designed. Vaughan concluded that, on present knowledge, the constraints of the scarcity of uranium supplies would restrict the nuclear contribution to his 1220 GWe, even if fast reactors with 40 years doubling time were operating in 1985, and 15 year breeders were being introduced from 1995 onwards. This argument illustrates how important the possible FBR is from the supply point of view, the reactor programme point of view, the full fuel cycle and the contribution to electricity generation. Hence the status of the FBR warrants a closer look.

In thermal reactors, energy is released by the fission or splitting of uranium$_{235}$ which is induced by bombardment by slow, 'thermal' neutrons. Some of the neutrons so released are absorbed by the more abundant, fertile uranium$_{238}$ isotope, converting it to the fissile plutonium$_{239}$, which may itself then be split and add to the total release of energy. But the amount of plutonium created is less than the amount of uranium used up or 'burnt'. To increase the amount of fissile material created, advantage can be taken of fast neutrons which produce, on average, more fast neutrons from the fission they cause, than are produced from fission caused by slow neutrons. Therefore a fast reactor is one in which fast neutrons are used to induce fission of plutonium$_{239}$, or a mixture of Pu$_{239}$ and U$_{235}$, together with some fertile U$_{238}$. The use of fertile Thorium$_{232}$ to produce fissile U$_{233}$ has the advantages which are being developed. An important criterion of merit of a breeder is the speed with which new material is produced, and the marker used is the time taken to double the amount of fissile material associated with the operation. The associated material is all that which is involved in the whole cycle, from the fuel in the core to its return, after cooling and reprocessing. The breeding gain can only be obtained whilst the fuel is in the reactor, hence to lessen the doubling time then either the fuel must spend more time in the core, or be subjected to a higher neutron bombardment, or the time spent in cooling and reprocessing be reduced. These alternatives have their problems, and the challenge of attaining a doubling time of less than ten years is very great.[288]

The first operational true breeder reactor was the British Dounreay Fast Reactor which has been working since 1959. A new generation of prototype FBRs has now been developed in France, Britain and the Soviet Union, and another is being planned in the USA. The construction of the French Phénix 250 MWe prototype began in 1968, was completed in 1973 on schedule, was generating in July 1974, and has been on full power for 80% of the time since then. The Superphénix, with a 1200 MWe rating, is scheduled for completion by 1982, and two more breeders were planned to be ordered between 1978 and 1980. The British prototype fast reactor (PRF), at Dounreay in Scotland, is now on full power, is more nearly a commercial design than the Phénix, and plans for a 1300 MWe commercial fast reactor (CFR), are reported as being well in hand. The Soviet 350 MWe prototype breeder at Schevchenko was reported as completed in 1972, but has had serious troubles. The USA did not go from their experimental reactor of 1963 (EBR-2), to a prototype, but preferred to build an advanced experimental plant, the Fast Flux Test Facility, (FFTF), which was completed in 1980. A whole new technology has had to be developed, for example, to handle the liquid sodium which is being used as the coolant. The mixed oxides of plutonium and highly enriched uranium being used in the British and French reactors require integrated working fuel cycle facilities, and reprocessing techniques for handling the highly active fuels are being developed simultaneously. The Phénix was planned to use 100% plutonium in 1976, rather than the 80% plutonium fuel used hitherto. The Phénix has a long doubling time of 50-60 years, whereas the PFR has a doubling time of 30 years, and the goal for the CFR is 10-12 years. The French say that they are not aiming for such times in the Superphénix but will wait for the third generation. The USA have started building a 350 MWe Fast Reactor at Clinch River (CRBR), to be completed in 1982. The USA is behind Europe in development of the breeder, but Europe has greater incentive to work fast; Britain has no indigenous uranium but does have plutonium from its thermal reactors, and France has a high dependence on imported petroleum.[274]

There are still a number of technical problems which have to be solved before fully grown commercial plants can be made available. This is not the place for a comprehensive review of these problems, but mention of a few, from different parts of the fuel cycle, may indicate how serious they are, and also that they are being tackled seriously. Fuel costs are low, but there are problems of swelling of the fuel under fast neutron bombardment and of 'creep' due to irregular swelling; both these phenomena mean that the fuel cannot be kept in the core as long as desired, so that the doubling time is increased. Research is in hand which may develop uranium carbide as the fuel instead of the oxides, which may help. Fuel fabrication techniques and facilities need improvement. The British have facilities to produce 5–10 t of fast reactor fuel at Windscale, near Manchester; the French have facilities at Cadarache, in Southeast France, which, when complete, probably in 1981, should produce 20 ta of oxide fuels. The 233 MWe Phénix has 4.3 t of mixed oxides as fuel, clad in stainless steel, whereas the 275 MWe Magnox Thermal reactors in the UK use 304 t of natural uranium clad in a magnesium alloy. ERDA plan to build new, more automated facilities to augment the 5 t annual total production from two plants. No one has any experience, except on a small test scale, of fast reactor fuel reprocessing. With its high 'burn up', the spent fuel has high residual temperatures and has to be cooled before the reprocessing can begin. This time in the cooling ponds may be six or even twelve months and is wasted time from the breeding aspect. The total time for the fuel to be out of the reactor at a plant being completed at Dounreay is a year with the fuel in the reactor for a year. Both the UK and France are considering plants with a 200-day cooling period which would process about 1 t/day, and so service 10–15 power plants and might be ready in the 1980s. The safety precautions with breeder reactors are different from those with thermal reactors, and there are differences of philosophy between the British, French and American nuclear safety scientists, as to the criteria to which they should be built. However, there appear to be every expectation that one crucial problem of preventing the aggregation of all the fissionable material into a critical condition can be solved. There is a major difficulty at this time in that the eventual costs of the fuel cycle are not known — a perennial problem in almost all aspects of the nuclear energy industry, because it is young and working at the frontiers of science in many fields. The cost of the fast reactor, whether of the pot type of the Europeans, with comparatively small self-contained reactors put into one large container, or a loop type breeder, will probably always be higher than for a thermal reactor, but the reactor is only part of the full cycle. From this sketchy incomplete look at the status of the fast reactor it should be appreciated that the difficulties may be daunting, but there are many people trying to solve them in many ways and in many countries.

7
Transportation and storage

The problems concerned with the transportation and storage of hydrogen, discussed in the last chapter, emphasize that resource/reserve production figures of the primary energy sources are subject to amendment, through alterations in technology and economics, all along the chain from resource-in-place to end-use. Transportation and storage are important considerations for all primary energy resources.

TRANSPORTATION

The increasing volumes of primary energy sources required have led to remarkable developments since 1940. The most dramatic developments have been in very large tankers and very large pipelines. These have also raised the costs to very large sums which have to be financed. At the same time, the small but multiple developments, with small transportation and storage costs because of local exploitation and consumption, as illustrated in the People's Republic of China and possibly applicable elsewhere, must not be forgotten. The four main categories of global, inter-regional and intra-regional flows of primary fuels and electricity for distant consumption are water-borne, pipeline, railroad and electrical systems.[162]

Water-borne systems
For the first ten years or so of the twentieth century there was little inter-regional movement because of local energy self-sufficiencies. By 1925 major flows of coal had developed but by 1950 oil had substantially replaced coal, even in intra-regional flows. By 1975 oil was dominant, providing 90% and coal only 7% of all inter-regional flows. The main movements of oil in 1974 were from the Middle East to Western Europe (505 Mt) and Japan (202 Mt), from the Caribbean to USA (123 Mt), North Africa to Western Europe (91 Mt), West Africa to Western Europe (63 Mt), and USA (40 Mt), and South-east Asia to Japan (47 Mt).[163]

Crude oil tankers now reach towards 500,000 longtons dead weight (dwt) but very large crude carriers (VLCCs) are usually classed as being over 150,000 dwt carrying capacity. Vessels of this size are constrained in use by the technology of size in construction, restricted channels and terminals, multi-port discharge requiring lightening at sea or entrepôt trans-shipment as at Bantry Bay, Ireland, by ownership/charter options, navigational safety and insurance and the apparent as well as real environmental risks. At the end of June 1973, some 41% of the world's total (dwt) tonnage was in VLCCs. At the end of 1974 some 4% was over 285,000dwt built since 1971. Japan's fleet in mid-1973 was 62% in VLCCs because the distance cost of imports to Japan provide the stimulus of economies of scale in carrier and in shipbuilding,[164] which have led to significant developments in computer-based design methods. The mv. *Globtik Tokyo*, (1963) is of 483,664 dwt, 379 m in length overall, breadth 62 m and draught 28 m. Such large ships need special onshore facilities for construction, maintenance, loading and offloading. Single buoy or single point moorings (SBMs or SPMs), central terminal systems (CTS) etc are all responses by the tanker industry to the obvious growing needs for more economical transportation, greater safety and minimum pollution. The drive to eliminate the dumping of oil waste at sea has led to improved tank cleaning, loading using the 'load on top' system, and provision of dirty ballast tanks which are offloaded at refineries equipped to handle oil/water separation. Loading and offloading with inert gas blanketing has reduced explosion risks. Such technological developments can be expected, but the important point about them is that without such developments this petroleum resource would not be available in the quantities and at the transportation cost currently possible. It was co-incidental, but none the less important, that the impact on Europe of the closing of the Suez Canal from 1967 to 1975 would have been much greater if, in 1967, and increasingly so later, there had not been tankers which could travel economically fully-laden from the Persian Gulf round the Cape of Good Hope to Europe.

There will be many more developments in tanker transportation and it is important to consider some which are being discussed because there are facets of great importance whilst the bulk of petroleum movements are by sea over large distances and over 50% from the Persian Gulf.[165]

Coal or coal slurries may again be used as bunkers. Nuclear powered ships may eventually be an economic as well as practical reality. Re-inforced tankers, capable of traversing Arctic waters, have already been proved feasible by the experimental voyages of the mv *Manhattan* in 1970. There will be improved navigational techniques, so important in restricted trade routes, like the English Channel, better non-corrosive non-fouling materials and new propulsion and engine designs. Some argue that propulsion design may restrict growth in size.[166] Some major shipyards in Japan and Europe are equipped to build tankers of one million dwt. On present technology however, tankers over 7/800,000 dwt need twin shaft propulsion. Hence it is suggested that tankers over 500,000 dwt are unlikely to be generally accepted by 1980. At end 1974 there were 164.4 M dwt of new building or on order with 57.9 M dwt between 205–285,000 dwt and 57.5 M dwt of 285,000 dwt and over. There have however been cancellations since then because of the reduction in consumption, due to mild winters and energy conservation.

Tanker availability and tanker rates do not depend solely on total tonnages. Ownership has a marked effect as does traffic pattern. Oil companies owned 36% of the total tanker fleet in 1961[167] and 33% at end 1974[163] and have most of the remainder on long-term charter. It is the small proportion of tankers on short-term or spot charter which create the volatility in tanker rates. The reduced demand has, since mid-1974, indeed led to a big fall in tanker rates as well as in much increased laying-up. Through 1967 to mid-1971, under 1 M dwt of tanker tonnage was laid up, but this rose to almost 5 M dwt in mid-1972, dropped sharply to under 0.5 M dwt in the third quarter of 1973 but rose even more sharply from mid-1974 to over 4 M dwt at the end of 1974.[168] Middle East to North-west Europe spot freight rates were over 250 on World Scale at end 1973 and under 50 on World Scale at the end of 1974. The re-opening of the Suez Canal in mid-1975 increased the over-supply of tankers by shortening the distances, and voyage times. Vessels of 250,000 dwt can use the Canal on the return journey to the Persian Gulf when in ballast, though they must go round the Cape of Good Hope when fully-laden. The opening of the Canal is an important shift in the traffic pattern, which will be changed if the Canal is deepened and widened to take bigger ships, and with the SUMED pipeline completed, linking the Red Sea and the Mediterranean. In March 1975 there was talk of the need for rationalization of tanker layups, in order to stabilize freight charges.[169] It was perhaps significant that this talk was initiated by an independent owner, for it is some of the independent owners who use their ships on spot, single voyage or short-term charters who play the tanker market, are the first to suffer from a fall but first to gain from a rise in freight rates. A new factor in the market, and probably of increasing significance, are the tankers owned by the national companies of the major oil-producing countries. They may well insist on cargoes being carried in their own vessels. They have the USA as precedent in this field. Libya, Kuwait, Algeria, Indonesia, Iran and Venezuela all have programmes for developing national fleets.[170]

In August 1974[171] it was estimated that, on current ownership and ordering, the Arab oil-producing states would have 7 M dwt by 1978. The Arab Maritime Petroleum Transport Company (AMPTC) is planning to spend $2,000 million over the next five years in addition to national plans. Hence the fleet could be a politically significant percentage of world tanker tonnage. This totalled, at the end of 1974, some 256 M dwt. In addition to the 83 M dwt owned by the oil companies, 163 M dwt was in private ownership, (54 M dwt under the Liberian flag, 25 M dwt Japanese, 23 M Norwegian) and almost 9 M dwt under government ownership.[163] By the end of 1978 actual Arab-owned tonnage[474] was 8.8 M dwt and, by mid-1979, some 9.1 M dwt of a total of 327.4 M dwt.

There will also be developments in the use of conventional tankers by carrying methanol, perhaps even reducing the expansion rate of transportation of liquefied natural gas (LNG) in specialized vessels. The technology for liquefaction, storage, marine transportation and re-gasification of LNG already exists. Supply and demand on a scale to justify plans and financing on a stable long-term basis are needed, as said earlier, to underwrite the economic feasibility of LNG projects, and for methanol projects.[172] Until the development of LNG tankers, long-distance transportation was impossible except by pipelines. When applicable under current economic and operating conditions, that development translated a resource to a reserve and led to a new industry.[173] This industry must develop to utilize a resource which has a premium value as a non-pollutant. There are many new techniques being developed in all phases of the industry.[174]

SUMMARY OF OIL TANKER POSITION AT END 1978

World Total Tonnage:	328.5 million long tons deadweight
By Flag:	Liberia 31.5%, Japan 8.9%, UK 8.6%, Norway 8.1%, USA 4.4%
By Ownership:	Private 60%; Oil Company 34%; Governments 4%
By size in '000 dwt:	205–285 43%; 65–125 15%; 285 + 14%; 125–205 10%
By employment:	74.5% in voyages from the Middle East

Source: *BP Statistical Review*, 1978.

Very large coal carriers have been talked about as possible developments which might bring coal back into greater significance in international trade in competition with high-priced oil. A 1974 study quoted[175] typical costs for moving coal from Blair Atholl, Queensland, Australia to Japan. The total cost of 58 cents/GJ, was made up of mine price 21 cents/GJ, overland transportation, 10.55, ocean transport 21, unloading etc. 5.275 cents/GJ. The total was said to be the equivalent, after allowing for bunkers, of '$4.10/barrel oil ... somewhat greater than the cost before October 1973 but much less than the prices of over $11/barrel oil in 1974'. Such calculations are interesting but even with 1979 prices almost double, do not mean that tens of millions of tonnes will move immediately.

Pipeline systems

The world pipeline system, outside the USSR and China, has been growing since 1966 at a rate of approximately 40,000 km/year, reaching some 1.72 million km in 1972. These lines were for natural gas 1.53 million km; petroleum products 50,000 km; crude oil 50,000 km and about 15,000 km crude oil-lines offshore. The dominance of natural gas-lines is most marked.[176] Increased volumes of trade since 1950 have necessitated increased size, as with tankers. Natural gas-lines of 1.22 m diameter have been laid in USA and Europe and even 1.47 m diameter in the USSR, whilst lines over 71 cm were 30% of the new construction in 1972, indeed more than 20% since 1967. One line of 2.5 m diameter is under design in the USSR but this may be exceptional. Watkins, 1975,[177] considered that the bulk of future demand will be for diameters within current capacity of modern mills. Steel is still the preferred material and there have had to be important improvements both in the steel and use of sizes in the difficult conditions now being tackled. The worst problems are those involved in carrying hot oil over permafrost areas and in long submarine lines, particularly in the hostile environment of the North Sea. Thick wall pipe of large diameter may be required for submarine traverses. Reinforced steel pipe and reinforced resin pipe are being developed, but still have problems. Much research and development work has been and is being done on main pipeline design, stress and strain and life-expectancy for the larger normal land-lines in Europe. Pipeline problems are enhanced under Arctic conditions. Thawing of the frozen ground could lead to earth slides or subsidence breaking lines. In some areas, as in Alaska, zones prone to earthquakes may have to be crossed. Furthermore flow must be ensured, otherwise if the flow is stopped the oil might solidify. In deep water offshore pipelining there are the problems of strain during laying and burial and, sometimes, after disinterment by seabed scouring.

These points stress the complexity of the problems which arise from the safety requirements to satisfy, not only the self-interest of the operator to avoid loss or shut-down, but also to ensure adequate margins to satisfy environmental conservation safeguards required increasingly by governments. Development of the oilfields in the northern North Sea have caused some Scotsmen to imagine a crazy mesh of pipelines covering the seafloor and all liable to be ruptured and spew out thousands of tons of oil. Pipelines are very expensive. The common carrier principle is well established in the oil industry, whereby as in North America, a line is obligated to carry other oil if so requested and practical. Economies of scale can be effected by two or more operators joining. Pipeline segregation by batching or, if the crude oils are compatible, mixing are common techniques. Hence it is in the operators interests to minimise the number of lines. Furthermore, it is in his interest not to waste the oil and elaborate precautions are taken to ensure that any break is detected and flow stopped.

Such safety considerations apply equally to coal slurry-lines as to oil lines.

Slurry-lines have been discussed for many years in the USA and mention has already been made of projects being planned. Development some years ago was inhibited by the swift action of the railroads in reducing tariffs and devising traffic systems which coped with the requirement. Practical experience over five years is available though from the Black Mesa line in Arizona. The line is 438 km long, 0.457 m in diameter and has a design capacity of 5 Mta. Extrapolating from this line, which worked to 99% capacity in 1972 and 1973, another 25 Mta line, 600 km long, is in the design stage. Proponents of such lines[175] stress that they have a low labour need, and so less liable to inflation, and have economy with distance. Hence they argue that a 20 Mta system over 1,500 km would be much more economic in the 1980s to move coal from the west of the USA to the mid-west than any other system. One study estimates a possible increase in coal slurry pipelines to 100 Mta capacity by 1985.[54]

There is no doubt that to some there is an attraction in envisaging a massive revival of coal in international trade, using slurry pipelines, and very large coal carriers. It may be opportune therefore to consider this scenario and it cannot be done merely as a transportation exercise. The report of the US National Academy of Engineering, 1974[54] did consider that the construction in ten years of $4 \times 1,600$ km $\times$ 25 Mta coal slurry pipelines was feasible in the USA. They stressed the advantages of the automated, continuous, hidden nature of the pipelines; their multiple land-use capability; their ability to operate one-way without empty returns and their relative economies. These economies are estimated at 4 mills/kWh at 4 Mta down to 1.5 mills/kWh at 18 Mta coal throughput. They say that this cost-line crosses that of unit train costs of 3 mills/kWh (i.e. 0.6 cents/ton mile) at a 6 Mta coal capacity. The study postulates an increase in western USA surface mining from 60 Mta in 1973 to 560 Mta in 1985, requiring 520 Mta new capacity because of 20 Mta depletion. This increase implies one hundred new 5 Mta surface mines, the recruitment and training of 45,000 new western coal miners and manufacture of one hundred new 100 cubic yard shovels and draglines. This task is mitigated by it being said that the equipment necessary for an expansion of 100 Mta in western mining is already on order and should be in place by 1979 or 1980. There are, however, admitted constraints to growth, even in the USA with its industrial strength and re-awakened desire for energy self-sufficiency. New, long-term purchase contracts and investments in new mines has been inhibited by uncertainty as to competition from alternative domestic energy sources and from imports. There has also been uncertainty as to the terms of new mining and restoration regulations and to restrictions due to water scarcity. Hence massive new capital of the order of $21,000 million would not be easy to find without government aid and an increase in coal price by $4 to $6 per ton. The equipment required is also a constraint. In 1974, in the USA, only two manufacturers of large power-shovels, 'and both are quoting deliveries in the first quarter of 1979. Today's total industry capacity has been estimated at about one shovel per month...'[54] There does not appear to be a reserve constraint in the Western USA. The coals are in thick horizontal seams, near the surface, low in sulphur and are non-coking. There is, given the equipment, a high recovery factor and a high productivity per man hour. This discussion of the effort required and the difficulties foreseen in the USA can be the background to a look elsewhere.

Outside the Soviet Union and China there are only Australia and South Africa with known reserves of economically mineable coals which might be available. Australia had some 10 Gt of brown coal and 3.8 Gt of bituminous coal recoverable by open-cut mining; South Africa had some 10 Gt of bituminous coal recoverable reserves of which only 5% are exploitable by surface mining, as reported in the 1974 WEC Survey. Neither, however, has the headstart for expansion as in the USA nor the manufacturing and manpower capabilities to mount, in the near future, the effort required. Sir Henry Cotton, Australian Consul General in New York was reported in the *Financial Times*, London, of 5 March 1980 as saying that Australian mines expect to be able to export 100 Mta during the 1980s and to increase exports to 200 Mta by the end of the century, when annual production should be 400 Mta. This implies that Australia will amend its former strong nationalistic policy and will accept a massive import of funds, machinery and manpower. Some major oil companies and some Japanese companies extended their coal interests in Australia in 1979 and 1980. Similar arguments would seem to apply in somewhat different circumstances to South Africa which is currently much more reliant on coal as its energy source than Australia. Even granting the immediate provision of the requirements there must be lead times of from five to ten years even to initial export. On these arguments it would seem unlikely that inter-regional energy flows will be transformed in the medium-term by massive transocean movements of coal fed by slurry pipelines on land.

Slurry pipelines in some countries will increase in importance and the historic growth, shown in table 55, can be expected to continue.

Table 55
Growth of slurry pipelines

	1950	1960	1970	1975
Maximum diameter (cm)	18	25	46	46
Maximum line length (km)	27	175	440	440
Total installed length (km)	27	325	1025	1260
Total capacity (all products) (Mta)	0.4	2.3	11	17
Total capacity (coal only (Mta)	0	1.3	4.8	4.8

Source: 'Transportation of Energy', J. E. Robb, *IX WEC Preprint*, Position Paper 5, 1974.

Pipelines, as a method of energy transportation, have a number of features which must be remembered. Pipelines have little flexibility; only in special circumstances can their function be changed, e.g. in the oil industry when crude-lines can be used to carry products and even the flow reversed. Pipelines are fixed installations and set the pattern of movement. They can, therefore, be a restraint on availability. The inertia of their investment can have a marked influence, as, for example, in the USA, and perhaps elsewhere, where the existence of the natural gas piped distribution system naturally encourages the replacement of the reducing supplies of natural gas by some other form of synthetic gas. In the USA there are 180,000 miles of natural gas transmission lines and 400,000 miles of distribution lines, representing a massive investment.[173] Pipelines are vulnerable to interruption. The Trans-Arabian Pipeline from Saudi Arabia to the Mediterranean and the lines from Iraq to the Mediterranean have been cut a number of times. On the other hand, the Trans-Alpine Line has had a quiet history, as have most others in the world and table 56

Table 56
Oil and gas pipelines

	1940	1950	1960	1970
OVERLAND				
Maximum diameter (cm)	75	90	100	120
Maximum length (km)	700	1,000	1,400	2,700
Total network ('000 km)	5	10	40	370
Network over 60 cm diameter ('000 km)	1	1	12	110
OFFSHORE				
Maximum diameter (cm)	45	60	100	140
Maximum length (km)	5	10	100	200
Maximum water depth (m)	20	30	50	100

Source: 'Transportation of Energy', John E. Robb, *IX WEC Preprint,* Position
Paper 5, Tables B—1 and B—11, 1974.

illustrates the growth of oil and gas pipelines on and offshore.

Many international frontiers are crossed by pipelines, though not always without controversy and delays. There are many oil lines crossing frontiers in Europe, North America and Asia. There are gas lines from Iran to the USSR, though a second line, IGAT II which was planned to go via the USSR to Western Europe from Iran was stopped by the Islamic regime in Iran in late 1979. There are lines from the USSR into Eastern and thence Western Europe; from the Netherlands to France and Italy. A gas line was first mooted in the late 1960s to carry gas from the Prudhoe Bay oilfield on the North Slope of Alaska through Canada to the USA below the 49th parallel. Agreement was reached on the route in 1977. In 1979 the cost was estimated for the 7,700 km line with a capacity of 90 million m³/day, as $14 billion. The capacity is almost the equivalent of 750,000 bd of oil and the completion was expected in 1984/1985. In 1979 agreement was finally reached on a line from the Southern Mexican oilfields to deliver to Texas, starting in January 1980, 8 million m³ of gas (300 million cu. ft. per day) at a price of $3.625/1,000 cu. ft., (roughly $1 million per day), with the price to be reviewed every three months and linked to world oil prices. The Mexicans held out for two years to get 'the right price'. But the line is of great importance to Mexico for, in addition to the revenue, it allows the oil production target of 2.25 mbd (roughly 112 Mta) at end 1980 to be reached without any waste surplus gas over the domestic demand. This is a classic example of the interrelation of oil production and associated gas.

Natural gas can be transported at about 750 psi, but hydrogen has to be compressed for pipeline transmission to about 2,000 psi. This may give a transmission cost for hydrogen of 3¢ per million Btu's per 100 miles, which is roughly double the cost for natural gas. Even so, one opinion[446] in 1979 was that when the distance exceeds 805 km the transmission costs for hydrogen are lower than for electric current, even at voltages of 700 kVA. Another opinion[447] was that for distances greater than 322 km it was cheaper to transmit energy through pipelines than as electricity by overhead cables. Such variations in opinion illustrate that the 1980s were entered without any practical experience on which to base firm prognostications.

Nevertheless the examples given suggest that in the 1980s further progress is probable in the continuing development of the energy pipeline systems which started in the 1930s in North America and in the 1950s in some other countries. However, the disadvantages of pipeline systems which have been mentioned must be remembered. Basically a pipeline can only compete with ocean transport when the pipeline cuts off a corner, e.g. the Trans-Arabian Pipeline (TAP-LINE). Yet that line carries lessons in political problems, including sabotage. But, of course, there are many places where pipelines are essential and there must be continuing progress in design, construction and operation.

Railway and road systems
Railways are effective carriers of solid coal over short and long distances and in many countries there is no alternative. Mention has been made of the effective reply of the US railroads on long hauls with unit-trains to the challenge of slurry pipelines, though they may eventually lose. Railways will continue to be effective carriers of petroleum products but only in small quantities, for pipelines have many advantages. Rail transport of crude oil is rarely economic, though for almost a hundred years Romania measured crude oil production in wagon-loads.

A much more important feature of railways is as a consumer of energy rather than as a transporter. The changes in fuel for rail engines from coal to diesel oil to electricity have had a marked effect, particularly in Britain and Europe, on petroleum product demand and refining pattern. One may well expect continuing electrification of railways where the density of traffic justifies it and particularly where a cheap source of electricity is available, such as from a massive hydroelectric scheme.

Roads will also continue to carry coal and petroleum products but on a local scale. The quantities of energy carried has hardly any significance in the total freight volume on the roads.[178] Roads are the extreme tips of the distribution system to the consumer. Their importance in the energy field is in their development as trackways for motor vehicles. This has, and will, until an adequate alternative is devised, increase the use of petroleum, has determined conversion/refining patterns, and by changing lifestyles, has had an influence on energy demand patterns beyond that of fuels.

In brief, roads cannot expect importance as major energy transport channels. Railways have lost passenger traffic to road and air, and energy resource freight to pipelines and electricity lines. They have not attracted investment and the condition of many has deteriorated. It is difficult to imagine how they can expect to increase and even maintain their present importance as energy carriers, even though, in the industrially developed countries they might recover both passenger and other freight traffic if enormous investment was available for wider tracks, new signalling systems and the development of much higher technology.

Electrical conductor systems
The generation of electricity consumes primary energy resources and its distribution over short distances is basically concerned with end-use demand. Long-distance transmission does, however, warrant brief consideration for it is one form of energy availability, albeit secondary energy, but its development has and will affect the deployment of primary energy resources.

John E. Robb gave an excellent review, for the Ninth World Energy Conference in Detroit, 1974[162] on long-distance electricity transmission and the following summary is based on that review. Long distances require high voltages because of line loss. The first high-voltage line in France carried DC electricity in the 1880s but from the first AC line in 1891 in England, AC circuits increased rapidly. 'In North America . . . lines tended to become longer as more remote hydroelectric sites were developed, then shorter on the average as more load was generated by thermal power plants, then longer again because of ascendancy of nuclear and mine-mouth thermal units plus remote Canadian hydro.' This quotation stresses the interaction of source and availability distance. High-voltage AC transmission systems exist for flows up to 5,000 MW(e) at voltages of 765 kv over distances up to 800 km. Flows of 7000 MW(e) or more are being planned over about 1,500 km and equipment up to 1,100 kv has been tested. Higher voltages are being sought because of better economics. Each line has a maximum capacity so, as load demand increases, either new circuits must be constructed or efficiency improved. Higher voltages overhead bring problems. Underground transmission lines of various types are in use but have their own problems. DC transmission systems 'appear to offer by far the most promise for longer distance underground transmission . . . but transmission cost is the big inhibitor to more DC use. . . '.

It is in the large countries, particularly the USSR, that there is an incentive to go beyond 1,500 km, where large cheap supplies of a primary energy resource are far from consumption and transmission as electricity may have advantages. Elsewhere there does not seem any good reason in the near future, why any trans-continental super-grids of super-conducting lines with massive flows will be developed. One attraction is that, by such grids, capacity installed to cover peak demand in one region could be used to cover peak demand at a different period in another region. This would only seem possible if, for instance, a massive solar energy electricity generating station in the Sahara could be developed to supply energy to Northern Europe and Southern Africa. Such dreams might materialise in a hundred years. Meantime it can be said that the technology exists for all practical transmissions in the foreseeable future. Improvements can be expected but they are unlikely to affect materially energy availability.

There will be schemes developed of the type in operation in the Nordic countries. This started in 1913 and benefits from both hydroelectric and thermal power plants. Transmission lines link Finland, Norway, Sweden and Denmark. Sweden is the focal point for the co-operation. As a result of heavy spring flood, Finland may have surplus hydro-power and sell to Sweden where the spring flood may be delayed. Heavy autumn precipitation in Sweden may allow power to be sent to Norway and save water in Norwegian reservoirs which can then be used in the next spring to give power to Sweden or to Denmark. Denmark only has thermal power plants and so can obtain hydro-power from Norway and Sweden when they have a high output, so saving oil and coal, and sell thermal power to them when they have a water shortage. A number of new connections are being considered by the Nordic Governments and Nordel, the Nordic agency for co-operation in electric power.[179,180]

By 1979 the European electricity pool tied together the supply systems of 24 countries.[513] Britain had a weak link with France by a small trans-Channel cable but tests were in hand towards the digging of trenches to bury four cables between Folkestone and Calais[514] some 5 feet below the seabed which would allow an exchange of 2 GW in the early 1980s. Public enquiries on both sides of the Channel would have to consider the environmental impact of the terminal transformers. Other plans would raise the possibility of exchange in Europe in the mid-1980s from the 1979 capacity of 55 GW (12% of total generating capacity) to 90 GW. The benefits of this pool vary with the different countries, but in general they were savings in investment in peak load capacity, all-year round economic advantages in trading in surplus and scarcity situations, support in the event of breakdowns or construction delays and the reduction in need for operating margins or idling generators. Eastern Europe has a power pool and there exist ties between USSR and Finland; Czechoslovakia and Austria; Bulgaria and Romania with Yugoslavia; and Hungary with both Austria and Yugoslavia. Technical as well as political problems face any major increased linkage in the 1980s, though on 27 June 1979, the USSR announced[513] that it planned to link its power grid with FR Germany through Poland, West Berlin and East Germany.

Finally figure 25 illustrates the comparative costs of transmission of energy in 1972.

COMPARATIVE TRANSMISSION COSTS OF GAS, ELECTRICITY & OTHER FUELS IN 1972

Fig. 25

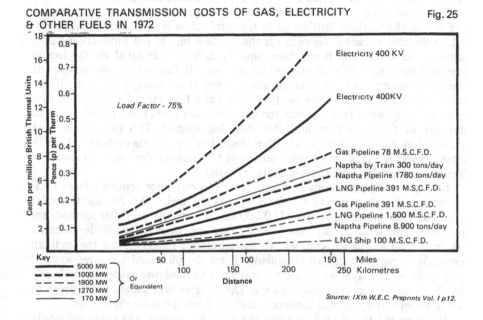

Source: IXth W.E.C. Preprints Vol. I p12.

STORAGE

The extension of the availability of a resource by storage is a basic concept and is manifested in small and large units. It is only the large units or aggregates of small units which will be discussed. Examples will be taken from the petroleum industry and then from the electricity industry with extension into some advanced concepts.

Coal is a solid energy source and can be stored unmined underground or in piles on the surface. These need not concern us. If coal is converted into liquid or gaseous hydrocarbons its storage will be according to the principles which govern their storage.

The main purposes of energy storage are to combat abnormal demand over the long-term (strategic), short-term (seasonal) or very short-term (diurnal or peak shaving) periods. The hydrocarbons are implicated in strategic and seasonal storage and electricity in the very-short term.

In the hydrocarbons industry the longest-term strategic storage is to shut-in a proved field. The USA Naval Reserves might be one example. The withholding of exploration or exploitation permits, as practised first by Venezuela but also by others up to present-day Norway, might be considered a type of storage. The reduction of output to a calculated optimum production, as by Kuwait and Libya, on the other hand, are energy conservation of 'good oilfield practice'. The recycling of fuel oil, naphtha from a refinery, or surplus gas from a separator at the oilfield, by the operating companies in Iran, started in the 1930s by the Anglo-Persian Oil Company, was both a form of storage and an enhanced recovery method. The production of an oil or gas field to match requirements, using the reservoir as the working storage tank is the universal industry practice. With unit control of one or more fields and integrated operational control of a refinery this can be sharpened to almost diurnal control as in Iran.

Integrated systems of reservoir/refinery storage have the obvious ceiling of the size of the smallest bottleneck in the whole flow system. Just as the practical production capacity depends on the smallest link in the production chain, whereas the theoretical production capacity may be calculated on the largest installed link, so is it with storage of any sort. If the re-injection equipment cannot overcome the reservoir pressure, there is no storage capacity. If the inlet and outlet to a 10 t storage tank only has a flow rate of 1 t/day then that tank can only store 10 tons after ten days and after being full will not complete delivery of those 10 tons until ten days has elapsed. This example may be simple, but the principle is so often forgotten by the theoretician. A 1979 example was the lack of pumping equipment to recover the crude oil stored in Texas as part of the US Strategic Storage Project.

Storage of hydrocarbons becomes more obvious once removed from their reservoirs. The storage may be above ground in steel tanks. Storage tanks for crude oil have increased to over 160,000 m^3 in some places. Other methods are possible. One scheme, suggested some years ago to make available potential oil from the Canadian Arctic, was to use submarine tankers under the ice to an entrepôt storage site in East Greenland. There the oil would be stored in open pits under a thin layer of plastic microspheres to minimise evaporation already small owing to the low temperatures. This scheme was suggested in order to allow movement by normal tankers in summer from the entrepôt storage.

Underground storage offers much the most economical and environmentally acceptable storage possibilities for both gaseous and liquid hydrocarbons. West Germany is using caverns in salt domes for its strategic storage. The creation and operation of these caverns with capacities up to 3.6 million m^3 have their problems but they are being solved in Germany, France and elsewhere.[181] Storage is also possible in horizontal strata, as developed, for example, for gas storage at Gatchina, near Leningrad, USSR.[182] In Belgium, natural gas storage in two abandoned coal mines at Fontaine l'Eveque and Anderlues has been thoroughly investigated.[183] Artificially-made caverns are used in granites, as in Sweden, in limestones, as at Lavera in Southern France, in chalk, as in the Paris basin, and in salt formations by leaching as in West Germany and elsewhere. The possibility of making storage caverns by nuclear explosions has been studied in France, where a great deal of thought has been given to underground storage, particularly for natural gas.[184] The cost of the storage depends on the type of rocks, the depth of the chosen formation, the ease of construction, the volumes required, the number of different types, for example, of petroleum products, involved and the ease of disposal of the excavated rock or saline water. There are many possibilities and the cost decreases usually as the volumes increase. In the USA at the end of 1973 there were three hundred and sixty reservoirs in twenty-six states with a total capacity of 170 km³ of which 133 km³ was in use.[185]

Many countries insist on a strategic reserve of petroleum, usually for both crude oil and products, in quantities to cover ninety days of normal demand. There has been an increased incentive to store more since the embargo on export to some countries by OAPEC in later 1973.

Short-term seasonal storage problems are usually overcome by changing the refinery outputs to produce more gasolines in summer and more fuel oils in winter. There has to be some seasonal storage to reduce the peaks of demand and this is accepted industry practice.

Very short-term or diurnal peak load shaving is a major problem for the electricity generating industry. Insofar as power storage is possible, then the efficiency of this secondary energy form is enhanced, substitution for primary energy source use is broadened and the competitiveness of nuclear energy for electricity generation is enhanced, for it is basically a continuous, high-load source. The possibilities of power storage must therefore be considered.

The first form is that of pumped-water storage. Water is raised to elevated reservoirs during off-peak periods, usually during the night, and allowed to fall and provide extra generation during peak periods. This substituted for otherwise essential extra equipment and extra fuel storage. The UK Central Policy Review Staff[186] estimated that an increase in UK pumped storage capacity from about 2.5 GW to 12.5 GW would allow annual fuel savings to reach over 1 Mt coal equivalent, by the year 2000. Overground pumped storage can only be installed when the site topography is favourable. If the site is in an area of accepted scenic beauty there can be opposition from environmentalists. The storage site need not be at the power plant. In South-East Brazil the possibilities are being considered of putting pump storage plants nearer to the consumption centres of Sao Paulo and Rio de Janeiro than the hydroelectric stations themselves.[187]

There are a number of advanced concepts for storage which are summarised in table 57.

In underground pumped storage, the lower reservoir and power plant are located in deep-mined caverns and the upper reservoir at surface. In this way the number of possible sites is increased and the objections of the environmentalists reduced. Compressed air storage uses a modified combustion turbine, uncoupling the compressor and generator at different times and incorporating the intermediate storage of compressed air. That storage can be in natural porous beds or caverns or man-made caverns. Battery storage research spans from the high temperature (580°C) lithium/chlorine battery to improvements in the lead/zinc battery. It is said[188] that of the advanced batteries 'the more attractive for the future are the alkali metal/sulphur, sodium/sulphur and metal sulphide batteries. Lower temperature (180°C) molten salt batteries and ambient temperature aqueous batteries may also be possibilities.

Hydrogen storage entails the production of hydrogen by electrolysis during off-peak periods and then using it in fuel cell, gas turbine or boiler for electricity generation. Superconducting magnetic energy storage (SMES) uses the principle that energy can be stored in an inductor of zero resistance for, theoretically, an infinite time; current is kept flowing in a closed loop within a strong magnetic

Table 57

Characteristics of various generating and storage systems

Type	Probable minimum economic size	Approx. capital cost	Expected life in years	Likely energy/ unit volume	Resource Considerations	Potential efficiency
	(MWh)	($/kW)	(Years)	(kWh/m^3)		%
1. Underground pumped hydro	10,000	200	50	2	Sites	65
2. Compressed air storage	200	230	20	4	Sites Fossil fuels	45 (75% storage)
3. Batteries	10	150	20	250		75
4. Hydrogen storage	10	300	30			50
5. Superconducting magnets	10,000	700	30	20	Nb, He	85
6. Superfly wheel	10	400	30	35		85
7. Combustion turbine	50 MW	120	20		Fossil fuels	24
8. Steam cycle plant	500 MW	350	30		Fossil fuels	37

Notes: Assumes discharge duration = 10 hours. Costs given in 1972 US Dollars.

Source: 'Assessment of advanced concepts in energy storage and their application on electric utility systems,' R. R. Fernandes, O. D. Gildersleeve and T. R. Schneider, *IX WEC Preprint*, 6.1–17.

field.[189] Superflywheels of various types have been conceived, the rotor being spun by an electric motor operating on off-peak power. Superflywheel power density is claimed to be superior to that of most other storage devices and the characteristics can be compared in table 57 with other forms including the combustion turbine. Thermal energy storage varies from storage on the consumer's side for off-peak water or air heating or cooling, intermediate storage as feed water at the generating station or high temperature storage as steam[190] in tanks or caverns, or as high temperature molten salt, or combining air storage and coal gasification.

The fuel cell is a battery/generator and there are a number of variations on which research is underway in many countries. The fuel cell has the advantages of low pollution and high efficiency. It has proved very useful in specialised circumstances but the euphoria of the early 1960s has disappeared. Much further research and development would seem to be necessary before there can be widespread commercial application.[191]

Research and development will continue.[500] Scientific meetings will bring out the results, on which the reader would be well advised to keep a watchful eye, even though the proliferation of literature makes any comprehensive cover very difficult, e.g. the Second World Hydrogen Energy Conference was held in Zurich, Switzerland, 21–24 August 1978.

In summary, one might say that this discussion has shown both the importance of storage to the availability of a resource and also the breadth of thinking which is being applied to the problems by many people in many ways.

8
Demand

Some general considerations
Mankind demands energy in four final forms:

- **heat** – space-heating and conditioning, food preparation, hot water supply, many different industrial processes.
- **mechanical energy** – static; electric motors, steam turbines, oil or gas engines, or mobile; transportation by land, sea or air.
- **chemical energy** – chemical synthesis processes.
- **radiant energy** – light and telecommunications.

This summary already illustrates that, just as the supply side of availability of energy resources is not merely a totalling of figures of so-called resources, reserves or production, the demand side is not merely a multiplication of figures of population and average use per capita. Man is the end-user and man exists in many different and dynamic environments, physical and societal. The physical differences are the most obvious but not always simple. The Eskimos in Inuvik in the Canadian Arctic now have refrigerators. The societal differences are obvious in the extremes, particularly in material wealth, but are less so in the middle and are often ignored when considering demand.

Countries may be divided into those with:

- negligible indigenous energy resources, who must import all their energy
- indigenous resources sufficient to supply all or part of their needs.
- surplus resources which by export can satisfy their non-energy needs.

Countries may also be divided into the:

- industrially developed (Japan is in the first category above, USA is now in the second).
- industrially developing (India is in the first category above, China in the second and Iran in the third).

Countries may also be divided into those with:

- centrally controlled economies.
- economies which are not centrally controlled.

The category of economies which are not centrally controlled is a diminishing one, for the increasing governmental interference/responsibility in energy matters in many countries is effectively lessening freedom of choice. There are many other broad divisions but within each there will be differences. Within the oil producing and exploiting countries there are differences in size of population, in size and maturity of energy reserves, their dependence on oil, their political ambitions and so on.

Demand can also be divided into domestic, commercial (with or without transportation) and industrial categories or by primary energy source (with or without the indirect, secondary or intermediate energy form of electricity).

As in the resource field there is a looseness of description which can be misleading. Demand, needs, requirements or consumption are often used synonymously because of an English language custom of avoiding repetition of the same word. Demand is essentially the quantity desired at a certain price or price range over a certain period under the conditions of supply and price at that time. Consumption is what has been consumed over a certain period under the conditions of supply and price at that time. Consumption extrapolated into the future becomes demand, but very often the parameters which shaped the past are not analysed and amended to the future. The arithmetical exercise of altering the percentage growth of the past is all too easy and the effects of changes in supply price and end-use need are so difficult to assess. Need or requirement implies a basic necessity of society. The UN Secretariat in a 1974 paper[193] define need for net useable energy as the amount, in one or more final energy forms, absolutely necessary to perform an activity and/or to assure adequate ambient conditions for living and work. Demand for net useable energy is taken as that actually consumed. Demand includes the energy wastage through negligence or profligacy in the final consumption stage. Gross demand, or gross useable energy consumption, includes energy losses which are part of the final process – these may be reduced but rarely eliminated.

Gross demand includes waste, no matter where, how or in what form energy is used and that use has an impact on the environment. It is in these two areas of conservation, of energy, and of the environment, that new factors have been added to more customary factors in forecasting demand. The difficulty of forecasting demand is always increased at a time of transition. The present day is a period in which the rules of the immediate past, of unrestricted economic growth potential, through abundant supplies of cheap energy, are being changed to the unformulated rules for a future, even the immediate future, of costly energy and lessened growth. The behavioural sciences have not yet progressed sufficiently to form a firm base as to how man will react to the new conditions. This is a simple truth, though either hotly contested or ignored by many economists and forecasters. However unfirm the base, demand must be considered, for it is the incentive for and justification of supply, and there are certain broad areas of special importance.

It has been common practice for many analysts to take the Soviet bloc and the People's Republic of China as self sufficient in energy, and consider only the World Outside the Communist Areas, (WOCA). This is no longer adequate. Some, for instance, have doubted whether USSR domestic oil production can satisfy its internal and export requirements for the next decade. On the other hand China may have the capability and political will to use oil and coal exports to pay for increased trade with others. Hence both these areas will be considered later.

The developing countries have become recently a more important feature of forward energy demand planning. This is partly because of their increasing political strength, in the UN Organization, and in such events as the 'North-South Dialogue' of the Conference on International Economic Cooperation. But there are also the pressures towards a 'New International Economic Order', as shown in many UN General Assembly Resolutions and resulting studies.[319] In these studies there are marked differences in treatment and detail of the developing countries, according to the origin of the studies. Increasingly, however, more consideration is being given to the special circumstances of the non-oil-producing and non-Communist countries of South America, Africa and South-East Asia. The NEPSG Report[320] recommended a US energy strategy which implied a heavy dependence on oil imports; they ignored the impact that this might have on the poorer nations of the world. Many, however, outside the USA appreciated that this was a major defect in the study, which was too parochial to be wise. A 1979 study[505] corrects this omission.

Discussion of energy demand involves many natural and man-made energy resources applied, usually through secondary products, to satisfy both human needs and human desires. Therefore there is danger in over-simplification of a very complex subject and its interactions with the whole of human society and the myriad life-styles encompassed. Yet in order to communicate, simplification is necessary, and numbers can take on a pseudo-validity. This is a danger which must be guarded against in the increasingly common method of using computer models. These vary in complexity and power of resolution, but all are dependent on the validity of the input.

The NEPSG used a 'competitive market' model designed for energy techno- logy assessment. The contributing members of WAES[322] used many different types of models based on their national and/or industry establishments. For the UN study of the future of the world economy[319] a model was constructed, mainly by Wassily Leonti of Brandeis University, USA. Many other models have been constructed and the International Institute of Applied Systems Analysis (IIASA), was established specifically to develop the 'technique of analysing an operation . . . in order to use a computer in seeking ways to im- prove its efficiency . . .' (Concise Oxford Dictionary, 1976). This Institute, with its national member organizations, including the USSR and East European countries, has had, since its foundation, an Energy Program to study the world's future energy supply and demand. Another model is, of course, that of Masaro- vic and Pestel, on which was based the second report of the Club of Rome. All these models are tools to help in the testing of hypotheses and the relative importance of chosen factors. They rarely resolve problems to the satisfaction of many beyond their own operators. Yet the conflicts of opinion are the battles from which, one hopes, greater understanding will emerge.

There are conflicting opinions on many aspects of energy demand and it is important to take note of some of them as otherwise there may be a danger of believing, rather than being wary of, some opinions when stated as if they were facts. There are conflicting opinions as to the strength of market inertia, and therefore as to how fast markets can change with substitution or conservation: some would consider as conservative the estimate that only 3% of capital stock could be adjusted each year to take advantage of energy-saving opportunities.[320] Some assert that there are 'absolute' constraints in energy production and end- use; others say that these constraints can be overcome at some cost which will be paid. The NEPSG[320] adopted this latter philosophy, and considered that, for the USA, '. . . we see little likelihood that costs will get so high that they become important influences on the shape of the political, social or economic future'. In contrast, the UN report,[319] noted above, states that, 'Accelerated development in developing regions is possible only under the condition that from 30–35%, and in some cases up to 40% of the gross product is used for capital investment . . . these levels necessitate drastic measures of economic policy in the field of taxation and credit . . . Significant social and institutional changes would have to accompany these changes'. One of the most important problems is the rela- tionship between energy consumption and the growth of gross domestic pro- duct, for many writers use GDP as an index to growth of energy demand, and often speak as though the relationship which they use is a fixed and firm fact. There are also strong views expressed on either side of any figure adopted for the 'price elasticity of demand', (i.e. the percentage by which demand is reduced when the price of a commodity or service increases by 1%). In the energy field, prices were held stable for almost twenty years, 1948–1968, and the shock of the four-fold increase in oil prices in late 1973 to early 1974, was taken at a time when supply-shortages were man-made and the position recovered quickly to an oil-surplus, so that experience of reaction to price change was short. Furthermore, there are conflicting opinions as to whether the past experience of the effect of income elasticity is applicable to the future. Will the increase in energy consumption with increased income, in the developing countries, whether of the individual or as the average of his income bracket, follow the precedent of the developed countries? Note has already been made of the need for greater capital investment for accelerated growth in the developing countries to which higher energy costs contribute by increasing the amount of work needed to produce a unit of income. Even in the USA, higher energy costs will, probably, slow down the increase in incomes, but how this will affect demand on energy is very debatable. These conflicts are minor compared to the almost irreconcilable conflict between more extreme views on broader matters.

On the one hand, some, like Professor Meadows, author of the first report for the Club of Rome, 'Limits to Growth',[321] take a pessimistic view. The main features of their scenarios are:

- world population will not rise from the present 4 billion (4×10^9), above 8 billion, perhaps not above 6 billion, because of reduced fertility in some areas and increased mortality in others, with war, pestilence and famine continuing their 'normal cycles'. Keyfitz of Harvard[375] expects a total world population in 2030 of 8 billion.
- possibility of massive climatic change through damage to the environment,
- the possibilities of free trade world-wide are low because of political instabilities,
- only a doubling of total global energy production above current levels can be expected, say to 2.5 kW/person.
- programmes requiring massive shipments of energy across national boundaries cannot be expected because of strong nationalistic security feelings,
- only sources regionally available, particularly the 'soft technologies', utilizing energy from the sun, winds, water, agricultural waste, should be developed, with deliberate efforts to achieve zero growth in the wealthier nations as soon as possible,
- the alternative 'hard technology' approach, with centralized energy systems vulnerable to interruption, will exacerbate international tensions and lead to the self-destruction of the technologies and infrastructures which are designed on the assumption of international order and altruism.

On the other hand, the more optimistic view[321] is that:

- the technological possibility exists to produce ample energy.
- the earth can support a rise in population to a plateau of about 12–13 billion,
- a rise in average energy consumption from current 1.8 kW/person to 3–5 kW/ person can be accepted as possible,
- technological and socio-political considerations should be distinguished; it is the politicians and the people who must ultimately decide between the alternative cases which the scientists can offer,
- hence those options should be given, with the implication that the politicians

will succeed in extending global cooperation and economic exchange sufficient to allow regional resilience to absorb impacts from outside without being destroyed.

These two views are not necessarily the most extreme views possible, but it is clear that global demand patterns based on either one would be very different.

Population growth

Population is an obvious determinant of energy use. The world's population is growing at about 2% per annum, that is, doubling every thirty-five years.[194] There is evidence and argument that this rate will lessen.[28] Population is subject to regional changes[195] from rural to urban areas.[196] In Japan and the UK, growth of the large metropolitan areas has slowed but movement continues to the smaller cities. The big migration in the USA to California is another example of massive movement.[197] Californian population can also illustrate the increasing use of energy in leisure as productivity increases and more leisure is available.

The population growth study in a US Report,[25] after considering mortality and fertility rates, international and internal migration, marriage patterns etc gave a very wide range, for the US population in 2000, of 252–295 million from a 1970 base of 205 million. In developed countries they estimated an 18% growth by 2000, assuming replacement fertility from 1970. In developing countries they gave a range of growth of 80% to 112% depending on a rapid or traditional demographic transition to the 'developed' category of nations. These are profoundly different figures. The 1979 World Bank Development Report estimated world population in 2000 as 6,000 million and feared that the increase in the urban population in the developing world was continuing at such a rapid rate that the problems already admitted would be even more pernicious unless strong action was quickly taken. Obviously there are many possible choices within the ranges and also many choices of multiplication factors when estimating future energy needs, demand and gross demand.

Energy consumption growth

As with reserve statistics there is little doubt that past consumption of the various energy resources has been recorded with widely-varying degrees of accuracy around the world. Hence the validity of some of the apparent trends in energy consumption is suspect.[200] The best example is in the simplistic trends used in the 'Limits to Growth' report.[198] This report created a furore because of its message of impending doom. The argument was that, if the current growth of population etc continues without change, the limits to growth on this planet will occur within the next hundred years, stressing that the ultimate limiting determinants are the stocks of physical resources. The report did have, as its authors desired, a very useful and widespread effect of making many people aware of the problem and of the need to think about the world's natural resources, their exploitation and its effect on the environment. An extreme viewpoint was taken but continuing discussion is very necessary.

A 1972 UN Study[199] set out a useful summary of historical facts on which table 58 is based. The world totals at the base of the table illustrate the total growth, with natural gas increasing over the twenty years faster than crude oil and solid fuels increasing at less than 1% over the last ten years. However, looking only at the solid fuel rates of growth over the 1950–1970 period in these three categories the disparity is 0.1%, 5.3% and 3.9% masked into the world

Table 58

Energy consumption by socio-economic groupings of countries, 1950, 1960, 1970

(terajoules by conversion from tonnes of coal equivalent at 23.28 TJ/t — rounded)

	1950 million TJ	1950 %	1960 million TJ	1960 %	1970 million TJ	1970 %	% Compound annual growth 1950/1960	1960/1970	1950/1970
I Developed market economies[1]									
TOTAL	43,460	100	59,810	100	99,360	100	3.2	5.2	4.2
Solid fuels	25,380	58	24,060	40	25,670	26	−0.5	0.6	0.1
Liquid fuels	11,290	26	22,160	37	47,550	48	7.0	7.9	7.5
Natural gas	8,270	14	11,990	20	23,530	23	7.2	7.0	7.1
Hydro & nuclear	850	2	1,610	3	2,650	3	6.6	5.1	5.9
Per Capita (GJ)	77,615		94,633		140,424		2.0	4.0	3.0
II Centrally planned economies[2]									
TOTAL	11,930	100	31,760	100	46,330	100	10.7	3.8	7.0
Solid fuels	9,820	82	25,300	80	27,850	60	9.9	1.0	5.3
Liquid fuels	1,790	15	4,460	14	10,610	23	9.6	9.0	9.3
Natural gas	170	2	1,750	5	7,290	16	21.0	15.3	17.9
Hydro & nuclear	50	1	250	1	580	1	16.7	9.2	12.9
Per Capita (GJ)	13,921		31,544		39,209		8.0	2.2	5.3
III Developing countries[3]									
TOTAL	3,250	100	7,060	100	13,590	100	8.0	6.9	7.4
Solid fuels	1,315	40	1,955	28	2,830	21	4.0	3.8	3.9
Liquid fuels	1,735	53	4,180	60	8,210	60	9.2	7.0	8.1
Natural gas	140	4	690	10	2,140	16	17.3	11.9	14.6
Hydro & nuclear	60	2	170	2	475	3	11.2	11.0	11.1
Per Capita (GJ)	2,980		5,122		7,845		5.6	4.4	5.0
WORLD TOTALS									
TOTAL	58,640	100	98,540	100	159,300	100	5.3	4.9	5.1
Solid fuels	36,530	62	51,310	52	56,315	35	3.5	0.9	2.2
Liquid fuels	14,880	25	30,800	31	66,350	42	7.6	7.8	7.8
Natural gas	6,360	11	14,430	15	33,010	21	8.6	8.6	8.6
Hydro & nuclear	960	2	2,000	2	3,660	2	7.7	6.2	6.9
Per Capita (GJ)	24,537		32,662		43,976		2.9	3.0	3.0

[1] North America, West Europe, Japan, South Africa, Oceania
[2] East Europe, including Yugoslavia, USSR, China, Mongolia, North Korea, Vietnam
[3] Africa, Asia, Latin America

Source: UN Report E/C 7/40 Add. 1 Sept. 1972. Tables 1 & 2.

total of 2.2%. The per capita energy consumption world-wide is almost 44 TJ in 1970, masking disparities of 140 TJ, 39 TJ and 7 TJ in the three categories. This table, in every item, illustrates strikingly the differences between the categories and within them the differing rates of growth. How much more then are the differences between the countries within those categories which have been lost in the aggregations? How great, therefore, the possible misunderstanding if the aggregations are used beyond the limits their applicability? Furthermore, as said earlier, even these figures of past consumption vary considerably in validity from country to country and type to type of resource. Lastly these energy content figures have been obtained, for illustration, by conversion at a standard rate of 23.28 TJ/t of coal equivalent to figures of coal equivalents derived by application of standard conversion factors which did not take into account the variations in energy content, according to coal rank or crude oil quality used, and efficiency in end-use. In other words, the consumption was gross consumption and the net useable energy could be very different even in the various categories, ignoring the disparities of the countries making up the categories. In brief this very useful table must be used only with full appreciation of the pitfalls of misuse.

The most comprehensive, recent study of the energy statistics of the world economy is that by Joel Darmstadter and others in 1971[116] Their figures differ only slightly from those of table 58, but they rightly stress that such figures are of greater value as a 'quasi-independent check on energy demand projections arrived at by disaggregated or sectoral methods'. They note that the same assumptions as to the overall economic growth may be common to both. This point accepts the possibility and the dangers of unconsciously circular arguments being accepted as confirmation – an all too common pitfall in demand forecast checking.

Darmstadter, in looking to the future, used an earlier projection, calculated in mid-1969 which may be taken as a classical model of expectations, at that time, for energy consumption up to 1980[202] for the USA. In this forecast, the rate of growth in energy consumption, 1965–1980, was 3.5% in comparison with the 3% historic growth from 1950 to 1965. The actuals are shown in the table below:

Period	1965–66	1966–67	1967–68	1968–69	1969–70
Percentage growth	+5%	+3½%	+6½%	+5½%	+4%

Period	1970–71	1971–72	1972–73	1973–74
Percentage growth	+3%	+7.9%	+5%	−4%

This run of figures illustrates several important points. The 1966–1973 figures averaged 5%, a high start to a fifteen-year forecast of 3.5% but not irreconcilable. The variations in annual growth are also clearly shown. The drop to a negative growth of 4% in 1974 over 1973 was of course not foreseen, nor the new situation caused by the events of October 1973. However, as a classical forecast in 1969, it is interesting to see that a 3.5% energy growth 1965–1980 carried on to 1985 would have given a figure for USA energy consumption of 83 million TJ, 3,564 Mt coal equivalent or 2,376 Mt oil equivalent.

A range of nine forecasts for the USA in 1985[54] was from 151 to 117 million TJ of which 129 million TJ was taken as the most reasonable; compare the extrapolated 1969 forecast of 83 million TJ. However, the 'reasonable' forecast of 117 million TJ was estimated as liable to reduction to 114 million TJ (51 million barrels/day oil equivalent) by conservation measures. On this point of conservation it may be noted that the National Petroleum Council[203] estimated that in the first three months of 1974, during the Arab embargo of oil to the USA, owing to consumer cooperation and mild weather, US oil consumption alone was cut by 6 million TJ (2.7 million barrels/day). Conservation and curtailment of activity were said to account for 37%, warmer than normal weather 16% and high fuel prices, lowered economic activity and unavailability of some products for 43%, and 4% was due to conversion to or use of alternate fuels and by reduced exports. Another study[204] argues, from a base case of 122 TJ million in 1985, two self-sufficiency strategies for the USA, with no imports of crude oil or natural gas, but with variations on domestic oil, natural gas and coal productions. These two cases give 1985 possible consumption figures of 113.3 and 117.6 million TJ and are discussed more fully later.

These variations for the best-documented and discussed energy system in the world illustrate the problem of demand forecasting. The 1975 US National Academy of Sciences Report on 'Mineral Resources and the Environment', says[25] (p. 12) that, unlike forecasts of the mid-1960s that tended to underestimate demand, because the effect of declining prices was not appreciated, there is now a tendency to exaggerate demand because of inadequacies in the data and the relatively primitive state of forecasting techniques. The report states (p. 306) that:

'we conclude that the demand projections forming the basis for current emphasis on large increases in energy supply are likely to be much too high for conditions that might be expected in the next twenty-five years . . . these popular demand projections are exaggerated by assumptions concerning the continued cheapness of energy and minerals that seem unrealistic for the foreseeable future. Because of the many shortcomings of current forecasting data and techniques, we are not able to quantify how much too high these forecasts may be.'

The report then goes on to argue the dangers of too high a demand target leading to a policy to match that demand. Subsidies for increased production from known sources then lead to increased consumption to use the cheap supply and so demand becomes self-fulfilling at distorted costs. They recommend a policy to encourage reduced consumption and reduce waste. It is such considerations which have created the need for a much harder look at the whole complex, inter-related energy systems of industry in order to measure the energy cost of each product and draw up energy budgets.

A simple example of an energy budget might be that it is said that every calorie of food consumed requires one calorie equivalent of fuel for growing and supplying the food and three calories equivalent for cooking.[115] However, such generalizations can be misleading and it is important to learn how this result was calculated and for what purpose. Energy studies usually have one or more of four main aims[205] in that they attempt to analyze:
– particular processes to deduce efficiency and recommend conservation measures
– consumption of energy on a large scale to forecast or reduce demand.
– consumption of basic technologies, such as food production, in order to

indicate future consequences of technological trends
- energy costs and flows to understand the thermo-dynamics of industry, possibly linked to a 'world modelling project'.

Chapman[205a] notes that extreme care must be taken with the method used. He instances that a detailed examination of the copper-smelting process by the heat inputs to fuel-heated furnaces compared to electric arc furnaces gives a two to one advantage to the latter. However if the subsystem of that calculation is raised to include the electricity supply industry, then the advantage is reversed. Hence energy studies can give useful results but they must only be used when it is clear what:
- subsystem of the world has been analysed
- energy inputs have been included
- calorific values are being used for primary fuels
- efficiencies have been ascribed to the energy industries
- conventions are being used to apportion energy costs within plants and industries.

These points are not known about the examples with which this paragraph began.

The largest energy-consuming industries are the energy industries themselves. In the UK the five energy industries, coalmining, oil refining, coke, gas and electricity production consume 30% of the total UK energy input.[205b] Hence it is important to know the energy efficiency of individual fuel industries in order to evaluate the comparative efficiency of say two processes, using one ton of coal or 1000 kWh of electricity. Furthermore it is essential to look ahead. As natural energy resources become harder to find they use more energy and are therefore inflationary. This suggests confirmation of the quotation given earlier from 'Minerals Resources and the Environment' that many demand-projects do not appreciate this point. Chapman uses an oil example. He states that the energy cost of a unit of Middle East oil is made up, as parts of the fuel energy obtained, by 4% in extraction, 5% in transportation to the UK. He states that the energy cost of an oil rig suitable for the North Sea represents about 10% of its total fuel output, add 4% for extraction and 4% for pumping ashore and the 'total energy cost rises to about 18% of the fuel output. Thus an oil industry based on North Sea oil may have an efficiency as low as 80% compared with the present 90%.' It is not clear, however, on what output the 10% energy cost of the fuel cost of the North Sea rig was based but the order of cost may be appropriate.

Published government statistics and the flow of products between the various industries of an economy can be used in a number of ways. Data on process costing are much more difficult to obtain.[206] Hence broad energy balances are popular. One such energy balance forecast has been made for South Africa.[123] Energy balances were constructed for each year from 1933 to 1972 using conversion figures from the UN 1963 study,[207] and projecting to 1980 and 2000. The balances were carried in megajoules and then reconverted to normal units. Assumptions were made on installed capacity of nuclear and hydroelectric power and it was assumed that all oil would be imported. No expansion of the SASOL oil from coal plant, was assumed though, as noted earlier, plans for a new larger plant were announced.[142] The result of this very neat exercise is to suggest that South Africa will be self-sufficient in energy, except for oil, and will have uranium available for export. Coal consumption and production in 2000 are given as just over 150 Mt but, from an estimated saleable reserve figure of

16,431 Mt, suggests a maximum production of 200 Mt in 2030 which implies an export potential.

In this relatively new development of energy balance calculations, there are probably some examples where indifferent statistics have led to some poor results. Perhaps one such is that it takes more energy to build and fuel a nuclear plant than it can ever produce and that a nuclear plant programme may never produce as much power as it consumes as long as it continues to expand. Such arguments are strongly attacked by advocates of nuclear power. W. Kenneth Davis[208] has calculated the energy required to make the materials and equipment for, and to build, an 1,100 EMW nuclear power plant. He included all fuel cycle requirements and the fuel, taking in mining, milling, enrichment and reprocessing both for initial and replacement fuel loadings, expressed in kWh of electricity. He assumed that all thermal energy used had a potential to produce electricity at a 33% efficiency. The results are summarised as:

Plant construction	0.74 EkWh $\times 10^9$
Initial fuel load	1.00 EkWh $\times 10^9$
Replacement fuel	0.50 EkWh $\times 10^9$
	per full power year

At full power the plant would produce 9.64×10^9 EWh per year. Thus all the energy invested would be repaid after two to three months of full power production, perhaps four to five months of actual initial plant operations.

In existing industrial plants of all types there is often remarkable ignorance as to the real energy efficiency. One grass roots attempt to obtain practical facts by personal inspection has been pioneered in the USA by Bruce C. Netschert. These studies are currently being used to assess the differences between demand and need and so justify curtailment or maintenance of offtake for plant during restricted supply periods. Such studies will undoubtedly increase in use. In both developed and developing countries, the possibility of supply shortages and the fact of higher oil prices has enhanced the national security aspects of energy supply. There is a real need for deeper examination in energy use in every process in industry.

At the other extreme from the individual process evaluation, to determine demand from energy budgets, are the broad brush estimates, dependent less on methodology than on crystal ball gazing into the political future. It is important to consider some of these because they so often influence government policies. Most of these forecasts concern both supply and demand and will be discussed later, but one example is given here to illustrate variations. Four estimates of world demand for OPEC oil, at differing prices, were as follows:

	Price, f.o.b. Persian Gulf	Million bls/d. rounded	
		1980	1985
OECD (1972 dollars)	$9.00	20	21
Ford Foundation (1973 dollars)	$6.25	33	29
Brooking Inst. (1974 dollars)	$7.00	28	34
W. J. Levy Inc. (1974 dollars)	$8.00	38	

Source: *Oil & Gas Journal*, p. 143, 11 November 1974.

By early 1980 crude oil prices were, in fact, very different, being around the $34 pb and some suggested would be $40 pb by 1985. Rising prices, conservation measures and sluggish growth had in fact caused world oil consumption to fall from 1973 to 1975 and the average increase from 1973 to 1978 was only 2.2% p.a. OPEC production in 1979 was 31.5 mbd, giving a rise of 4.7% over OPEC 1978 production, despite the drop in Iran's production and despite the high rise in prices from around $13 pb in December 1978. As noted in the Supply Chapter, OPEC production in the 1980s could be only of the order of 30 mbd.

Improvements in utilization or conservation in end use.
There is currently a many-pronged attack to improve end-use and so conserve energy. There is a lot of literature accumulating on the subject. This is an area where the bases for extrapolation of consumption into future demand must be carefully noted.

There are two types of energy conservation. The one differentiates between 'need' and 'demand', by eliminating waste, whilst the other differentiates between 'demand' and 'gross demand' by lessening losses. The first type might also be termed the 'voluntary type' for results can be gained by using less — for example using electric lighting only when illumination is essential, as rarely in an empty room. Such actions can be encouraged. Higher prices and exhortations, specific propaganda to use energy saving appliances and reduced advertisement of wasteful appliances can be reinforced by taxes on high consumption. Policing by inspection and punitive taxes on certain fuels are also possible to increase the 'voluntary' actions. The second type to reduce losses rather than waste usually requires additional capital and effort to replace inefficient equipment with a type using less energy. The two types may not always be easily distinguishable. Their importance varies according to the general and particular area of application.

Transportation is one area where obvious savings are possible. Private automobiles are the epitome of the 'mobility revolution' of the middle half of the twentieth century. They have transformed society's life style in the industrially developed countries but they are wasteful of energy — in the USA automobiles use 13% of total energy consumed. The concept of the social necessity for individual vehicular mobility in the automobile will be hard to shake, yet the deeper springs of behavioural pattern are already being studied by the Center for Human Mobility, Indiana University, USA. A 1974 paper by the UN Secretariat attempted an estimate of possible savings.[193] The study was based broadly on the estimated savings during the fuel shortages of the 1973–4 winter in the northern hemisphere. The contention is that a 20%–30% reduction in demand could be affected in the medium-term. A combination of both types of conservation would be needed. The elimination of waste by the more reasonable use of the automobile, which was said to account for 10%–20% reduction in the short-term, should be supported by re-design and operation on a less energy-intensive basis.

Considerable research and development is being done in many countries to make the gasoline automobile more energy effective despite anti-pollution devices which tend to decrease efficiency:
— electric road vehicles, as in West Germany[211]
— high speed trains, as in France[212]
— urban movement in USA[213]
— guided systems[214]
— of many kinds, noting differences between countries, e.g. 90% present London

commuter traffic is by public transport. A mini-train seating twelve, computer controlled to give high peak load capacities was planned for Sheffield, England for the late 1970s.

One study in the USA[54] suggests that change from present 30:70 to 50:50 ratio of small to large cars by 1985 would save at 1973 mileages, some 1.5 mbd of oil, and with 0.5 mbd from improved engines would give a total saving in the USA by 1985 of 2 mbd, almost 100 Mt. In the UK a study by the Central Policy Review Staff in a report to the Prime Minister of July 1974[186] rejected severe speed restrictions, car pooling and switching from the private car to mass transit as economy or conservation measures. They objected to them not because of the undesirability but their impracticability, as the savings would not forseeably outweigh the social strain. They did, however, favour smaller cars with more miles to the gallon as a major conservation measure.

Space heating and cooling is another end-use area where considerable savings could be effected. In the USA savings in 1985 could, realistically, be one million barrels/day of oil equivalent (mbdoe) and a further 1.1 mbdoe by improved building insulation standards.[54] It is, however, almost impossible to generalize. There are great differences, country to country and even within the larger countries, on the existing insulation level, the temperature difference within and without buildings, the accepted minimum external temperature of the design of buildings and the number of actual and potential district heating schemes or heat-pump usage. The suggested savings in the USA may be of the right order for the USA. In Europe it would be difficult to calculate potential savings. In contrast to the USA the houses are a much more polyglot collection, using all forms of space heating, wood, coal, natural gas, electric fires as well as coal, oil or gas central heating and there are wide variations in individual tastes. Undoubtedly voluntary conservation could and will effect considerable savings. Attempts have been made to assess the possibilities. In France space heating is said to consume 25% of total energy and, being mainly coal and wood, creates a pollution problem.[216] A 1974 Norwegian study concentrated on the common type of electric space heating at marginal cost. It argued that this is a cost incremental to the provision of other more necessary uses and so can be delivered at half the cost for the other residential purposes alone. This argument might be classed as promotional propaganda aimed at the second type of conservation by the installation of new devices.[217] As some consider that electrical space heating is energy wasteful it is interesting to note that a 1974 US study,[209] states that 'electric heating as an alternative to individual oil-fired heating, in actual practice, can provide substantially the same overall conversion efficiency of basic energy resources. Furthermore, the types and quantities of resources potentially available to run power plants are much greater.'

District heating for space heating is being considered in all countries. In Switzerland the Federal Office of Energy commissioned a study.[218] In the USSR one study[219] estimated that 15–17% of total fuel necessary to cover the heat loads of buildings can be saved by a centralized heat supply system. They give an additional saving of 0.6 to 0.7 t of a reference fuel (7000 k cal/kg) or a calorific value of 29.3 MJ per kW of installed electrical capacity of district heating turbines and up to 1 t/kW of installed capacity, when centralization is taken into account. In the USSR, about 80% of the heat energy produced is consumed in urban areas, only 20% by rural communities. Of the reference fuel 470 Mt were consumed for heat in 1973 and they forecast perhaps 700 Mt by about 1985. Large district heating schemes contribute 21% of the heat load of the whole country and 27%of that of urban settlements. This illustrates the size of the construction since 1946, typically in clusters of tall apartment blocks. These are ideal for district heating schemes, with the combined heat and electricity plants outside the urban areas. A Finnish study[220] comparing alternate fuels for metropolitan Helsinki, recommended a nuclear plant but stressed the rapid escalation of costs to provide district heating as the distance to be covered increased.

Other areas for energy conservation in end-use obviously exist outside transportation, space heating and the examples given in earlier chapters, particularly chapters 6 and 7. In the UK the National Industrial Fuel Efficiency Services (NIFES) was established in 1954. They proved, even in 1965, when energy was cheap, after an exhaustive study, that a total capital investment of £2.5 million in new and modified plant in industrial establishment could save 300,000 t coal equivalent/year. The payback period was two years in the sample taken.[221] The CPRS study[186] cited earlier, noted possible annual savings in the UK of 10 Mt coal equivalent, by widespread use of efficient variable speed drives instead of the constant speed characteristic of the standard AC electric motor, of 4.5-5 Mtce by pyrolysis of all refuse and general industrial waste and about 12 Mtce by using diesel generators or back-pressure steam turbines with waste heat recovery. The Finnish National Fund for Research and Development supported a scheme to pilot plant stage, using the waste heat from an oil refinery cooling water for desalination of seawater by vacuum distillation.[222] This scheme had the added attraction of reducing the potential pollution of that cooling water.

Substitution and interdependence.
Finally, under both conservation and demand, one must consider substitution of alternate energies and the growing interdependence of the different energy industries.

The steel industry is one of the higher consuming industries. Examples from Poland and Britain can illustrate both past and future substitutions. In the steel industry, as a whole, there has been a decline in specific energy consumption. The reasons are better quality of input materials, including higher quality coke, more efficient fuel use and improvements in blast furnaces and other processes. In Poland the use of coke fell from 812 kg/t in 1966 to 640 kg/t in 1972 but the use of natural gas increased from 32.4 m^3/t to 61.5 m^3/t and fuel oil from 2.1 kg/t to 13.6 kg/t. The coal inputs in open-hearth furnaces had declined from 70% in 1958 to about 6% in 1972. The USSR has recently increased its price for export crude oil and Poland's domestic oil production is declining whilst its coal resources are considerable. Poland therefore will combat high oil import prices by maximising efficiency of use of that oil, reducing the specific consumption of coke and introducing new technology of coke-making from steam coals. In the UK[119] considerable effort has resulted in a commercial plant to produce a good quality coke by coal-blending techniques. This conserves the prime coking coals of which supplies were declining yet were considered to be essential to good quality steel. British consumption of coke in blast furnaces and iron foundries had fallen from 11.1 Mt in 1967 to 9.9 Mt in 1972. Coke constitutes two thirds of the energy purchases by the British steel industry. 1974 use was 500 kg/t and oil injection was 50 kg/t. Future plans were for 190 kg/t for coke, 150 kg/t for oil and 700 m^3/t for reducing gas. In the short term plans of the next five years, to 1980, the UK expectation was to continue with coke replacement technology and hope for some indigenous natural gas and oil, with increasing supplies in the medium term of five to twenty years. Plans also foresaw increasing use of electric arc furnaces in steel making, with some electricity from nuclear plants and, in the longer term, perhaps eventual integration with nuclear power plants. These would supply heat as well as electricity and hydrocarbons would provide only the chemical reducing agents and not base heat.

The possibilities of substitutions on a wider scale can be illustrated[204] from a study in the USA, referred to earlier, of two alternative cases of self-sufficiency in 1985 against a base case on projections of 1960–1968 trends and unrestricted imports at US$6.50/barrel. The two self-sufficiency cases imply the following conservation practices, which are interesting in themselves.:

- **space heat** – 10% increased efficiency in all fuels by improved insulation in commercial and residential buildings.
- **air conditioning** – 10% increase in average performance coefficient of equipment.
- **water heat** – 10% increase by improved insulation on all equipment.
- **industrial process heat** – 10% increase in all fuel efficiency through waste heat recovery and 10% decrease in demand by use of more efficient processes.
- **iron production** – 25% reduction in coke requirements/ton of iron.
- **ground and air transport** – 10% reduction in fuel/passenger and freight mile, through better load factor, smaller cars and partial shift to more efficient modes of transport.

In all these cases hydroelectric, geothermal and nuclear productions are held at 4.5 million TJ, 1.8 million TJ, and 15.8 million TJ respectively. The big differences are that in the two self-sufficiency cases, with no imported petroleum, case A postulates a high domestic petroleum production and case B a high domestic coal production.

Table 59 shows the relevant figures with the 1974 US National Academy of Engineering Study [54] estimates of what was considered practical.

The NAE figures for hydroelectric, geothermal and nuclear power were 3.20 million TJ, 0.44 million TJ and 18.3 million TJ, indicating a greater reliance on nuclear power. There is also greater caution in the development of shale oil, liquefied and gasified coal of case B. Both estimates are of demand for they have eliminated waste. The NAE study argued for a saving by 1985, through elimination of waste, of 8.3 million TJ (3.8 mbdoe) and 12.3 million TJ from use of energy-saving equipment, minus 15% for partial overlap to give a total conservation potential of 17.5 million TJ. The differences between these four estimates are important for they indicate:

- the variations in substitutions which were considered feasible,
- differences in preferences for practical speeds in development,
- the wide range of demand figures used,
- lack of any guidelines from government.[225]

There are three main categories of substitution:
- direct use of waste heat in the thermodynamic cycles involved in generating electricity (low temperature processes);
- direct use of the heat output of reactors in chemical, industrial processes, supplemented by electricity production (high temperature processes), e.g. steam reforming of methane for use in reduction of iron ore;
- direct or indirect use of nuclear heat supply to produce synthetic liquid or gaseous fuels by the intermediate step of producing hydrogen.

223

Table 59
Summary of alternative energy resource consumptions, USA – 1985
Hoffman et al. 1974

	Base case		Case A		Case B		NAE 1974	
	Million TJ	mbd*	Million TJ	mbd*	Million TJ	mbd*	Million TJ	mbd*
OIL								
Domestic	25.3	11.55	33.7	15.4	25.3	11.55	27.4	12.5
Imported	28.7	13.11	0	1.0	5.3	2.5	1.1	0.5
Shale Oil	0		2.1		0			
NATURAL GAS		Bcfd		Bcfd		Bcfd		Bcfd
Domestic	23.7	60	34.0	90	26.8	70	31.76	81.5
Imported	3.5	10	0		0			
COAL		Mta		Mta		Mta		Mta
Direct use	16.9	790	15.4	720	28.7	1330	21.9	950
Liquefied	0		3.8	210	5.3	290	1.3	50
Gasified	2.2	120	2.2	120	4.1	230	2.4	260
TOTAL	122.4		113.3		117.6		107.8	

Conversion Factors: 1 tonne US coal (East: West, 7·5) = 21.5 TJ, 1 tonne W. US coal = 18TJ
1 barrel US oil = 6TJ, 1 barrel oil/day = 2,190 TJ/year
1 cu.ft/day natural gas = 385,000 J/year

Sources: Future patterns of interfuel substitutions etc. K. C. Hoffman et al, IX WEC Preprint 1.3–14, 1974 and US Energy Prospects: an engineering viewpoint, Nat. Acad. Eng. 1974.

Research is being done on many fronts in these categories. No obvious successes have yet been commercialised.

Romania can provide an interesting example of substitution.[6] From table 60 it is apparent that the cheapest system is the gas-fired thermal power plant, generating 8,308 GWh.

Table 60
Alternative thermal power station costings Romania
Four 330 MWe stations

Type of fuel	Updated overall expenses million lei	Cost price of electric, power lei/kWh	Total investment million lei
Lignite	9,488.7	0.1796	5700.4
Imported crude oil	10,193.8	0.2614	4424
Natural gas	7,181.2	0.1425	4424

Source: Energy Development and its Social Impact, O. Groza et al,
IX WEC Preprint 1.2–15, 1974.

Minister Groza et al, point out however that the same volume as the total flue gases in the gas-fired station would have produced 1.5 Mt of nitrogenous fertilizers. On 1971/1972 international market prices of US$11/1000 kWh electricity delivered and US $150t for nitrogenous fertilizers, it was calculated that the value of the natural gas, as converted into electric power, would be US $87 million, and as converted into chemicals would be US $225 million. Hence substitution of natural gas was not considered truly economical.

Two major 1978 reports can illustrate a number of points made above. The WAES report[322] considers that the 'soft technologies' are unlikely to contribute significant quantities of additional energy in 2000 on a global level, although they would be of particular importance in particular areas. This is, indeed, the most generally accepted current view. It is also spelled out in the second report which is discussed below, or rather in the section on unconventional energy. Hence both reports rely on the 'hard technologies' even in to the first quarter of the twenty-first century.

The WAES Report described different scenarios using variables in two periods of the time span to the year 2000. These variables for WOCA with which they were concerned may be summarized as:

1977–1985 — economic growth rate 6% or 3%,
— oil price per US barrel rising to $ 17.25, constant at $ 11.50 or falling to $ 7.66,
— national policy response, either vigorous or restrained,
1985–2000 — economic growth rate 5% or 3%,
— oil price per US barrel rising to $ 17.25 or constant at $ 11.50,
— gross additions to oil reserves 20bb/year (billion barrels per year) or 10 bb/year,
— OPEC oil production limited to 45 mbd (million barrels per day) or 40 mbd,
— principal replacement fuel, coal or nuclear energy.

Using these criteria the WAES participants made a detailed analysis of energy use in 13 of the countries represented in the study group, and estimates were made for the remaining countries within WOCA. Variations between countries, even on the basic WAES assumptions, can be illustrated by comparing the high and low annual growth rates used for Italy at 5.1% and 0.7%, with those for Norway at 3.2% and 2.5%.[340] Finally these estimates were reconciled to give a consistent WOCA pattern. In the reconciliation the initial assumptions had to be changed, e.g. the total annual WOCA growth rates, 1985–2000, were lowered to a high rate of 4.0% and a low rate of 2.8%, from the initial 5% and 3%. The resulting scenarios for WOCA ranged in total delivered energy demand in the year 2000 from almost 160 mbdoe (million barrels per day of oil equivalent) to almost 207 mbdoe.

The report[424a] by the Conservation Commission (CoCo) of the World Energy Conference (WEC) was based on reports to the Commission by teams working on the major aspects of world energy supply and demand. The Executive Summaries of these reports were discussed at the 1977 WEC in Istanbul. The final report of the Commission was the result of still further study and reconciliation of the supply and demand reports. Although energy demand models had been used, the final report favoured a broad brush approach, stressing the complexities which arose when attempting to include all the econometric elasticities and aggregations/disaggregations applicable. The demand report from the Energy Research Group, Cavendish Laboratory, Cambridge, under Dr. Richard Eden, used three different models, dealing with energy demand, oil demand and inter-fuel substitution, and adopted the scenario approach. The Commission, however, proposed an 'alternative scenario' in their report. In brief they considered that the Cavendish report underestimated the future rise in internal energy prices, overestimated the long range demand of the industrialized countries and underestimated that of the developing countries. Table 61 illustrates the resulting differences. However these figures should not be taken by any reader further out of context or used as more than a broad guide, without reference back to the original sources.

Table 61
Projected global energy demand scenarios, 1972–2020
(exajoules per annum)

	CoCo Alternative	Cavendish	
		High growth energy-constrained	Low growth oil-constrained
1972	250	250	250
1980	328	320	315
1990	447	430	414
2000	613	580	554
2010	782	741	686
2020	1000	945	847

Source: *World Energy, Looking Ahead to 2020*, Conservation Commission of the World Energy Conference, Table 1, p. 240, IPC Sci. & Tech. Press, 1978, Guildford.

Comparison of these figures with most other reports is difficult because of dissimilarities both in total areas and their subdivisions, in the periods studied and often because of lack of background data and assumptions. However, from chapter 5, appendix 4 of the CoCo report[424a] the centrally planned economies (CPE) were allocated in the year 2000 some 167 EJ in the 'alternative scenario,' and, from Exhibit 21 of the Demand Executive Summary[328] can be roughly measured the allocations for the CPE for the Cavendish high growth, energy-constrained and the low growth, oil-constrained scenarios of 200 EJ and 160 EJ respectively. Deducting these values from the world data for the year 2000 in Table 61, the WOCA data become 446 EJ, 380 EJ and 394 EJ, which can be compared with the WAES high and low estimates of 455 EJ and 352 EJ, after converting the latter figures at 2.2 EJ = mbdoe. In summary all the CoCo estimates quoted lie within the very wide range which WAES quoted.

The United States of America
Since 1974 there have been many forecasts and energy future scenarios for USA demand. Estimates published[323] by the US Department of Interior in 1976 are illustrated in tables 62 and 63 and can be used as markers for comparison with others, such as those in references[320,328,329,331,364,488,505]. The most important trend is that targets have been reduced with time. Project Independence,[505] in the Nixon Administration in 1974 used 109-115 EJ as the 1985 target; the 1976 projection in the Ford Administration used 108 EJ for 1985 and 123 EJ for 1990, and on the same annual increase of 2.7% would have reached 160 EJ in 2000; the 1977 NEPSG[320] early in the Carter Administration gave 150-130 EJ for 2000 and 160-204 EJ for 2010; the Committee on Nuclear and Alternative Energy Systems, (CONAES) in 1977, gave only 99-143 EJ for 2010; the Energy Information Administration (EIA) annual report to Congress, 1978, gave[505] 96-102 EJ for 1985 and 106-115 for 1980; the Exxon report[488] in December, 1978, fell in the lower part of these ranges with 99 EJ for 1985 and 108 EJ for 1990. The Exxon report[239] of December 1979 gave 90 EJ for 1985 — a drop from the 1974 target for 1985 of some 8.6-11.3 mbdoe.

The Carter Administration presented its energy policy soon after the publication of the NEPSG report at the beginning of April 1977, and there seemed to be close similarities in principle. Indeed the lack of reference to the world outside the USA was even more noticeable. The Plan, unlike the NEPSG study, was an overt political document by the President of the United States, who has foreign affairs responsibilities. There were no references either to OPEC, the International Energy Agency, or the Conference on International Economic Cooperation, or to the common need of all countries for increased research, development and demonstration (RD&D) in energy matters. There was no reference to the problems of other countries due to high energy prices, yet the Plan implied continuing reliance on oil imports, which would be increased if the coal or nuclear plans faltered. On conventional arguments, developing countries may be expected to be increasingly dependent on oil for a number of reasons. The higher levels to which oil prices could rise, with the USA a buyer in a seller's market, would therefore be a very great burden to the poorer countries. It was noticeable also that whilst the emphasis of the Plan was on the reduction of imports, which would avoid this contingency, yet there was no mention of favouring supplies from those countries which have traditionally been considered more 'secure' countries. Some might say that this omission was an oversight, others that the Plan was for domestic audiences, whilst others might think that

Table 62
USA energy consumption 1973–1990
(percentages of world consumption)

	1973		1980		1985		1990	
	Exajoules	%	Exajoules	%	Exajoules	%	Exajoules	%
Coal	13.97	(19.25)	18.1	(22.4)	22.4	(25.3)	26.4	(27.9)
Petroleum	33.92	(27.26)	43.0	(30)	48.5	(30.6)	52.0	(29.5)
Natural gas	26.46	(52.72)	21.6	(38)	21.1	(32)	20.9	(28)
Hydro & Geothermal	3.15	(22.7)	3.6	(23)	4.1	(24)	4.7	(25.4)
Nuclear	0.95	(39.0)	4.5	(37)	12.0	(35)	19.0	(30)
Totals	78.44	(29.8)	90.8	(29.4)	108.1	(29.7)	123.0	(28.8)

Table 63
USA energy production, 1973 & 1990

	In industry units		Domestic production as percentage of consumption	
	1973	1990	1973	1990
Coal	534 Mtce	1030 Mtce	108%	105%
Petroleum	457 Mtoe	681 Mtoe	58%	57%
Natural gas	641 km³	507 km³	98%	98%
Hydro & geothermal			97%	100%
Nuclear			100%	100%
Total			83%	83%

Conversion Factors:
Coal   100 Mtce = 3 EJ
Crude Oil   100 Mtoe - 4.4 EJ = 2 Mbd
Natural Gas   100 km³ = 4 EJ

Sources: Actual data for 1973 UN
Forecasts — world — US Dept. of Interior, Bureau of Mines — USA —
'US Energy through the year 2000' (revised), and modified to conform
with UN energy Conventions — see Ref. 323.
1990 figures in industry units converted from the Quads (10^{12}Btu) by
using the conversion factors used by the source for 1973.

it was appreciated that if the plans for the other energy sources failed, then only the Middle East could be the suppliers, and not Canada, Venezuela or even Mexico. Since the oil embargo of 1973/1974, imports from Nigeria and the Middle East had almost doubled in 1975 and 1976, whereas imports from Venezuela and Canada had fallen by 38% and 54% respectively. Domestically, of course, any US energy plan has the problem of reconciling Federal interests with those of the individual States as voiced by Congress; and Congress tends to protracted investigation. After a law has been passed, there are many Federal and State Agencies involved in its interpretation and there is a propensity to test some of these interpretations in the law courts, so that the impact of any plan on demand takes time to become obvious.

When the National Plan was presented to Congress there was every expectation that Congress would take some months considering it and would probably modify considerably its already very complicated provisions. The main themes of the Plan were: conservation and a shift to coal; thermal, or 'conventional' nuclear reactors to be used as a last resort; commercial development of the breeder reactor to be halted but research to continue together with the long-term research on nuclear fusion; RD&D in the use of the renewable energy sources, particularly solar and geothermal sources, to be increased as the very long term basis for energy policy. On the demand side the general principle was to rely on price to effect conservation, by letting the prices of oil and oil products rise to world levels. The case for this policy was argued strongly in a 1979 study.[505]

The conservation measures proposed seem complicated and would take time to bite and the swing from oil to coal may not be easy. Indeed the President of US Steel is reported[363] as saying that this would be the greatest problem facing American business. He estimated that the costs would include $20 billion, to open new coal mines by 1985, $8 billion for railroads and $70 billion for public utilities to convert to coal. Furthermore, except for tax provisions there was no encouragement for substitution of oil by other energy sources nor for the production of more indigenous coal, oil or gas supplies; indeed some say that on balance there is discouragement.[327].

In May and June 1977, there were many critics of the Carter Plan who said that unless Congress changed the Plan there would be no chance of the targets being met, whether for coal production at 1 Gt (1,000 million tonnes) in 1985, or the savings and increased efficiencies on the conservation side. In August 1977, however, the House of Representatives made only a few changes, which tended to stiffen some clauses, e.g. no sales of 1980 automobiles unless they ran at more than 16 miles/US gallon (19.21 miles/UK gallon). In September the Senate attacked many aspects of the Plan. In mid-November it was problematical as to what compromise would be worked out by the Joint House/Senate Committee which was the next stage, but President Carter on 7 November said that he would veto any Bill which did not fit his energy strategy.

In the world sense, there appeared to be some indication that the Administration was beginning to appreciate the wider importance of the US energy policy. A discussion paper was reported[365] as circulating in the State Department which stressed the importance of peace in the Middle East to ensure the flow of oil to the USA. On 27 October Defence Secretary Brown was reported[401] as saying that, even if the US acts to improve its energy situation 'I am disturbed at the potential for strife in an oil-starved world of the nineties'. There also seems to be a growing appreciation that action must be taken in the near future. A report called *Geopolitics of Energy* was published by Senator Henry Jackson, Chairman of the Senate Committee on Interior and Insular Affairs, which stressed the long lead times for the development of known resources. Lead times are quoted in most papers on energy supply, particularly in recent times, yet they are repeated here because they are so much more important to any demand discussion, when as is becoming increasingly obvious, demand will be increasingly governed by supply.

These lead times emphasize that it is now that decisions must be taken to affect the supply in the middle to late 1980s. If the Carter Administration is re-elected for a second term to 1985, there may be little chance for a change in basic thinking, and any new major projects after 1985 could only be effective after 1990/1995. The day-to-day operations, not just 'long-term plans' in the USA are therefore of great importance to the world because of long lead times.

Table 64
Some lead times for development of known energy resources

	Years
Offshore oil from lease to peak production	9—14
Underground coal mines	3—6
Coal-fired power plants	5—8
Hydro-electricity dams	5—8
Uranium exploration* and mining	8—10
Nuclear power plants, (Cf. 6 years proposed in the USA)	7—10
Coal gasification	10—15
Conversion of tar sands and oil shales	5—10

(* presumably this means the delineation of a known ore body DCI)

Source: *Geopolitics of Energy,* US Congress, Common Interior &
Insular Affairs, Sen. H. Jackson, as reported in *The Times,*
18 August 1977, London.

Despite these signs of increasing appreciation of the problem, there was much talk but little action for many months. At the annual meeting of the American Petroleum Institute in November 1978, the oil industry called for an end to price controls, but in February 1979, it was said that controls would continue. The Department of Energy in December 1978, was firm on a billion barrels as the target for a strategic stockpile, but by end January 1979 had dropped the target to 750 mb, and in September 1979 had to review the target because only 92 mb had been collected. In February the Carter team were reported as saying that energy consumption in 2000 need only be 20% higher than in 1979, and might be as low as 15-10% higher, so suggesting a range of 92 EJ-106 EJ. Then in mid-July 1979, President Carter came out with a revised energy package.[516] The stated aim was to reduce current imports of oil of 8.5 mbd to 4 mbd by 1990. This target was given despite a recent Department of Energy forecast that oil imports would be 14 mbd by the late 1980s, even allowing for an increase of coal production to 1,000 Mtce. There was therefore some scepticism as to the reality of the Administration's target. However, the President took two immediate steps, being within his own authority. He imposed a quota of 8.2 mbd for oil imports for 1979, which was not really much above the anticipated actual, and he took the controls off heavy oils. The latter were estimated would raise the price of some 10 billion barrels (bb), of heavy oil, mainly in California, from $2.50 to $15-16 per barrel, with the hope of increasing production by 200,000 bd by 1985 and 500,000 bd by 1990. The other measures in the package required legislation. The major points were:
— an Energy Security Corporation to develop 2.5 mbd of oil substitutes by 1990. (A Bill already before the House of Representatives envisaged synthetic fuel production of 0.5 mbd by 1984 and 2 mbd by 2990, yet a Rand Corporation report earlier in July had hinted that even 1 mbd by 1990 would be difficult.) The targets for 1990 were
— coal liquefaction and gasification 1.0-1.5 mbd
— shale oil   0.5 mbd
— biomass   0.1 mbd
— unconventional gas   0.5-1.0 mbd
to a total between 2.1 and 3.1 mbd. Some $88 billion (b) would be allocated plus drawing rights on an issue of $5 b. low denomination bonds.

— an Energy Mobilization Board to expedite construction on energy projects and cut red tape. This Board was expected to encounter strong environmental and local interest lobbies and soon both Senate and House were seeking to ensure that they would have a right of veto against any of the Board's actions;
— new incentives for development of heavy oils, shales and unconventional gas;
— utilities were to cut oil consumption by 50% by 1990, saving 0.75 mbd;
— major residential and conservation programmes to save 0.5 mbd by 1990;
— a $2.4 b scheme to insulate the poorest citizens against high energy prices;
— $16.5 b. in the next decade to improve public transport and car fuel efficiency.

The crux of the package was that the profits which would accrue as a windfall to the oil companies as prices rose would be taxed to provide $146–$270 b. over the next ten years, which would finance the new programme. At the end of October 1979, the Senate Finance Committee was considering a Bill to collect $141 b, but the House had already approved a Bill to collect $272 b. Hence the Administration was thought probable to enter the 1980s with a compromise between the two. Nuclear power was not mentioned in the President's original presentation of the package to the nation, but the comment was included in a speech the following day that it was 'a hard fact' that nuclear power was contributing 13% to electricity generation, but definitive future plans would have to await the report of the Kemeny Commission on the Three Mile Island incident. This was published on 30 October 1979.

This description of the attempts by the Carter Administration to formulate an energy policy and then get it implemented illustrates both the complexity of the energy problem and the complexity of the American system of government. Energy has become so important and topical a subject to so many people from the environmentalist to the farmer, that any proposal or regulation seems to get increasingly complicated and cumbersome in trying to be just to all. The American governmental system is such that domestically any plan has to reconcile Federal interests as seen by the Administration with those of the States as seen by their representatives in both House and Senate. This tends to protracted investigation by Congress. After the law has been passed it is subject to interpretation by a host of Federal and State Agencies. The proposal for an Energy Mobilization Board accepted that streamlining had become necessary, yet the Board had not been established by the end of November 1979. Then there is a propensity to test the interpretations of any new law in the Law Courts. Therefore it is obvious that in a democracy such as the USA it can be difficult in peacetime for any change in any demand pattern to be effected quickly. Indeed some Americans considered that only acknowledged crises can effect any real change, particularly if the full force of price mechanism is not allowed to act to adjust to an energy-scarcity situation, and no early loosening of controls to equalise oil prices within the USA to work prices was included in the package.

DEMAND FOR URANIUM

The uncertainties which beset the demand aspect of the nuclear energy industry can also be classified as those inherent to the downstream side, those which depend on collaboration with the upstream side, using the jargon common in the oil industry, and those uncertainties inherent in factors outside the industry.

Inside the industry, the first item is the demand as forecast by the industry for nuclear power and then how this may be affected by industry, such as by its stockpiling policy, reactor type pattern, variations in fuel cycle processes and facilities, and progress towards more advanced reactors with lower calls on uranium.

The 1976 OECD Report was based on the submissions of the countries who are members of the NEA and IAEA, augmented by estimates made by the staffs of those organizations, in order to fill the gaps and obtain a picture for the world excluding the USSR, China and Eastern Europe, but including Yugoslavia. It is assumed throughout this discussion that the centrally controlled countries so named will be self-sufficient and in no way affect the supply or demand for uranium in the rest of the world. Table 65 gives the annual and cumulative requirements for uranium; the high estimate is based on the estimates submitted by members, without any allowance for the use of recycled plutonium; the low estimate was calculated by the NEA/IAEA staffs on a lower growth rate in electrical energy consumption, on the recycling of plutonium in thermal reactors from 1981, and on a stated reactor pattern.

From table 65 it can be seen that the cumulative requirement to the end of year 2000 is roughly equal to the total reserves and estimated additional resources up to a cost of $30/lb U_3O_8 as detailed in table 25. This does not imply that the uranium requirements are covered without additional discoveries within an economic range, for the rate of production of these resources is not entirely at choice, but circumscribed by differing factors, e.g. the dependence of the bulk of South African production on the rate of gold production. Furthermore, as each plant has a 30-year nominal life from commissioning, for the capacity installed in 2000, a further 6–9 Mt has been committed of which 4–5 Mt would be required after 2000. If a ten-year forward reserve was accepted as adequate insurance, as OECD and Canada use, then the suggested requirements in 2000 of 236 kt to 313 kt would mean that 2.36 Mt to 3.13 Mt would have to be included in the proved reserves at that time. The task as foreseen by the OECD is therefore even more formidable than at first sight with productive capacity being planned ten years ahead because of the long lead times for new mines.

Table 65
Estimates of world* uranium requirements and world* nuclear power capacities

	1975	1980	1985	1990	2000
1. World uranium requirements in kilotonnes of uranium					
Annual a. High estimate	18	53	101	168	313
b. Low estimate	–	48	82	130	236
Cumulative a. High	18	192	594	1,295	3,826
b. Low	–	178	513	1,066	2,974
2. World nuclear power capacities in gagawatts electrical					
a. High estimate	68.6	194	530	1,004	2.480
b. Low estimate	–	179	479	875	2,005
Total under supply constrint, after Vaughan[270]	–	–	370	590	1,220

* Both estimates exclude USSR, China & Eastern Europe

Source: 1976 OECD Report, Paris, December 1975.

Stockpiling

Many consider that stockpiling is necessary and this could add to the amount to be found and mined. The present position is that the OECD took a total of 80 kt as known stocks, but the information was minimal; the USA Government stocks were said to be 34 kt in enriched uranium and 21 kt as natural uranium; USA consumers were said to hold 15.2 kt, and USA producers 3.3 kt; Canadian Government stocks were the only others reported as of significant size and were 5.5 kt. In total the 80 kt is over three times 1975 production, but these stocks have been accumulated over time for a number of different reasons, mainly resulting from the fall in the market after 1959, (see figure 23 and discussion), and not following a deliberate policy, with provision for maintenance and increase with rising production. Stockpiling may be considered desirable, or even essential, to provide against three main contingencies, (i) possible interruptions on the supply side (ii) temporary increases of demand on the process side, e.g. in enrichment plants, (iii) export restrictions or war. Mandel argued[187c] that three months supply could cover the first contingency and a further 21 months could cover the others. A stockpile to cover the first point could be located partly at the producing end and partly at the consuming end; stocks for the second point would need to be at the enrichment or other plants; for the third contingency the stocks would need to be with the consumer. Five important points are involved in the matter of stockpiling uranium:

— the mine has no other market than the power plant, and the plant has no other substitute fuel; hence both producer and consumer have a mutual interest in stabilized and secure supply, with which stockpiles might assist;
— the interests of the producer and consumer diverge if there is fluctuation in price, as has happened in the past, when selfishness or 'commercial acumen', accentuate the swings; long-term contracts of the right kind, or/and jointly financed stockpiles might help;
— uranium ore can be stored easily and safely;
— the cost to cover 2 years supply to any plant would not be a major item compared to the capital cost of the plant or to the cost per kWh of electricity produced;
— the size of the aggregated 2-year stockpile is however a very significant figure to add to the reserve and production requirements.

Mandel calculated[287c] that by 2000, some 100 kt should be held at the producing end and 700 kt at the consuming end, so adding 800 kt, or 18%, to the 4.4 Mt figure which he used for the cumulative annual requirement. Petit[287f] argued that the difference between the high and low OECD estimates for 1980, (see table 65), is only 14 kt, but the difference, even on the low estimate, between being with and without a two year stockpile is 96 kt; a stockpile to cover production in 1986 and 1987 adds 191 kt to the cumulative demand at year end 1985, or an average of 10 kt over the five years from 1980, which is equal to the total production in 1976, only four years earlier. In brief, the stockpile policy, even to cover only two years' production, is of greater moment to planning production, within a ten-year lead time, than whether or not the high or low estimate of demand is taken.

Demand by area

There was a dramatic change in forward thinking in the nuclear industry between the compilation of the 1976 OECD Report,[275] discussed in the first edition of this book, (first supplement, p. 46), and the compilation of the views of the members of the Uranium Institute which formed the basis for the February 1979 Report of its Supply and Demand Committee. The differences are illustrated in table 66.

Table 66
Estimates of installed nuclear capacity by area outside the communist areas

	1980		1990	
	A	B	A	B
USA	82	62	385	157/192
Canada	7	6	41	20/22
EEC	56	45	292	151*/190*
Other Western Europe	23	14	86	12/18
Japan	17	17	84	45/60
Australia/New Zealand	–	–	2	–
Total OECD	**185**	**144**	**890**	**386/481**
Africa & Arab M. East	–	–	7	3/4
Latin America	4	2	35	11/17
Asia excl. Japan & Arab Middle East	5	4	72	21/31
Others	–	1	–	15/21
Total WOCA	**194**	**150**	**1,004**	**436/555**

A. 1976 OECD Report [275] — High Estimate
B. 1979 Uranium Institute Report [515]

* including Spain

The 1976 OECD figures in the table were their high estimate, taken deliberately to give a maximum uranium demand. However the low estimate for WOCA were 179 GWe and 875 GWe for 1980 and 1990 respectively. These are far higher than the Institute totals in the table and all the higher than the range for 1990 of 410-530 GWe which the Institute members in general favoured as the best overall view. The Institute figures reflect the constraints and uncertainties which they described as inhibiting expansion:
— lower economic growth than expected after the first sharp rise in oil prices in 1973-1964, and apparent easy oil supply position in 1978;
— opposition to nuclear power as a principle;
— problems in some countries of technical development and construction capacity;
— uncertainty of fuel supplies, increased by interruptions in supplies owing to the non-proliferation policies of some suppliers;
— uncertainty as to future levels of stockpiles and access of the consumers, the plant operators to them;
— uncertainty as to the market for enrichment services which would alter the stockpile policies and so supply availability.

Demand by reactor type
Mandel[287c] illustrated the influence of the reactor type pattern on the demand for uranium by taking a base case for 1990, with a pattern not very dissimilar to the OECD pattern, but with a total of only 960 GWe, and then in his second case effecting a 25% reduction in demand for uranium by amending the reactor pattern. This exercise is displayed in table 67, which also shows the OECD data.

Demand based on enrichment plant capacity

The Uranium Institute's Supply and Demand Committee considered[515] that the customary forecasting of uranium demand from anticipated nuclear reactor installation needed to be checked, particularly in the short term, with the anticipated feed requirements of enrichment plants. These are highly capital intensive and by 1979 had been built to meet the earlier higher level of fore-

Table 67
Exercise illustrating effect of
reactor type pattern on demand for uranium, 1990
(percentages of total installed generating capacity)

	LWR	PHWR	SGHWR	GGR	AGR	HTR	FBR
Base Case	90.7	5.4	2	0.8	0.7	0.4	–
Case 2, reducing demand by 25%	70.0	10.0	2	0.8	0.7	16.5	–
OECD Pattern	85.7	6.7	2.1	0.3	0.6	3.0	1.6

Source: 'Uranium Demand and Security of Supply; a consumer's point of view', H. Mandel, International Uranium Supply & Demand Symposium, Uranium Institute, 16 June 1976, London.

cast demand. Uncertainty increases beyond 1985 as to what the actual operating plant capacities would be but their estimates are summarized in table 68, together with the estimated demand based on the reactor capacity of the previous table, and a summary which stresses the uncertainties.

The three ranges of demand for uranium are all based on a constant tails assay of 0.20%, but, as noted in the chapter 6 on conversion, the tails level markedly affects the demand, e.g. the low/high range in cumulative requirements based on the estimated enrichment plant capacities in the year 1990 varies from 598 kt U with 0.16% tails to 1,168 kt U with 0.30% tails, cf. the quoted 705–790 kt U range with 0.20%.

The Institute's supply estimates suggested a possible increase in annual output from 1979 of 41 kt U to 78/108 kt U in 1990, with the higher figure being a maximum and requiring adequate incentives. The cumulative requirements (1978-1990), would be 760/790 kt U. This estimated production must then be related to the proved reserves given earlier as being at 1 January 1977 as reasonably assured at $30/lb U_3O_8 at 1,650 kt U. This implies that, with the higher requirement of the 'summary', at 0.20% tails assay, at 800 kt U, there would still be remaining from the 1977 reserves some 850 kt U with a reserves : production ratio of 8.6 : 1. This would be a most unsatisfactory ratio and indicates that much more uranium will need to be found in the 1980s, even on this anticipated forecast which was so much lower than the 1976 OECD demand forecast.

Pricing

Prices in a free commodity market are a compromise between the costs to the producer and the value of the raw material to the manufacturer/converter in relation to the price he can get from the end user. In the nuclear energy industry it was long considered that the cost of uranium would be a low and stable item,

Table 68

Estimated enrichment plant capacity, uranium demand forecasts (with no recycling or stockpiling) and summary of estimated demand (with constant stockpile 1980, 1985 & 1990)

	1980	1985	1990
ESTIMATED ENRICHMENT CAPACITIES in SWU/year			
World outside communist areas	27	38.2–38.6	45–59
USSR Exports	3	3–4	? 3–4
Total	30	41–43	48–63
URANIUM DEMAND BASED ON ABOVE at tails assay of 0.20% and low/high estimates			
Annual requirements, kt U	44	58–63	71–92
Cumulative, 1978–1990, kt U	112	379–388	705–790
URANIUM DEMAND BASED ON ESTIMATED INSTALLED REACTOR CAPACITIES at tails assay of 0.20% and low/high estimates			
Annual requirements, kt U	28	50–61	75–99
Cumulative, 1978–1990, kt U	81	282–315	597–725
SUMMARY OF ESTIMATED DEMAND RANGES at tails assay 0.20% and with stockpile kept constant at 100 kt U			
Annual demand	36 ±8	57 ±7	87 ±12
Uncertainty	±22%	±12%	±14%
Cumulative demand	97 ±16	335 ±53	698 ±102
Uncertainty	±16%	±16%	±15%

Source: *The Balance of Supply and Demand, 1978–1990*, Uranium Institute [515] 1979.

always dwarfed by the high capital costs of the conversion plant. The price was controlled in the earliest days and, though it dropped after the fall in demand from the 1959 production peak, the general feeling was, perhaps, similar to the cheap oil price fixation of the 1960s. However, the price per pound of oxide rose over eight-fold over four years and prices have become important to producers, consumers and governments.

In the 1960s and early 1970s, most of the supplies of yellow cake were by long-term contracts on a fixed price plus escalation; prices for mid-1970s deliveries were reported as $7-$8/lb, with increases of 60¢-70¢/lb per year of later delivery. Some prices were said to be below $5/lb oxide in 1972; the Tenessee Valley Authority invited bids from 53 suppliers for 33 kt to be delivered over the period 1979-1990, and only three companies were interested enough to reply and these asked $12-$16/lb, with escalation from 1973. Similar prices were obtained in early 1974 and the upward movement movement had started. Since 1973, long-term contracts, with price escalation, have been replaced by long-term contracts with the price to be renegotiated one or two years prior to delivery; an escalating floor price is often included, as an insurance for the producer, and, in some cases, down payments are made to finance exploration or expansion. For short-term deliveries, fixed price contracts in 1974 were for $12/lb for immediate delivery and, for 1976 delivery, $16.50/lb + $1.50/lb yearly escalation; in mid-1976, prices were reported as $40/lb for immediate delivery + $2.50 for each later year of delivery.[287a] These examples of some contracts and prices are intended merely to stress the variety of a shifting practical factor in the nuclear energy industry.

The price must be high enough to provide incentive for further exploration, development and production for the producer. That price cannot be calculated on a cost plus 'reasonable profit', defined like a utility, for this is a high risk game. Exploration costs are comparatively low, but are increasing as the richest and more obvious deposits are found; appraisal and development costs increase as more careful delineation is required to attract investment capital for exploitation; costs for both machinery and labour are rising; but the biggest problem for the producer is to estimate demand. Ideally the producer would like to prove reserves to satisfy the high demand estimate, and equip and develop his mine in such a way as to meet the low estimate, but able to expand to meet the high estimate if necessary. The richness of each body is unique and this richness can vary within the body, so that optimum development may entail variable annual costs and, certainly, different costs betweem mines. In a totally free market the richest and cheapest mines would put the low grade, most expensive mines out of business, but the government interest may be to keep all mines working. Hence the problem of price-fixing returns. Should the costs of the poorest mine be covered and the richest reap the benefit, or should the richer mines subsidize the poorer mines on a median price? On the other side, the consumer/converter wishes to pay the price he can afford, and yet have security of supply for the 30-year life of his plant. As said before, the consumer is just as locked to uranium as the producer to the nuclear plant. The consumer is also befogged as to the future requirements beyond the plant in existence or nearing completion, and even these could be put in jeopardy by actions such as the Californian Project 15, discussed later. Yet he has to look ten years ahead in order to plan his next plant, just as the producer has to plan his next mine. Can the consumer share with the producer some of the risks, by contracting on a base demand, with graduated flexibility and advance payment on delivery?[287c]

Indeed this is happening already with some contracts. Can stockpiling be jointly financed to stabilize both the supply and the price structure? Can, or should, the governments of producing countries fix prices, as have the oil producing countries? Should governments finance or subsidize stockpiling, as has been done by some coal-producing countries, and was done by the USA and Canada after the fall in demand for uranium in 1959? At what point is the advantage of a high price in encouraging exploration, for a finite and depleting resource, counteracted by making the end product uneconomic in relation to its competitors? This last question is the most difficult. The problem is not new or unique to the uranium industry. Some economists would answer with their own, current theory, but this problem is new to the uranium industry, and in fact also to the economists, for this is a new situation and the old catchwords or formulae may not work. Answers to these questions are particularly important at a time when the positions of the competing energy sources are also in constant flux.

The present generation of thermal reactors achieve a burnup of about 5,000 MWD per tonne of heavy atoms, without plutonium recycle. On this basis 1 lb of U_3O_8 has the same energy content as 28 barrels of oil. This might suggest that the breakeven price of U_3O_8 would be about \$300 per lb, when the price of oil was \$11/barrel. However, this oversimplifies the matter, both because uranium is a replacement of only the heaviest and least valuable part of the barrel of crude oil, and because, while oil is burnable almost as it is, uranium can only be burnt after much expensive processing. The calculations can be done in various ways, and limiting prices have been quoted between \$100 and \$150/lb of oxide. Above this price an electric utility, with spare capacity in both oil-fired and nuclear powered stations, would find it more attractive to buy oil. It is noteworthy that a further increase by a factor of 2.5 would bring uranium up to the lower of these limits. These limits may be some sort of yardstick for thermal reactors but certainly do not apply to fast reactors.

Many calculations have been made, and many more will be made in the next few years in order to get 'some feel' for the 'proper price', but it would seem difficult, if not impossible, to get agreement, on any broad, theoretical basis, between all the producers, consumers and governments who are involved. Yet the industry, on both sides, obviously appreciates that something must be done in order to gain stability. The pricing problem is only one of the many areas of uncertainty in the nuclear energy field.

Safety
One of the greatest and most difficult of the imponderables facing all energy industries in 1980 is the extent to which public concern about the safety of the processes of the conversion of nuclear to useful energy will inhibit its growth.

In the USA, in the first half of 1976, the intensity seemed to increase in the arguments with which both laymen and scientists had been questioning the safety of nuclear reactors, and with which other scientists, from industry, government, and the academic world, had equally been expressing confidence in the multi-layered safety measures taken and had pointed to the good record over the past 15 years. The argument had taken a serious turn in January 1975, when Dr. Hans Bethe, supported by 34 other scientists, made a pro-nuclear case which was countered, in August 1975, by the Union of Concerned Scientist with 23,000 signatures. The 'Second Gathering of the Citizen Movement to Stop Nuclear Power' was held in Washington in November 1975. The National Academy of Sciences established a Committee on Nuclear Power and Alternative

Energy in December 1975, with the primary aim to clarify the debate by getting wider agreement as to the established facts and the necessary further research. It was typical of the controversy that the Committee was criticized as being composed of people partial to the protagonists. Some people say that, 'although nuclear power will eventually have a part to play, yet technology is not ready for major acceleration'. At the close of the 1975 session, the USA Congress extended for ten years the Price-Anderson nuclear insurance law which limits the nuclear industry's liability in the event of a catastrophe to $560 million, which a special study had considered might be exceeded on a two million to one chance. However, the movement against nuclear energy had become more organized, and in March 1976, swung the National Council of Churches to its support. The organized drive was particularly strong in California, and the heat of the debate was increased by the resignations, early in 1976, of three senior scientists, who had been working for the General Electric Corporation, in protest about more nuclear plants. One anti-nuclear platform might be summarised as that nuclear reactors mean the mass production of poisonous substances, that nuclear energy means nuclear bombs, that nuclear energy is not necessary because it is not cheap, and the funds necessary for development could be used better for the early exploitation of clean solar and geothermal energies. The rebuttal of these charges by the nuclear protagonists has been cautious, for the sake of honesty and scientific integrity, and is therefore less emotive and vote catching. However, when the Project 15 initiative, which would have stopped all further construction of nuclear plants and phased out existing plants, was put on the ballot papers for a vote on 6 June 1976, in the State of California, the protagonists did put forward their arguments. The people rejected the proposition by a 2:1 majority. However, one cause was that Governor Brown had recently brought in some strong safety regulations. The antagonists had not given up the fight. State-wide votes were taken in Oregon and Colorado in late 1976, and in over a dozen more States in 1977 on similar issues. These votes, and the results of legislative measures under discussion in over 28 States, were crucial to the USA energy position in the 1985-1990 period. The drop in electricity consumption, after the 1973/1974 fuel price rises, cut the revenues of the public utility companies and forced the deferral for a year or more of plans for some 160 GWe of new generating capacity, and the cancellation of plans for a further 10 GWe. The coal industry has still not overcome strong environmentalist opposition to strip mining, particularly in the west, nor the bureaucratic restrictions on mining methods and environmental restrictions on pollution through utilization, so that new coal-fired power stations are not being planned. The Government has discouraged the construction of any oil-fired stations, because they would need imported oil, and of any gas-fired stations, because the declining domestic gas supplies are needed for premium use in households. Many energy analysts therefore see an enforced reduction in electricity consumption in the 1980s, if new plants are not scheduled immediately. The 1976 nuclear contribution to USA power needs was 9%, and, even allowing for recent deferrals, the 1985 contribution could be 18% of an estimated total of 1,786 Mtce;[271] this assumes that existing plants continue and current construction is completed. In late 1979 this was doubtful.

The European Economic Community (EEC) is less homogeneous even than the USA. The United Kingdom has officially adopted a cautious continuation of expansion of nuclear power, through additional thermal stations, even though there are controversies about types of reactors and the role of the fast reactor.

Denmark has refused to embark on a nuclear programme but can benefit from the nuclear capacity installed in Sweden and Finland, through the Nordel interchange system of electricity supply management. The Netherlands deferred a decision on three nuclear plants in January 1976. Belgium is aiming to have 6–9 GWe of nuclear capacity by 1985. France is the strongest proponent of nuclear power in the EEC, though because of the lower recent electricity consumption, as in most countries, the 1985 nuclear contribution of 79 Mtoe in the Spring 1974 plan, was reduced to 55 Mtoe in the April 1975 plan; this is still a massive increase from the 1975 contribution of 4.6 Mtoe. FR Germany in 1974 revised its 1973 plans upwards to 20 GWe in 1980 and 45–50 GWe of nuclear power in 1985, and FR Germany can illustrate both a diffusion of responsibility for energy and an active government educational programme, which is not matched elsewhere.

The German Federal Ministry for Research and Technology (BMFT), is responsible for research and development of nuclear power; the Federal Ministry of the Interior is responsible for licences and is advised by the Institute of Reactor Safety, which also advises the *Länder* or States, who have ultimate control of all plants; the Federal Ministry of Economics has the responsibility for the creation of an energy policy for the whole Republic. A novel, *Die Explosione*, serialized in a weekly magazine created much discussion early in 1976. The BMFT published a booklet on nuclear issues, have had a study made of public opinion and are encouraging public debate. The Federal Government is thus taking a positive attitude, rather than a low profile, but is officially impartial. The anti-nuclear lobby say that the Government is pro-nuclear, and have raised the same arguments as in other countries, to which the nuclear industry has not attempted any technical replies.[277] The West German Reactor Safety Commission in May 1976 favoured the nuclear power plant which the BASF Group proposed at their Ludwigshafen chemical works, to provide both electricity and steam, but the Commission did recommend that the nuclear plant be moved five miles in order to avoid reciprocal accident dangers with the chemical works. One might say that FR Germany is taking a pragmatic view at the moment, but favours nuclear developments, for the Germans are active in many joint ventures and are exporting nuclear plant and technology.

Spain could be said to have chosen nuclear power as the main source of its electricity generating needs by 1985. Spain has some local uranium, but only a 1976 capacity of about 200 t of yellow cake each year; this is planned to be 800 ta by 1979, from reserves of about 10 kt, with an equal amount expected to be confirmed in the near future. However, a £100 million exploration programme is being launched aimed at reducing by the mid-1980s the present 50% dependence on imports. The power programme, announced in May 1976, and costed at £8,000 million, calls for 22 GWe of installed nuclear capacity by 1985; this is a massive expansion from the 1.14 GWe of 1976 capacity at 9% of generation to 54% in 1975. Spain has been importing both plant, technology and skilled manpower, mainly from USA Westinghouse, but is taking firm steps to train its own scientists and technicians, and to build up its own manufacturing capability so that eventually the whole fuel cycle will be under Spanish control. Indeed there is a proposal that the 20 ta spent fuel, which is now being reprocessed abroad, should be stored until a viable commercial plant to handle 1,000 ta could be build in the 1990s. Nevertheless, Spain is willing to cooperate internationally and is a partner in the multinational Eurodif enrichment project.[293]

Britain was a pioneer in the application of nuclear physics, and public antagonism has been comparatively mild, though at times vociferous. The cautious expansionist policy of the Government has recently received quite a lot of support from other industries and the trades unions, all of whom favour a broad-based energy policy in which nuclear power would have a major role. At the National Energy Conference, organized by the Secretary of State for Energy, in June 1976, there was a strong feeling that Britain should not lose its place amongst the pioneers in the nuclear field, and, indeed, that the fast reactor programme should be pursued with vigour, whilst simultaneously increasing thermal nuclear capacity. There are critics of this attitude. Their views are worth summarizing, for they are typical of many in Britain and it is noteworthy that, whilst the safety factor and the unknown hazards are part of the criticisms, yet other local points are made. The arguments are, in brief, that the potential energy gap of the 1990s for Britain is illusory and therefore the huge cost, safety hazards and irrevocability of the fast reactor programme are unwarranted and uneconomic; that the demand for electricity will not rise as fast as some expect, for five reasons; energy consumption per capita in Britain remained stable for 50 years and only rose from about 4 tce in 1950 to 6 tce in 1975; over half the growth in final energy consumption, 1960–1975, occurred in road transport which could be controlled in future and is irrelevant to the fast reactor; domestic energy growth has only been 5% pa since 1960 and there is no cause for any rapid expansion; increased efficiency and conservation will lessen the impact of industrial growth; and, fifthly, that there is no obvious, great new outlet for electricity. The critics also argue that the serious technical problems should be solved in time but that the leap from the 350 MWe Dounreay plant to the 1300 MWe is a leap into the unknown which others might take first; that as long as Britain stayed amongst the front runners, but not in front, she could benefit from the mistakes of others and could employ the huge sums involved much better on R&D on North Sea oil, coal conversion, new energy technologies and improved thermal reactors. To each of these arguments there are counter arguments, and so the discussions carry on.

There are similar differences within and between other countries. Sweden, for instance; after a long public debate and referendum, decided in March 1974 to move ahead installing more nuclear capacity but step-wise, keeping all options open. Iran, Brazil, South Africa, Israel, Egypt, South Korea and others wish to use nuclear power whilst others hesitate. The arguments for and against the use of nuclear energy are going to continue for many years, but, wherever the debate may be held, there are common elements of which, one hopes, some will die as more becomes generally known about the subject. The debate is often emotional and therefore there are confrontations between arguments which are not comparable in type, (the technical is faced by the philosophical), in time span, (as next year's energy budget is countered with a hypothetical position in 2000), in space, (as the local situation is contrasted with the global), and so on. Rarely is the debate sufficiently comprehensive to cover all aspects of the problem. Often confusion is increased by the refusal of one side or the other to accept the validity of any knowledge, whether it is advanced as theoretical or practical. The technical issues raised are usually those of the risks of catastrophe, of long-term radiation from wastes, of terrorism and war blackmail, and of an eventual police state, if all the necessary safety measures were taken against these risks to have 100% surety.

The risks of catastrophe are not limited to the breakdown of a plant by accident or sabotage, but include proliferation of knowledge and capability to make war weapons, and this aspect warrants mention because it has affected the policies of governments and could be a barrier to the availability of the nuclear resources. Canada withdrew its aid from India when, in May 1974, India extracted plutonium from the CANDU PHWR, which Canada had virtually given to her for experimental purposes, and exploded a nuclear device. The agreement, announced on 30 May 1976, that France would build a nuclear plant for South Africa raised immediate and loud opposition from other African states, because of the potential military significance of South Africa gaining the means towards nuclear weapons. These risks have been the subject of much discussion, particularly between the seven 'nuclear exporters', the UK, USA, USSR, France, FR Germany, Canada and Japan, and they have been joined by the Netherlands, Sweden, DR Germany, Belgium and Italy, with Czechoslovakia and Poland expected in April 1976, to join soon.[273] The three most sensitive technologies are the enrichment of uranium, the reprocessing of spent fuel and the refining of heavy water. The first two can yield fissile material from which weapon grade fuel can be made (see figure 24), and the third is the key to the operation of some reactors. The first step in enriching uranium to nuclear fuel standards is difficult but the next step to bomb standards is comparatively easy. As put to the British House of Commons on 31 March 1976, by the then Foreign Secretary, Mr. Callaghan, four assurances would be required before Britain would export nuclear technology: no nuclear explosives would be be made for peaceful or other uses; adequate protection is taken against theft and sabotage; re-export of the technology would be under the same assurances; plant will not be replicated within twenty years. Contracts such as between France and Britain with Japan for the reprocessing of spent fuel do not involve export of technology. The International Atomic Energy Agency, which already had various safety inspection and approval responsibilities, approved, in April 1976, two new contracts under the new rules, namely the 1975 cooperation agreement between FR Germany and Brazil, and the reprocessing plant being supplied by France to Pakistan. It was reported in the British Press on 4 August that the USA Congress had been assured that the plants being proposed for Egypt and Israel would be subject to inspection by the IAEA.

There is little doubt but that regulations for precautions to be taken, such as those just mentioned, have been strengthened because of the debates in public and private, even though these often quickly progress into the subjective areas of the quality of life, reaching into morality, religion and the principles of society, indeed they reach these areas much faster than do debates on other energy resources. Illustrative of this intensity in the nuclear debate is that a principle of democratic government is often at stake; should decisions be made by the representatives elected to govern, who have a responsibility and capability of becoming informed on complicated matters before taking a decision, or should the decision be taken out of their hands by a referendum of all the people? Examples have been given of different views in different countries and of the different ways in which they are expressed. There is no doubt that this area of public opinion, stemming basically from doubts of the safety of harnessing nuclear energy, creates one of the major uncertainties of its future. Governments may sometimes have to steer a middle course between the extreme views of the protagonists and the antagonists, and, like any course set in weather of changeable winds, they may have to change course as they 'tack', and these changes of course add to the uncertainties.

Competition

Nuclear energy cannot be considered in isolation, and the economic competition from the other energy sources must be appreciated, even though it is very difficult even in local terms to evaluate, because there seem to be as many cases for evaluation as there are evaluators. The following tabulation[290] therefore should be accepted merely as one local example, which will not be applicable elsewhere, for the whole tax structure and costs in each country are unique to that country. However this table might be taken as showing that in Britain the rising costs of uranium may affect the comparative advantage of nuclear energy but there is some leeway.

Generating costs in pence per kWh in UK Central Electricity Generating Board stations commissioned since 1963

	1972–1973	1974–1975
Coal-fired	0.49	0.74
Oil-fired	0.40	0.88
Nuclear	0.48	0.48

International collaboration

The costs in every phase of the nuclear energy industry are becoming too much for many countries to handle alone, and there is increasing cooperation between nations.

Exploration is the least expensive phase and yet there are a number of joint ventures which illustrate the international character of exploration and the interest of oil companies. A Japanese consortium and Denison Mines of Canada plan to explore in the Elliot Lake area of Ontario; The Arabian Oil Company of Japan was said in February 1976, to be planning to explore in Niger; Spain's Instituto Nacional de Industria will join Soquem, a Canadian corporation, to explore in Quebec; Japanese and French interests are exploring jointly in Mauretania, with the Mauretanian Government having the right to a 49% share in any commercial uranium mining and milling operation which may result; Ente Nazionale Idrocarburi hope to start mining in the Alps in 1979; Getty Oil has a large interest in the Jabiluka area in Northern Territory, Australia; the major international oil companies have uranium prospecting interests in many countries; a Canadian/Argentinian consortium will handle the planning of a mining and milling complex in Mendoza Province, aimed at an annual production of 600 t U from mid-1979, which should satisfy Argentina's needs for at least 15 years.

Collaboration in the reactor and enrichment fields is well advanced in Europe. Euratom, the European Atomic Energy Agency, was set up at the same time as the EEC in March 1957, as a further instrument of joint collaboration by the six members of the European Coal and Steel Community, which had itself been established in 1952; the UK, Denmark and Ireland simultaneously joined all three organizations on 1 January 1972. France and FR Germany agreed to pool efforts in the HTR and FBR fields, in an agreement which was said, in February 1976, to cover R&D, concept studies, the planning, construction and utilization of the reactors, and the manufacture of components for and of the full fuel cycle. This agreement for FBR's supplements the cooperation of France, FR Germany and Italy in NERSA, (Central Nucléaire Européenne Rapide Société Anonyme), and the French Commissariat à l'Energie Atomique (CEA) will set up a joint subsidiary with Interatom, which will include the company in which the Belgians and Dutch have a 16% share. This merging of interests is a cost-sharing exercise; the sodium-cooled 300 MWe breeder demonstration plant at Kalkar cost DM 2,490 million, instead of the DM 1,700 million estimated at the time construction started in Spring 1973; the thorium HTR at Ventrop was estimated to cost DM 1,480 million, instead of DM 903 million. Such sums warrant collaboration. There are a number of other well-established joint ventures, such as between the UK, FR Germany and The Netherlands in the centrifuge enrichment project, Eurenco; Iran joined the Eurodif project (the gas diffusion enrichment scheme), and had, it is said, made a $1,000 million loan for a 10% stake in the first plant at Tricastin and had taken a 20% share in the study for a second plant. Iran was also said to be interested in the fuel fabrication expansion of British Nuclear Fuels Ltd and in the possible South African enrichment plant.

National competition

One important international venture should be the Joint European Torus project, (JET), which is intended to be an important step towards the development of fusion power in the EEC. A jointly managed programme included research by an international design team, at Culham, England, on the magnetic confinement of plasma. There is general scientific agreement that the next step should be the JET experiment, costing £60 million over the 1975–1980 period. The EEC Commission, after a survey of possible sites, recommended Ispra, Italy, which has under-utilized EEC facilities and abundant electric power available. Britain, France and FR Germany objected in February 1976, and favoured sites in their own countries, and a site in Belgium was also mentioned. The political decision as to the location for the experiment by the EEC Council of Ministers was delayed, which led to budget problems, to the team losing members to other work and so a loss of momentum at a time when the project seemed to be leading the world scientifically, in the face of competition from work on similar lines in the USA and Japan. Culham was eventually chosen.

This example of national interests delaying a project which is an obvious regional project emphasises both the nationalist and governmental interest in the whole nuclear industry. In the USA the nuclear industry might be said to be almost in a state of siege from 1976 to 1980; the Government being loth to take any controversial step and lose votes in an election year. In Britain and France the attempts to let many private firms compete for the available business have given way to reorganization into single big units in order to concentrate both expertise and capital in reactor supply. Governments still maintain control in the three sensitive areas, mentioned earlier, and appear to have preferences as to where they will export. It is, however, the big sums involved and the long lead times before any returns are possible, which inhibit financing by private enterprise and make government involvement necessary.

The nuclear energy industry is very big business; world export trade was some $3,600 million in 1974 and the stakes are rising; the deal in June 1975, between FR Germany and Brazil was alone said to involve $55,000 million; Iran was willing to devote $25,000 million to acquire nuclear capacity to provide all its electricity needs in 20 years time, but more recently it was estimated that more than that sum would be required to provide 52% of its anticipated needs in 1993/1994; part of the high cost is due to problems of locating plant near abundant water supply in an arid country. Nevertheless the programme was being

pushed and was a measure of the Shah's conviction that petroleum is too precious a commodity to squander and that a nuclear facility may have many advantages. Iran spread its orders widely, to France, FR Germany and the USA in order to benefit from the keen competition. The French Government took over the 45% in the Framatome organization from the US-based Westinghouse company for national security but also commercial reasons; France believes that it has a chance of an early lead in the future market for fast breeders. The Americans have a large share of the present reactor market and were said to have built or be building 70% of the industrialized world's nuclear power stations,[281] and be earning $1,500 million in annual export sales. American interest in exports has been quickened by the slow down in their own domestic market, but the competition is keen, with strong sales pressures from the industrialized countries on the developing countries; 45 countries were said to have announced plans in 1976 for 257 plants.[318] It has been suggested[281] that there are two types of nuclear power customers in the developing world:
— those able to afford a nuclear weapons capability and abundant power;
— those resigned to not being able to afford the bomb, but convinced by sales talk, including advice from the IEA, that nuclear power is their cheapest alternative energy source for the next fifty years.

Into the first category was put Iran, Brazil and the Argentine, with Pakistan and South Korea encouraged by India's 'bomb'. Iran had been mentioned as having had an interest in the whole fuel cycle from exploration to enrichment. Brazil persuaded the Kraftwerk Union to tie its eight reactor deal to facilities for enrichment, fuel fabrication and plutonium separation, and as Brazil has indigenous uranium resources she will have a full fuel cycle. No country would admit that its prime purpose is to make atomic weapons, though it is sometimes noted that one advantage of the CANDU reactor is that it makes more plutonium than any other thermal reactor. In the second category are the poorer countries and it is perhaps difficult to see the real advantages to them of the installation of small reactors, which sacrifice economies of scale, except that same national pride which, at one time, encouraged many to build small uneconomic oil refineries. There seems to be little chance that nuclear energy will be cheap to install, for many decades. However, in 1976 it seemed almost inevitable that small and large countries would be adding to the number of nuclear power stations around the world and that nuclear energy would be a major factor in the late 1900s. In 1979 this seemed less likely, as the revised forecasts showed.

Summary of demand for uranium
The recent major changes in the resource and production position of uranium are the increases in Australia's reserves, the new discoveries in Canada, the reappraisal of South Africa's potential, the changes in USA reserves and the attempt to calculate the USA ultimate resources. These events illustrate that there are three groups of factors which affect the measurement and availability of these resources; those which arise within the supply side of the industry, such as improvements in exploration, mine construction and extraction; those which arise within the nuclear industry as a whole, because the users are also involved, such as attempts to stabilize production trends; thirdly, those factors outside the nuclear industry, such as the policies of governments, and the availability of money, manpower and materials, in competition with the other energy industries. The same three groups of factors affect the demand side.

Any forecast of the demand for nuclear power can be affected by industry policies such as stockpiling, reactor preference, variations in fuel cycle processes and facilities, and progress towards improved reactors; pricing and financing, with various forms of contract, are basically between producer and consumer, although governments are often involved; whilst variations in public opinion, in government policies and in competing energy supplies are outside the industry but have a most profound effect on it. The nuclear industry is only part of the total scene, and is affected, as are the other energy industries, by general world economic conditions, like the 1974/1975 lessening in energy demand, and the resulting difficulty in maintaining any momentum in expansion. There would seem little doubt, except amongst the extreme anti-nuclear lobby, that nuclear energy will increase its contribution to world energy input with more thermal reactors being built in the next twenty years; fast reactors may be introduced in the late 1980s, becoming the dominant type soon after the turn of the century; fusion reactors may start their climb to dominance, perhaps from 2020, to be joined by solar and other renewable energy resources which will then fulfill the bulk of the energy needs.

It is a salutary thought that the 'uranium age', or more precisely, the need for large tonnages of uranium as the fuel for thermal nuclear reactors, will probably last only fifty or sixty years, will be increasingly supplemented with thorium, and plutonium from the breeders, until the fusion process is controlled and exploited, when the energy source may move to the heavy hydrogens. In total the 'uranium age' that is using uranium as a non-renewable resource may be even shorter, at say 1990-2050, than the 'petroleum age', from say 1930-2000, with the decade overlap, similar to that in the 1930s which petroleum had with the 'coal age'. If this is so and assuming a continuing lead time of 10 years from discovery to production, and a mine life of 20-30 years, then it is only the existing mines, and those located in the next ten years or so, which will have a full and profitable life, for prices may decline with the reducing demand after about 2010. This scenario assumes that new rich deposits will be found in accessible and politically stable areas in sufficient quantities to allow this progression, without undue strain, and, as always in such looks ahead, without catastrophes or wars. The very large additional resources from low grade ores, would never be called into use, just as peat has only a very restricted use now as a carbon resource. This argument therefore reinforces the theme that it is the proved reserves which are important, and the immediate, and by no means an easy task is to find these reserves.

INTERDEPENDENCE/INTEGRATION OF ENERGY DEMAND

The natural gas industry developed as a separate industry in the USA because, at the beginning, gas was a by-product of the oil industry. Gas was costed as such. Distribution to individual customers obviously required detailed engineering and retail sales techniques. Crude oil producers were not interested. In Europe manufactured gas also grew into a separate industry from the coal industry. It grew closer to the oil industry when processes to use petroleum products were developed. Now organisations like the British Gas Corporation, the nationalised body, is exploring for and producing natural gas. It is also the mandatory first purchaser of all natural gas in Britain on land and offshore, and has a monopoly of distribution and supply. Manufacturers of gas from coal did produce benzole and other anomalies from the coal tar residues but the British Gas Corporation has not yet moved into petrochemicals. In the USSR, until after 1971, the oil industry ministry also covered natural gas, but there are now two ministries. In theory, of course, the centralized economies plan the production and utilization of all energy resources to the maximum benefit of all. A Soviet paper to the Ninth World Energy Conference in Detroit, 1974, stated,[108] 'The optimal planning of energy economy involves the choice and introduction of such kinds of primary energy resources and ultimate energy carriers, the production, transportation and utilization of which enable to meet all the requirements of energy consumers at minimum cost' and went on to explain how this was attempted in the USSR. The Soviet Union has an abundance of energy resources but the biggest deposits, as we have seen, are far from the consuming centres. Increasingly sophisticated planning techniques and all its administrative strength will be needed to optimize the effectiveness of each source.

The interdependence of the energy industries, and the need to consider integration of those industries and the rationalization of demand between them, is becoming increasingly appreciated worldwide. Governments of all political systems and both public and private industry are taking a keener interest in all energy. Australia is one country which is moving towards close management of its energy resources. In the USA the 'Project Independence' concept and subsequent establishment of many Federal energy agencies and the initiative in setting up the International Energy Agency are examples of closer integration. Mention has already been made that many energy companies and particularly the oil companies, have been diversifying into other energy industries. This is particularly easy, of course, in the exploration aspect because of the similarity in techniques whether for coal, gas, oil or, even, uranium.

The petrochemical industry has an interesting position in this field. The quantity of petroleum being used as raw material, or building blocks, in the petrochemical field is only about 5% of total petroleum use. However, oil refinery/petrochemical plant complexes are a natural development in energy conservation and nuclear power/petrochemical plant developments other possibles. The oil-producing countries are showing a growing interest in upgrading their petroleum resources with petrochemical complexes. Many ambitious schemes have been announced. One total investment estimate for the Arab states alone was for about £22,000 million in firm decisions and a further £31,000 million under study.[215] The petrochemical industry has shown itself very prone to sharp swings between scarcity and over-supply. It is not easy to foresee what the effect on petroleum supplies would be if the oil producers were to use their oil as leverage to obtain petrochemicals outlets.

From all sides one must expect that many barriers between the individual energy industries will disappear.

WESTERN EUROPE

In Western Europe there is a great diversity of economic growths and energy demands. As a whole the area is a net importer of energy, predominantly of oil, but also of natural gas and coal. Britain and the Federal Republic of Germany have significant coal resources. Britain and Norway have petroleum resources from which production will make them self-sufficient in oil for the 1980s, and Norway for much longer. The Netherlands, Britain and Norway have natural gas resources, but the production from the Netherlands' giant field of Groningen is due for a decline from 1980. France has a little coal, natural gas and oil and is moving fastest in the nuclear energy field. The Scandinavian countries have a strong hydro-electricity element in their energy supplies, which through the Nordel system even benefits Denmark. To these differences in their energy bases are added the differing states and potential of their economies. Since the shock caused by the oil drama of 1973 every country has been attempting in its own way to define an energy policy.[337] For some, their policies must be conditioned by membership of one or other international grouping, the European Economic Community, or the European Free Trade Area (EFTA), or the Scandinavian Group, or being aspirants to join the EEC. These groupings add to the proliferation of scenarios and forecasts and to problems of comparable statistics. In some statistical aggregations Turkey and Yugoslavia are within 'Western Europe', in others one or both may be excluded. In this section, which seeks only an approximate figure for demand as a base on which to consider supply, particularly of petroleum, a broad brush approach will be taken. Earlier examples will have inured the reader to the variations in scenarios and their assumptions. However figure 26 sets out five variations on the theme of economic growth rates.

Comparison of this figure and figure 28 which illustrates Japan's growth possibilities, can make several points. Western Europe's growth in GNP is lower. The decline in growth rates in Japan from that of the 1960s is much greater than in Western Europe's scenarios. There is much more variation in opinions as to Japan's growth than in the future of Western Europe. In Western Europe energy consumption grows more slowly than the economy because all analysts expect savings through conservation and improved efficiencies, with greater scope for such saving in Western Europe than in Japan.

The range in total primary energy demand for 2000 in Western Europe (excluding Yugoslavia and Turkey) in the two CoCo scenarios H5 and L4 is 92 EJ and 88 EJ with the appropriate figures for 2020 as 130 EJ and 120 EJ. The total for 2000 suggested by Frisch is 3,800 Mtce (114 EJ) but he notes that two French companies have suggested higher figures. The full range of the three quoted figures is thus 88 EJ to 114 EJ (2,933–3,800 Mtce, or 2,000–2,600 Mtoe).

The EEC Commission set out an objective for energy consumption in 1985 in December 1974 which they revised in mid-1977. The bare bones of these targets in Mtoe were:

	End–1974	Mid–1977
Total primary energy consumption	1450	1280
Oil contribution	695	640
of which imports	515 (36%)	500 (39%)

The revision reduced the total demand but roughly maintained the amount of oil which would have to be imported. The OECD in 1977 offered a range of total energy consumption in 1985 from 1,704.4 Mtoe to 1,619.2 Mtoe. These examples show the considerable divergence of opinion, even in 1977, of what may be the total energy consumption in Western Europe in 1985.

A rough sketch of the total European Community energy resource position is given in table 69.

WESTERN EUROPE – SOME ECONOMIC AND ENERGY GROWTH POSSIBILITIES
Average Annual Percentage Growth Rates   Fig. 26

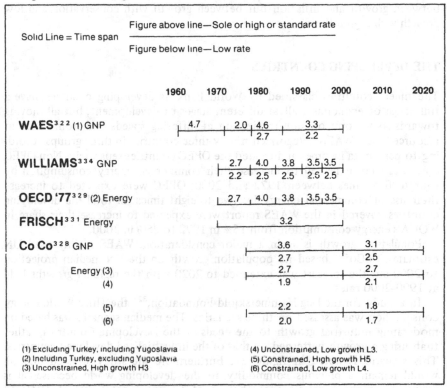

| Solid Line = Time span | Figure above line—Sole or high or standard rate |
| | Figure below line—Low rate |

(1) Excluding Turkey, including Yugoslavia
(2) Including Turkey, excluding Yugoslavia
(3) Unconstrained, High growth H3
(4) Unconstrained, Low growth L3.
(5) Constrained, High growth H5
(6) Constrained, Low growth L4.

Sources As referenced

Table 69
EEC energy reserves — 1979

	Annual use	Recoverable reserves	Reserves/ production ratio	UK reserves as % EEC reserves
Oil, Mtoe	548	2,329	4.25	90%
Natural Gas, km³	192	3,022	16	24%
Hard coal, Mtce	284	82,000	290	55%

Source: Hoare, Govett Stockbrokers, reported in *Financial Times*, 19 July 1979, London.

In 1979 there were a number of downward revisions to total EEC energy demand targets. The EEC agreed a reduction of oil imports for 1980–1985 of 472 Mta, cf 500 Mt of mid-1977. One study[523] offered a traditional scenario of 1,157 Mtoe in 1985 and 2,038 Mtoe in 2000 which conservation could convert to a 'reduced' scenario of 932 Mtoe and 1,361 Mtoe respectively (@ 1 t oil = 1.69 tce). Such an exercise is highly relevant to a statement made to the OECD meeting in July 1979 by the US Treasury Secretary that the choice was not between growth and inflation but between growth with conservation and low growth with high inflation.

THE DEVELOPING COUNTRIES

The ninety countries classified by World Bank as developing countries have a full range of economies, all at different stages of development, but all moving towards some form of industrialization at differing speeds and from different resource bases. WAES categorized ninety-three countries in three groups according to per capita incomes and treated the OPEC countries separately. Non-OPEC countries were expected to increase their commercial energy consumption by four to five times between 1972 and 2000; OPEC were expected to increase their use of commercial energy by five to eight times. In total the developing countries covered in the WAES report were expected to increase their share in WOCA's energy consumption from 15% in 1972 to 25% in 2000.

Population growth is again a major consideration. WAES mainly used UN estimates. CoCo[328] based its population growth on the UN median projection to 2000 but the forecast was extended to 2020 with the regional growths held at 1990–2000 rates.

In a study for the Dag Hammerskjold Foundation,[325] the Third World energy consumption was assessed on three scenarios. The median scenario was based on moderating industrial growth to the needs of the Developing Countries, rather than using a growth patterned on that of the industrialized developed countries. This scenario is noted in figure 27. Furthermore the study stressed that oil would remain a precious commodity to the developing world because oil is adaptable to the constraints of absence of transport networks, of scattered communities and diversity of use. Indeed, in their projections for the Third World (excluding China and Korea) oil, in all three scenarios, provided from 35–43% of the 2000 energy demand and 32–39% in 2025. The median scenario gave 40% in 2000 and 36% in 2025 of totals of 157 EJ and 455 EJ (7.1 Gtce and 20.7 Gtce).

In a more recent study M. Frisch takes a more optimistic attitude to world energy growth than the WAES or draft CoCo reports, (see figure 27). From consuming 10% of world energy in 1974, including the Communist areas, the Third World would be consuming 20% in 2000. The new study again emphasized that the Third World, even without the Middle East, could be dependent for 40% of its energy on oil and 55% on oil and gas. The growth rate assumed in the WAES and draft CoCo studies are indicated in figure 27, as are the figures from the OECD 1977 report[329] although, unfortunately, it does not go beyond 1985.

There are many ways of using the mass of information which has been condensed into figure 27. One way would merely be to ignore the data, reflecting that there are such variations that no conclusions can be drawn. Another way would be to inspect the figures, and, noting the many different labels to the figures, again ignore the data, but pray and wait for the time when a concensus is reached on standards. Unfortunately that time may never come, and is certainly a very long way away. Even if one concentrates on the energy demand

DEVELOPING COUNTRIES — SOME ECONOMIC GROWTH POSSIBILITIES
Average Annual Percentage Growth Rates   Fig. 27

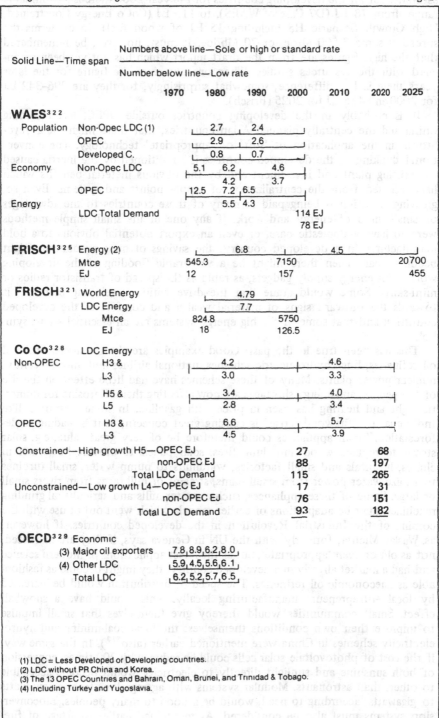

Solid Line—Time span

Numbers above line—Sole or high or standard rate

Number below line—Low rate

		1970	1980	1990	2000	2010	2020	2025
WAES[322]								
Population	Non-Opec LDC (1)		2.7	2.4				
	OPEC		2.9	2.6				
	Developed C.		0.8	0.7				
Economy	Non-Opec LDC	5.1	6.2	4.6				
			4.2	3.7				
	OPEC	12.5	7.2	6.5				
Energy			5.5	4.3				
	LDC Total Demand				114 EJ			
					78 EJ			
FRISCH[325]	Energy (2)		6.8		4.5			
	Mtce	545.3		7150				20700
	EJ	12		157				455
FRISCH[321]	World Energy		4.79					
	LDC Energy		7.7					
	Mtce	824.8		5750				
	EJ	18		126.5				
Co Co[328]	LDC Energy							
Non-OPEC	H3 &		4.2		4.6			
	L3		3.0		3.3			
	H5 &		3.4		4.0			
	L5		2.8		3.4			
OPEC	H3 &		6.6		5.7			
	L3		4.5		3.9			

Constrained—High growth H5—OPEC EJ   27   68
             non-OPEC EJ   88   197
         Total LDC Demand   115   265
Unconstrained—Low growth L4—OPEC EJ   17   31
             non-OPEC EJ   76   151
        Total LDC Demand   93   182

OECD[329] Economic
    (3) Major oil exporters   7.8,8.9,6.2,8.0,
    (4) Other LDC   5.9,4.5,5.6,6.1,
    Total LDC   6.2,5.2,5.7,6.5,

(1) LDC = Less Developed or Developing countries.
(2) LDC without PR China and Korea.
(3) The 13 OPEC Countries and Bahrain, Oman, Brunei, and Trinidad & Tobago.
(4) Including Turkey and Yugoslavia.

Sources: As referenced

estimates for all the developing countries for the year 2000 there is a formidable range, from 78 EJ (D7 Case of WAES), to 115 EJ (CoCo Energy Constrained, High Growth Scenario H5 including 15 EJ of wood fuel). In oil terms the spread is some 3,500 Mtoe, or 70 Mbdoe. It must, however, be remembered that the high figures are from the draft report which has not yet been reconciled with the resources studies. When one examines the figure for the later years there is less difference, somewhat surprisingly, for they are 296–342 EJ for 2000 and 455 EJ for 2025 (Frisch).

It is probably in the developing countries outside OPEC, but including China and the centrally planned Asian countries, that there might be a 'revolution' in the application of 'soft' or 'appropriate' technology. The conventional thinking in the developed countries is conditioned by the inertia caused by existing plant and investment and by the obvious material benefits which have accrued from the centralization of supply points and systems. Even so, growing attention is being paid in many of these countries to the advantages of small units of energy and work. If any one of the small simple methods were to have a domestic base, or even an export potential obvious to a bold manufacturer in a developed country, the savings of mass production might become clear. Then there might be a very rapid flooding of the developing world with energy supply gadgets, as rapid as the spread of transistor radios or mini-skirts. Some would argue that the drive within developing countries is towards the outward signs of material wealth and comfort of the developed countries, and that some of the big energy systems are an essential status symbol.

This has been true in the past. Good examples are the insistence on local oil refineries, huge hydro-electric schemes, national airlines and, more recently, nuclear power plants. Many of these schemes have had little effect on the life of the people, and energy shortages are now affecting them. Kerosine for domestic light and heating has risen in price with gasoline. In some countries, like Indonesia, the firewood 'crisis' is causing great concern, as it is leading to deforestation. Small appliances could therefore be of very great value, e.g. solar stoves to replace wood and dung fires, solar panels to provide hot water in clinics, hospitals and small factories, windmills to pump water, small turbines to utilize water power from small dams, streams or mill-races from rivers small or large. Some of these appliances, such as water-mills and agricultural grinding machines, may be adaptations of earlier models which went out of use with the coming of the Industrial Revolution in the developed countries. If however, as Waslav Micuta, formerly with the UN in Geneva says, if they are marketed, not as old or even 'appropriate', but as the latest applications of modern science and had a market already in a developed country, they might become as fashionable as uneconomic oil refineries. The speed of distribution would be increased by local entrepreneurs manufacturing locally. This would have a snowball effect. Small communities would thereby give themselves that small impulse to improve their own conditions themselves; the local coal-mining and hydro-electricity schemes in China were mentioned earlier (also[517]). In the same way, if the cost of photovoltaic solar cells could be reduced by 200 times, conversion of both sunshine and sunlight directly to electricity would become accessible to others than astronauts. Modular systems with aggregations from a few watts to gigawatts, according to need, would be a boon to many peoples. Bioconversion systems must also be considered. As mentioned earlier, shortage of firewood, leading to deforestation in some countries, does not give much hope in those places for 'energy plantations', even though they are very fertile. On the other hand even the most primitive methods of making gas from bio-wastes can be of great help, as so well shown by a 1977 paper[402] from Nepal.

For any of these 'soft technology' systems, market penetration would be very patchy, as Professor Peter Auer stressed.[403] Nevertheless, many studies, particularly of economic theorists, may be perhaps too cautious in the amount of energy which could be supplied by the year 2000 by such sources, particularly in developing countries.

There is no doubt, however, that in most developing countries the main economic and industrial growth will be through a fast growth from a low base of heavy and light industry. Even the UN, which is sympathetic to grass roots aid, says in the study, *The Future of the World Economy*:[319]

> an accent on heavy industry is essential for industrialization and economic growth on a broad regional, though not necessarily on the small country basis. This opens up vast horizons for cooperation and specialization between developing countries.

Unfortunately it seems very doubtful on present knowledge that such cooperation will be manifest in many places. Many of the developing countries, particularly in Africa, are so young that strong nationalism seems a prerequisite for survival. Every country is very differently endowed with resources so that development will be very varied, yet complementary actions, though valuable, will be difficult to arrange. However, the UN study suggests that the prices of natural resources will more than double between 1970 and 2000, with agricultural prices rising 14% on average and the prices of manufactured goods declining by almost 7%. These changes would, however, be slow and the benefits would go to the oil and mineral exporters, rather than to the non-oil, non-mineral producers.

Frisch[331] was bold enough to look to the very long term of 2075, based on his study published by IIASA in May, 1975. This set a limiting level of annual energy consumption of 10 tce/capita at about 2050; world population was levelled at 10 billion, split as to 1.7 billion in the industrialized countries of the West and East, and 8.3 billion in the Third World including China. Table 70 sets out some of the assumptions and illustrates the dramatic shift in consumption which is shown in the first three-quarters of the twenty-first century.

This is only an exercise but some 6–7 Ttce (teratonnes of coal equivalent, 10^{12} tce) would be needed to satisfy the energy demands so calculated over the next hundred years. In an exercise described earlier, it was suggested that the world's conventional fossil fuel reserves were possibly of the order of the following, in tonnes of coal equivalent:

	Coal	Oil	Natural Gas
Proved and possible reserves	0.6 Ttce	0.5 Ttce	0.8 Ttce
Resource Base	30 Ttce	2 Ttce	1.36 Ttce

If the energy need over the next hundred years is indeed of the order of 6 to 7 Tt, then the resource base of the world's coal obviously has to be dug into deeply, unless other energy sources are rapidly exploited.

To return however to more practical matters, there is no doubt but that the big problem for the medium term is the growing scarcity of the hydrocarbons, whose resource bases, indicated above, are small in comparison to future needs.

Table 70
A very long term exercise in world energy consumption
1950–2075

	1950		2000		2075	
AVERAGE ANNUAL GROWTH						
Industrialized countries	4.3%		approx 0%			
Third world *	7.9%		2.8%			
Total world	**5.0%**		**2.5%**			
CONSUMPTION per capita						
Industrialized countries	2.7 tce		13.2 tce		12 tce	
Third world *	0.1 tce		1.7 tce		8 tce	
Total world	**1.0 tce**		**4.3 tce**		**8.6 tce**	
TOTAL ENERGY CONSUMPTION	Gtce	EJ	Gtce	EJ	Gtce	EJ
Industrialized countries	2.4	69	19.4	560	20	580
Third world *	0.2	5.8	8.1	230	66	1,907
Total world	**2.6**	**58**	**27.5**	**795**	**86**	**1,485**

* Third world includes China and Korea.

Source: Frisch[331]

SUMMARY

There are obvious wide divergences in opinions in recent studies as to the bases on which future demand should be discussed. Accepting the many problems, it would seem that many expect the demand for energy in the year 2000 to be of the order of 12.5 Gtoe (550 EJ), and 20 Gtoe (say 900 EJ) in 2020. From the 1976 consumption of 6.4 Gtoe, the growth rate to 2000 would only be some 2.7% p.a. This is considered to be low unless there is a very marked change in the impact of conservation, and oil production is restrained.

In 1976 about 53% of the total world energy consumption was in the USA, Western Europe and Japan. The USA has not had any obvious energy policy for some years. Even in December 1979 Congress had not cleared the Carter Administration Plan which was presented in July 1979. The USA was therefore still continuing to increase its petroleum imports and if this policy continues it could have serious effects on the supplies available to the rest of the world. Nevertheless it would seem that a rough average of the authorities quoted would be for a total USA energy consumption in 2000 of about 3 Gtoe and for 2020 about 4.4 Gtoe.

In Western Europe the full range of estimates for 2000 is about 2.0 to 2.6 Gtoe. The only estimate quoted for Japan envisages a total energy consumption in 2000 of about 1.125 Gtoe. This implies an annual average growth to 2000 from the median of the 1985 estimates at about 5.25% p.a. If this rate continued through to 2020, Japan's total energy demand then would be about 3 Gtoe.

The aggregates of these three major industrial areas would then be about 6 to 6.5 Gtoe in 2000 and about 10 Gtoe in 2020, which would infer that they would retain their share of world energy consumption, but this is unlikely.

Opinions as to the probable future demand of the developing countries are very varied indeed, except that all analysts seem convinced that their growth rate will be greater than that of the industrial countries. The mean of the two CoCo demand scenarios noted in figure 27 is about 2.3 Gtoe in 2000 and 5 Gtoe in 2020. These can only be a very rough guide because in taking a mean between two scenarios the internal consistencies are ignored. Further speculation on demand must await the consideration of the USSR and China and the possible supply potential to be discussed. This rough, almost 'back-of-the-envelope' summary emphasises the many possible variations which must be considered.

9
Supply and demand

The interaction of supply and demand of the classical economic models on the concepts of Malthus that agricultural land and other natural resources are scarce, or of Ricardo that resources are used in the order of declining quality, and so to the 'law of diminishing returns' as population grows, is no longer tenable.[28] Since the capital stock has increased faster than population, increased capital per worker acts contrary to diminishing return per worker. Natural resources are not fixed on volume. Increased knowledge has brought forward more and different resources, and not necessarily in the order of declining quality which hindsight might set. 'Pervasive changes in knowledge, technology, social arrangements, economic institutions, market sizes and the overall sociophysical environmental context, have occurred in the advanced nations. The classical economists omitted socio-technical change from their economic growth models or underrated its significance or powers'.[28]

One basic technical change is that the contained molecules and the naturally occurring molecules of all natural resources are more important than their gross volumes, mass or surface area. The homily that 'all barrels of oil are not equal' is one way of stressing that it is the quality or characteristics of an energy source tht matters. Hence it is real costs of resource products rather than gross volumes that measure availability. In the past, availability on this criterion has increased with time in the advanced nations. An example can be taken from an economic study of the USA from 1870.[226] This cites the real cost or net product cost of fuels in terms of man-day input per unit of net output. The study calculates that this had fallen by 1957 to about a fifth of what it was in 1900. The difficulty is to decide whether the factors producing this phenomenon will continue and will also apply to nations in earlier stages of development. A continuing factor is that it is in the very nature of man to try to improve his living standards, in the broadest sense, intellectual as well as material. Even if the accepted standards change there is no reason to suppose that the next generation will not inherit a more productive world. The conversion of petroleum to synthetic fibres has enormously increased the world's capacity to clothe people. It is no longer necessary to dedicate millions of square kilometres to cotton or flax-growing or sheep-grazing. More examples of technological change like this one can be expected and affect the relationship between energy input and gross national product.

In the twenty-five years before October 1973 there was some fluctuation in energy input but broadly there was cheap energy predominantly petroleum, and many economic concepts grew in what was only a temporary normality. It is important therefore to look at the post-war period and pick out some of the features which may improve an understanding of that environment.

The present time is an accelerated transition from the cheap to the expensive energy era, from apparent abundance to apparent scarcity and the results of that acceleration are unclear. This is true on both supply and demand. Energy conservation, both in elimination of waste and reduced use, helped by two mild winters in the northern hemisphere, had reduced demands so that early 1975 saw a 'surplus', a cutback of production through lack of market. This is a position of which many more people were aware because of the October 1973 events than had been aware of equally important earlier situations.

256

1950-1973

Conditions in the energy field were remarkably stable in the 1950s and 1960s. Petroleum took over from coal as the almost universally major source. Energy costs declined. Energy use increased faster than GNP. There was growing dependence on the Middle East for oil imports. The planners in the oil industry and some economists appreciated that such conditions would change. Petroleum geologists were urged to think of themselves as energy geologists and think more of economics. Oil companies diversified into other energy industries, mainly coal and nuclear. In Europe the Suez Crises of 1956 and the Six Day War of 1967 caused temporary upsets but were soon forgotten as the flexibility and spare capacity of the international oil industry quickly overcame the problems.

The international oil industry before 1940 consisted of a handful of major companies who could use the oilfields as storage tanks to absorb changes in demand and avoid wild fluctuations of supply and price. After 1950, more American companies became international and although in competition, often fierce, with the original major companies they added to the internationalism which gave the industry strength and flexibility.

The raising of prices in the early 1970s and growing call by the oil producers for equity participation, started to change the climate in which the industry worked but still many thought that the transition would be smooth and gradual. A number of steps were taken in anticipation of change. There was increased exploration outside the Middle East. A harder look, but little action, was taken at alternative sources. Plans were laid for increased liquefied natural gas ocean shipments from North Africa and Nigeria. There were reappraisals of the rundown of indigenous coal production and of the building rate of nuclear power stations. However, the attacks by the environmentalists on many facets of the energy industries were also reaching greater heights and much mental energy was being forced into these problems. Naturally, the increased costs of environmental conservation measures added to growing concern about the inflationary effect in the world economy of rising petroleum prices.

October 1973-March 1974

In October 1973 at the outbreak of renewed war with Israel, the oil ministers of the Arab oil producing countries announced reduced production levels, a greater than average reduction to 'unfriendly' countries and a boycott of 'hostile' countries. From October to December all OPEC countries announced price increases. In the Persian Gulf this meant an increase from about US $3 to US $11.65 per barrel in 'posted prices' on which royalty and taxation dues were calculated by the governments for billing the operating companies. Effective sale prices rose from about US $2 to US $8, with more on some spot sales. Prices have increased since. These actions led to much scrambling, to unilateral deals, to many arguments, often acrimonious, and a sudden, world-wide appreciation that 'nothing would be the same again'. The transition from cheap to expensive energy came in six months, not in six years. The time for the development of alternative sources to set a real and not an imaginary, theoretical ceiling on OPEC prices had moved from the 1990s to the 1980s. This task seems beyond the capability of private industry on its own. It seems inevitable that governments must give adequate incentives to research and develop new sources to substitute more efficient and economic alternatives for inefficient fuels, to conserve more energy and set environmental standards which are reasonable and tenable. The role of governments was forced by the 'oil crisis' to change to greater involvement in the problems of pricing, of international financing of energy, and of the disparities of high prices on developing countries because of the shift in monetary wealth to the oil producers.

A most important feature of the October 1973 events was that antagonism to the Jews gave to the Arab core of OPEC a cohesiveness which was strengthened by the success of their use of oil as a political weapon. This success also ensured the adherence to OPEC of the non-Arab members who saw and felt immediate benefit. This cohesiveness, lacking for all the earlier years of the OPEC organization, was in glaring contrast to the scattered aims and actions of the importing countries. Only by early 1975 did it appear that the consuming-industrial countries were moving from the chaos of late 1973, towards rational thinking of the problem as a world problem. The long-term world problem of rising demand pressing on shortage of supply is still there. This will be solved in a way which will depend on the solution in the short-term of the problems raised in October 1973. This was that the supply of the largest, most transportable, convenient and pervasive source of energy suddenly came under monopoly control with a massive shift of wealth by the action of that monopoly.

The flexibility of the international oil industry and its determination to honour its contracts with its customers, lessened the impact of the reduction of supplies. Yet the structure of the industry effectively changed in October 1973. However, contracts still held and the whole of the physical structure of oilfields, pipelines, refineries, tankers, distribution points and storage down to gasoline filling stations and the enormous investment involved in these facilities still functioned.

1974-1979

The investment in the infrastructure of the oil industry is not only money but also men. It is their knowledge and expertise which makes possible the functioning of an extremely complicated, intermeshing world-wide mechanism with great variations in conditions in every phase of operations on both the supply and marketing sides. The role of the major international integrated oil companies has changed, with much of the oil supply, of which they had virtually unrestricted control, now out of their hands, and governments in consuming countries becoming more involved in the industry. The international oil companies may be taking up a less clear-cut role as agents but this will seem widely different in different parts of the world. The skilled operators are still within the industry.

These facts are undoubtedly appreciated by most producing countries. Hence the eagerness of some to integrate their own operations down the stream to retail sales and to gain world-wide operating experience. The influence of these national oil companies and their reactions with the national oil companies in some consuming countries are difficult to assess and predict. One very major uncertainty lies in the psychology of these operators. If they have the caution of the civil servant of the western world, of the accountant or standard banker, they will favour long-term contracts which will petrify trade and lose the flexibility of entrepreneurial activity and competition. It seems doubtful if this will be the dominant theme, for the merchants of the Persian Gulf have been in trade for centuries. The various nations joined together in OPEC all have their own basic philosophies which vary between each other and are very different from those of the consuming countries. Indeed the components of OPEC are more diverse than those of the Economic Community of Europe and yet they showed greater cohesion from October 1973 to 1975. In March 1975 two events in the Arab world occurred which seemed then might affect at least the OAPEC core of OPEC. The step-by-step policy of the US Secretary of State to reduce the Arab-Israeli confrontation seemed to fail and King Faisal ibn Saud, who was said to have a moderating and strong influence, was assassinated.

Walter Levy in the 1974 July issue of *Foreign Affairs* noted the main elements in the attitudes of the oil producers as being:

— desire for rapid and consistent economic development to ensure a viable and secure economy when the importance of petroleum declines as their reserves decline,
— the price rise in 1973/74 was justified by the high cost of alternatives and the need to encourage research and development for them,
— high prices encourage conservation and optimum use, and highest value use should be the basis for oil prices,
— high prices compensate for past low crude prices and the past loss of benefits due to export of crude rather than upgraded, higher priced, refined products and, also, past losses through the non-utilization of flared gas,
— the rich developed countries have a responsibility to aid the poor, less developed countries, who have been hurt by higher prices.

These points might well be considered the results of a rationalization, but undoubtedly these are some, if not all, the thoughts guiding past and future actions. However, they assume different priorities in different countries. Some producers adopt belligerent attitudes and others more reasonable ones. Nigeria, for instance, took over British Petroleum's main assets in Nigeria in August 1979 because of their interpretation of the attitude of the British Governmen to the 'rightful' aspiration of the majority of the people of Zimbabwe (Rhodesia). Saudi Arabia showed great moderation and understanding of the need for a stable world economy from 1973 to 1979, though it was said in November 1979 that they were somewhat disillusioned at the slow response of the industrialized countries.

Table 71
Oil exporters surpluses in 1975 prices, 1973—1980
(US $billions)

	1973	1974	1975	1976	1977	1978	1979	1980
A.								
Exports	58	140	116	139	140	119	150	166
Cumulative current surplus	1	69	100	137	159	148	167	189
B.								
OPEC Current Account								
Oil revenues (net)			97	119	131	124	173	
Non-oil exports			6	7	9	9	11	
TOTAL Exports			103	126	140	133	185	
Current account surplus			30	35	27	5	44	

Sources: A. *Financial Times*, 30 Aug. 1979, London,
B. *Financial Times*, 20 Sept. 1979, London.

The monies which have accrued to the oil producing countries since 1973 are by no means always very clear to the layman because of the bewildering way in which the data are displayed to him. The problem is illustrated by the two sets of figures in table 71, and the more detailed figures given in table 72. These alternative presentations are given to show yet another example, this time in the financial field, of the reason why very great care must be exercised when confronted with data in the energy field.

Table 72 sets out OPEC oil production, exports (including products where applicable) and government revenues for 1972, 1974, 1977 and 1978, and there are a number of points to note when examining the figures:
- the plateaux of total production and exports with a decline in 1978;
- the dramatic jump in total revenues, 1972–1974, meaning a shift of over $75 billion from the consuming countries, mainly the industrialized, to OPEC;
- the increase of over $38 billion of 1977 over 1974, though exports were equal, due to the rise in the average take per barrel[406] from $9.31 in January 1974 to approximately $12.70 at end 1977, with a two tiered structure of $11.57/$12.16 from January to June 1977;
- the dominance of Saudi Arabia;
- the second place which Iran lost early in 1979, as its production dropped below 4 mbd;
- the different production and export patterns over the years of the individual countries, particularly the steady growth achieved by Iraq, which was the most individualistic in its pricing and sales policies to maintain that growth;
- the decline in most countries, 1974–1977, because of slack demand;
- the decline in production in Kuwait and Venezuela;
- the decline in exports in Venezuela as its domestic demand increased.

OPEC countries have marked individual needs and though over the period 1973–1978 there was a cohesion which some observers had not expected, some of the strains of cohesion were showing in the disparities in prices and policies which Sheikh Yamani of Saudi Arabia called chaotic in October 1979. OPEC consists of the small producers like Ecuador and Gabon who tag along and then three very different groups.[228] Indonesia and Nigeria have large populations, in total 205 million people, with concomitant needs for funds to develop their countries. The four countries of Iran, Iraq, Algeria and Venezuela have only 70 million people in total but they have strong, though different, drives for development and considerable funds. Then, thirdly, the five countries of Saudi Arabia, Kuwait, the United Arab Emirates, Qatar and, though politically different in the 1970s, Libya, have a total population of 12 million but 53% of OPEC 1978 exports and revenues, totalling some $64 billion. These are the so-called 'low absorbers'; the countries considered in 1974 as the ones whose 'petrodollars' would be the least absorbed domestically and expected to flood the world monetary system and create chaos. This did not happen. The main reasons were:
- the corrective forces were stronger and the monetary system more flexible than expected;
- the high oil prices increased inflation and deepened the existing sluggish world economy by reducing demand;
- new supplies of non-OPEC oil, from particularly Alaska, the North Sea and Mexico, reduced the call on OPEC oil by about 4mbd;

Table 72

OPEC oil production, exports[1] & revenues — 1972, 1974, 1977, 1978

	Production mbd				Exports mbd				Revenues billion US dollars			
	1972	1974	1977	1978	1972	1974	1977	1978	1972	1974	1977	1978
Saudi Arabia[2]	6.01	8.5	9.2	8.3	6.03	8.5	9.1	8.1	2.8	22.6	42.4	35.8
Iran	5.05	6.0	5.7	5.2	4.82	5.7	5.0	4.6	2.4	17.5	21.3	20.5
Iraq	1.46	2.0	2.3	2.6	1.40	1.8	2.3	2.4	0.7	5.7	9.6	9.8
Kuwait[2]	3.28	2.5	2.0	2.1	3.22	2.4	2.0	2.1	1.4	7.0	8.9	9.2
Libya	2.24	1.5	2.1	2.0	2.22	1.5	2.0	1.9	1.6	6.0	8.9	8.6
Nigeria	1.82	2.3	2.1	1.9	1.76	2.2	2.0	1.8	1.1	8.9	9.6	8.2
UAE[3]	1.20	1.7	2.0	1.8	1.05	1.7	2.0	1.8	0.5	5.5	9.0	8.0
Indonesia	1.08	1.4	1.7	1.6	0.95	1.2	1.5	1.5	0.5	3.3	5.7	5.6
Venezuela	3.30	3.0	2.2	2.2	3.07	2.8	1.9	1.6	1.9	8.7	6.1	5.6
Algeria	1.06	1.0	1.1	1.2	1.01	0.9	1.1	1.1	0.6	3.7	4.3	5.0
Qatar	0.48	0.5	0.4	0.5	0.48	0.5	0.4	0.5	0.3	1.6	2.0	2.0
Gabon	0.12	0.2	0.2	0.2	0.11	0.2	0.2	0.2	0.02	—	0.6	0.5
Ecuador	0.08	0.2	0.2	0.2	0.07	0.1	0.1	0.1	0.03	—	0.5	0.4
TOTALS	28.18	30.7	31.2	29.9	26.19	29.6	29.6	27.7	13.95	90.5	128.9	119.2

(1) including products where applicable, (2) including half of Neutral Zone, (3) including Abu Dhabi, Sharjah and Dubai

Sources: 1972 — 'OPEC: oil Report', Dec. 1977, *Petroleum Economist*, London.
1974—1978 — *Petr. Economist*, Vol. Xlvi, No. 6, June 1979.

- energy conservation within the industrialized, oil consuming countries began, slowly, to reduce demand;
- the OPEC rapidly increased their imports of goods and services in 1975–1977, but then appreciated that this frenzied attempt to westernize had caused ridiculous wastage and had swamped their absorptive capacity. The example of Iran in early 1979 having fallen into political and economic revolution, at least partially due to over-rapid 'westernization', added further reason for caution and moderation in industrial expansion;
- that caution was said to be being expressed through greater diversification into longer-term and less liquid investments, yet this was not very obvious from a summary[518] which tabulated the investible OPEC surpluses in the years 1976, 1977, and 1978 as being in total $37.2 b, $33.5 b, and $13 b, deployed in those years, respectively, in USA 32%, 27% and 10%, UK 12%, 12%, and -10%, in other countries 33%, 59%, and 65%, and in international organizations 5%, 1%, and 1%. This data does stress rather the changes in investment policy, but a much deeper study would be needed to learn more than that simple fact.

Table 73 shows how these shifts have affected the world current account balances in some selected areas.

Table 73
World current account balances, 1973–1978
(goods, services and private transfers in billion US dollars)

	1973	1974	1975	1976	1977	1978
OECD Countries[1]	19	−15	16	−4	−8	22
USA	9	7	21	8	−13	−13
Japan	–	−5	−1	4	11	17
UK	−1	−8	−3	−1	3	4
OPEC Countries	7	68	35	40	33	9
Non-OPEC developing countries	−11	−35	−45	−31	−28	−33
Oil importers[2]	−9	−30	−38	−26	−25	−29

(1) excluding Greece and Turkey,
(2) including Greece, Turkey, Romania and Yugoslavia.

Source: 'Energy, the next twenty years', Table 4−1, p. 171, a Report sponsored by the Ford Foundation, administered by Resources for the Future, Ballinger, 1979, Cambridge, Mass., USA.

Some points to be noted from this table are:
- the wave in the OECD total caused by the strong US position in 1975;
- the remarkable resilience of Japan, despite its great dependence on oil imports;
- the UK position due to its own oil and gas productions;
- the declining surpluses of OPEC, which are very different from the 1974 forecasts, but 1979 was expected[518] to see increased surpluses, with higher prices, perhaps to a total of $44 b. Wide variations in prices actually gained, particularly on the 'spot market' preclude accuracy. However, one estimate was that Middle East and North African OPEC members might win 1979 surpluses of Saudi Arabia $3.2 b, Qatar $1 b, Kuwait $12.3 b,

UAE $5 b+, Iran $2 b, Iraq $9.3 b, Libya $2.5 b, Algeria minus $0.7 b, compared with a 1978 deficit of $2.3 b. This list again stresses the differences between members.

Currency fluctuations and inflation have dramatically distorted the true cost of oil to different countries. *Petroleum Intelligence Weekly* illustrated the point (*Financial Times*, 21 Nov. 1979, London), for the USA and UK, using 1972 dollars for the true cost of Arabian Light Crude oil, as follows:-

	1972	1975	1976	1978	1979 (3rd Qtr.)
Official price	1.90	10.72	11.51	12.70	18.00
UK	1.90	7.37	8.50	6.45	7.10
USA	1.90	7.30	7.49	7.08	8.97

Such variations are again matters which must be appreciated when OPEC aid to the non-oil developing countries was a matter of considerable discussion in the late 1970s and a factor in the spread of petrodollars. In actual disbursements these were estimated by OECD[519] as being in 1973 $1.36 b, in 1974 $3.4 b, averaging in 1975 and 1976 $5.4 b, in 1977 $5.76 b, but falling in 1978 to $3.76 b because the Arab Gulf States stopped some $2 b aid to Egypt after the Egypt/Israeli Treaty, and Iran's main aid programme stopped with the Revolution. The bulk of OPEC aid went to Arab neighbours; in 1976 about 71% went to Egypt, Syria, Jordan, India and Pakistan; aid to other developing countries was more directly aimed at specific projects but with less in practice than in discussion.

Iran

Iran was the first Middle East country to rebel in 1951 under Premier Mossadegh against the reputedly imperialistic control of the major oil company system and nationalized its oil industry. As the new system developed, 1954–1978, Iran under the Shah moved quickly towards a conscious energy policy and the deployment of its petroleum revenues to build up its economy and, increasingly with time, its power as a nation. Iran has a relatively large population of 30 million. It is a country of diverse topography, climate and agricultural potential and has other natural resources than oil, namely iron ore, copper ores, hydroelectricity potential and a coal production of over 1 Mt. The Shah encouraged the development of the National Iranian Oil Company (NIOC) into an international oil company, with joint ventures with other companies in exploration in the North Sea, in refining in South Africa, and even marketing in the USA was being discussed in 1978. Natural gas exports to the USSR began in 1970 with the completion of the pipeline IGAT I, and other schemes were under discussion, such as a second gas export line, IGAT II through the USSR to Western Europe, and LNG export schemes from a floating LNG liquefaction plant in the Gulf. A substantial petrochemical industry, started in 1961, was being built up and new investment of $8 b planned for the 1978–1983 period. Hydro-power schemes were developed and construction started on four nuclear power plants. Industry was encouraged to manufacture locally and so reduce the volume of imported manufactured goods, like cement, cars, shoes and sugar. An engineering capability and a steel industry were established. Iran's economy was set on a high growth trend; in 1962–1967 growth was almost 10% p.a.; in 1967–1972 growth was almost 13% p.a.[231]

The Shah appreciated that petroleum is a scarce and limited resource. He aimed to produce and market to provide a maximum income with which to build a strong modern, mixed agricultural and industrial economy by 2000. His oil pricing policy was aimed at the minimum of the cost of alternative fuels. At the same time he strongly opposed the profligate use of petroleum and favoured conservation in the broad sense, the maximum use for its chemical characteristics and therefore the substitution of alternatives for primary energy generation, both within Iran and elsewhere. In 1974 Iran's oil revenues jumped to over $17 b. An ambitious programme was begun of industrial construction, aid to developing countries, and the building up of a powerful military and naval force to offset the withdrawal of Britain's forces to west of Suez.

The Shah, however, became increasingly autocratic. He did not appreciate that there was growing resistance among the people to being dragged so fast towards a materialistic, industrialized, western state. Discontent overflowed late in 1978, oil production was reduced, (exports fell from 5.6 mbd to 1.5 mbd in October); the Shah retired into exile in January, 1979; an Islamic Republic was declared in February and the clergy, working through revolutionary councils, assumed real control. Virtually all expatriates left the country; construction stopped; oil exports had stopped at end of December and production for a time was below domestic need. Oil exports were resumed early in March. Gas supplies to the USSR were restarted in early April after a three months suspension. A target of 2.8 mbd of exports was set for six months, but there was confusion as to quantities available and as to price as policies and practices were somewhat disorderly. Continuing political problems caused fluctuations in production, which was reported as down to 3 mbd, with exports at 2 mbd in early September 1979.

In November 1979 it was far from clear how Iran would weather the storm into which its political, social and economic life had been thrown. It was clear however that a major contribution to world oil trade had been reduced and that the example would affect other producers to moderate their expansion and review their national interests. It seemed more than likely that a production of 4 mbd, 200 Mta, would satisfy Iran's reduced growth for some time, perhaps years, and that therefore world supply estimates for both oil and gas would require revision downwards.

The revolution in Iran shattered the complacency which appeared to be growing early in 1978. Reduced demand for oil had again created an apparent surplus; the rapid trend to industrialization, even in the 'low absorbers', but markedly in Iran, was recycling petrodollars, indeed making the oil-producers into borrowers; the world monetary system had not crashed; minor disturbances in the Middle East and North Africa had not stopped the flow of oil; the Israeli/Egypt pact might bring lasting peace. The overthrow of the Shah created first concern as to the stability of the OAPEC members, particularly Saudi Arabia, and so security of supplies. The other producers, led by Saudi Arabia, however, increased their supplies to help to overcome the 5 mbd short fall, of which the consumers themselves covered about 2 mbd by drawing on stocks and cutting demand. Nevertheless Saudi Arabia, in particular, stressed that this was a temporary measure and production would be reduced when Iran's exports recovered. This crystallized the appreciation of the consuming countries that production control by OPEC, and indeed by other producers, like Mexico, would be an increasingly dominant factor in world oil availability, with all its implications, not least to the USA.[456] Prices were raised in March

1979 to $14.55 for the marker crude, Saudi Arabian Light, with the new clause, granting members the right to levy surcharges at their discretion. This clause, together with the higher prices on the Rotterdam spot market, caused by bids for cargoes from temporarily embarrassed customers, caused chaos, as mentioned earlier.

1980 onwards

An OPEC meeting in December 1979 was expected to improve the general immediate situation, but the longer term was very uncertain. The 1979 events had again demonstrated the power of OPEC, and the cutting of supplies to Israel and South Africa by the Khomeini regime in Iran had restated the political side of the weapon. OPEC seemed to have a number of options open. Should they fix a production schedule for a stated period and let demand determine price? Should they fix a price schedule and manipulate production as necessary to maintain that schedule over a shorter or longer period? Should they fix a price and production schedule and raise both if demand would carry them? Should they separately set and collectively approve revenue schedules for members which would be kept by individual members playing price and quantity as they decided? These were only some of the possibilities. Hence the early 1980s will be a period of continued uncertainty, except that supply will be lower than possible demand, and a look at Venezuela, a founder member of OPEC and in very different circumstances from the Arab members, will be interesting.

Venezuela

Venezuela began producing oil at the beginning of the century and completed a slow process of nationalization of the industry on 1 January 1976. A cautious conservationist policy for 15 years had restricted production and oil revenues, but also had led to a decline in production from a 1970 peak of 3.7 mbd, (193 Mta @ 52.1) to 2.2 mbd, 115 Mta, in 1978. Renewed exploration effort since 1976 had, however, found new oil and particularly gas, and particularly offshore, of which development was expected to help to reverse the decline. In mid-1979 there were said officially[521] to be 2.6 Gt (18 bb @ 7 b/t) of proved oil reserves, 1.7 Gt of probable reserves in old areas, over 4 Gt possible reserves in new areas and at least 10 Gt in the heavy oils of the Orinoco basin, making a total of 18.6 Gt or 136 bb of possible recoverable oil. In 1976 Venezuela was financially very sound with an external debt of only $3 b, and so sixth in comparative indebtedness in Latin America, with oil revenues in 1978 at $5.6 b. Domestic demand for oil products was only 12 Mt in 1977, so that even though this could be expected to rise with domestic development, the export availability should remain considerable. Higher prices in 1979 would raise 1979 revenues and as most sales contracts expired at end 1979, it was expected that they would be re-negotiated on higher terms. Ambitious plans for non-oil development were formulated under President Carlos Andres Perez. In his term, 1974–1979, gross domestic product grew at an average 9% p.a., but his successor, President Louis Herrera Campins, declared[522] in July 1979 plans to stabilize the economy with fiscal discipline, cut imports, hold GDP growth to 5% in 1979 (compared with 7.3% in 1978), and 6% in 1980, hold unemployment to 5%, inflation to 10–12%, and halve the current account deficit; expansion would restart in 1981 but in a more controlled manner. Here, then, is another country intending to move forward into the 1980s at a much slower pace than was anticipated, even in 1977.

Western Europe

The discovery of natural gas at Groningen, Netherlands, in 1959 had a considerable impact on Europe's energy prospects, for, in addition to supplying gas eventually as far as Italy, the find led to the discovery of gas in the southern North Sea in 1965 and the discovery of oil in the northern North Sea in 1970. The rapid indication of very large resources altered dramatically the energy supply and demand pattern of the UK, Norway and so Western Europe's total pattern. The extent to which that oil will satisfy demand, for how long and from what date depends on the size of the reserves proved and on actions of governments.

Professor Odell of Erasmus University, Rotterdam, is the greatest optimist as to the size of reserves, as mentioned in chapter 4 regarding North Sea gas. In December 1974 he published the results from a simulation model exercise.[247] He had calculated an expected rate of discovery and had allowed for increases by development extensions by time. He gave a range of cumulative recoverable oil as 11–19 Gt, from which a mean curve of production was estimated to peak about 800 Mta in 1990. Assuming a target of 75% of his own forecast of Western Europe's oil consumption, the annual production curve would cross the demand curve at 400 Mt in 1982 and exceed it until, on the down curve, it crossed demand again at about 600 Mt in 1995. Odell argued that if the production which he forecast was controlled, between 1985 and 1995, to conform to Western Europe's demand, then that demand could be covered into the first quarter of the twenty first century. Whilst admitting the difficulties of such production control, he said that this would 'see the whole of Western Europe through into the post-oil age without any further undue dependence on supplies of foreign oil'. The actual results and thinking therefrom to 1979 had not lived up to these hopes. In the UK sector the UK Department of Energy in its annual reports increased the proved reserves from existing licenses rapidly from 895 Mt in 1974 to 1,360 Mt in 1976 but then gently to 1,397 Mt in its 1979 report. The estimates of probables fell almost 400 Mt from 1977 to 1979 and the total possible recoverable reserves in existing fields from 3,200 Mt in 1978 to 2,511 Mt in June 1975 to the end of 1978. The downward revision after such a short time of even the conservative estimates of total probable and possibles in existing licences, despite the increase in number of those licences, is an important indication of future expectations. In the 1979 report the Department estimated the reserves in future discoveries under existing licences, (including the 1979 Sixth Round), as 350–800 Mt, and at 550–1,000 Mt for the remainder of the UK Continental Shelf. The total remaining recoverable oil on the UK Continental Shelf is thus estimated to be 2,300–4,300 Mt, with 106 Mt to be added to give the original recoverable oil in place. The discoveries in Norwegian waters must be added to the British reserves, but even if they were equal, which few expect, the totals are much less than Professor Odell suggested before exploitation began and the considerable exploration done subsequently. It would seem highly unlikely if the late 1980s to early 1990s could see total European offshore production more than of the order of 200 Mt, and this is guessing into the more distant unknowns of the deeper shelf and further north in Norwegian waters.

An important aspect of productions from a number of fields is that the production pattern is not that of a single field. This principle was illustrated in the chapter on production by stressing that smooth curves often hide individual component vagaries. It was pointed out in November 1973[250] that, with discoveries spread over the years and the time lag from discovery to commissioning of production, it was more likely that the production curve would be a wavy plateau rather than the smooth bell-shaped curve which so many favour. The illustrating exercise on this point is still valid. Taking proven reserves at 1.6 Gt and current exploitation plans, it would be reasonable to peak production at about 100 Mt in 1980 but still be producing 10 Mt in the year 2000. If a further 1.6 Gt of reserves were proved in six equal annual increments from 1974, with a five-year lag to commissioning, then the peak of the total production would be delayed but higher, and the decline of the first 1.6 Gt offset by the new field productions, giving a peak of over 150 Mt 1986–1992 and some 50 Mt in 2000. If an additional 300 Mt of reserves were proved in the years 1981–1983 and 1985, allowing for greater exploration difficulties, then the rate over 150 Mta would be stretched from 1986 to 1997 and production would be about 100 Mt in 2000. This exercise illustrates that the timing of discoveries and, particularly of initial production, is a major determinant and, because of the production technique from platforms, offshore production increases are more stepwise than many onshore fields. Both these points are often under-estimated and sometimes ignored by tyro forecasters whether or not they are using computers and models.

Government attitudes and fiscal actions, whether concerned with royalties, no-risk participation or taxation in any form, are just as equally important in determining production rates. Other examples have been given in earlier chapters and are no less important concerning North Sea oil as in those examples which included the influence on North Sea gas. The rules under which the first exploration permits were allocated seemed a viable system but an experiment of auctioning by sealed bids on the Canadian pattern was tried which disturbed some operators. Then there was political talk, after the early discoveries seemed to indicate substantial potential, that the UK Government must ensure the maximum benefit to the country. There was much discussion in 1974, particularly after a Government policy paper in July 1974, and it is highly probable that some companies delayed their investment plans. Even after the British Government put down their final plans in April 1975, there was still uncertainty about the role of the proposed British National Oil Corporation (BNOC). Under the Labour Government (October 1974–April 1979), BNOC was given a privileged position with both operating and advisory responsibilities. In three years its staff increased to over 1,000 and had taken an interest in some overseas ventures. BNOC had remarkably rapidly become a major operator with assets, helped by the increased oil prices, estimated in mid-1979 to be about £2.5 billion. The Conservatives in Opposition had complained about BNOC's dual role, and in the early months of office had started to reduce its privileges and take from it its advisory role. However the possibility of it having to sell some assets was changed to raising funds by the forward sale of production and a possible public sale of bonds to encourage private participation. The British state oil enterprises therefore entered the 1980s in a somewhat fluid position.

The beneficial impact on the British economy of exploitation of its indigenous hydrocarbons was undoubted. One estimate[528] in mid-September 1979 was that the net gain in 1978 to the British economy was some £2.7 b from oil (and £2.6 b from gas), which might rise to £5 b in 1980, £10–£13 b in 1985 and £13–£20 b in 1990, if Government refrained from detailed interference with the industry. Nonetheless, the Department of Energy in a submission[529] also in mid-September 1979, revised downwards its 1978 estimates of both indigenous supply and demand in the next twenty years, but leaving a deficit in 2000. Table 74 illustrates this opinion.

On the supply side it was said that coal production could be as low as 80 Mt unless the National Coal Board's Plan could be implemented without any inordinate delays, which for each new project could mean a loss of 5–10 Mt each year. A gloomier view was also taken of oil production, perhaps to half the 150 Mtce of the 1978 estimate, say 880,000 mbd of oil equivalent. The lower demand figures were said to have taken account of effective energy conservation measures.

Table 74
UK energy balance in the year 2000
(million tonnes of coal equivalent)

	1978 Estimate	1979 Estimate
SUPPLY:		
Coal	170	
Nuclear/hydro	95	
Natural gas	50—90	
Indigenous oil	150	
Renewable sources	10	
Total supply	**475—515**	**385—410**
DEMAND	**450—560**	**445—515**

Source: Dept. of Energy, UK[529].

The share of natural gas in satisfying Europe's primary energy needs is illustrated by the increase in the European Economic Community from 16.7% in 1976 to 17.1% in 1978, being equivalent to about 164 Mtoe.[438] The 1978 use of gas varied from 0% to 45% of the energy needs of the various countries; some percentages were 18% in the UK, 45% in the Netherlands, in comparison with 26% in the USA and 26% in the USSR.[436] EEC production was expected to increase in 1979 by 2% but imports by 30% to meet an increase of 7% in demand. Some 85% of Europe's 1977 gas came from domestic sources but by 1985 one estimate[437] was that domestic supplies would only provide about 66%, particularly if the IGAT II pipeline from Iran via the Soviet Union was built. However, the early 1978 political turmoil in Iran cast doubt on when if ever, that line would be built.

Italy is of interest because of imports supplementing its indigenous natural gas. The first gas discoveries in Italy were made between 1944 and 1950 and the industry developed with less influence from the pre-existence of a manufactured gas industry, for Italy has only small coal proved reserves of some 33 Mt.[137] In 1950 natural gas provided 2% of the gross national energy consumption, rose to 8% in 1955, 10% in 1960, fell to below 8% in 1965, but rose again to be about 10% in 1972. Part of this was from imports of LNG. Imports from Libya began at end of 1971 and reached 3 km³ in the winter of 1972/1973, under a twenty-year contract at this level from Esso. A second twenty-year contract for 6 km³/year from the USSR was scheduled to start in early 1975. A third twenty-year contract for import of gas from Holland was made in 1970 for 1974. Finally a fourth contract for piped gas from Algeria was initiated in 1973, scheduled to arrive in 1978/79 and, after four years, reach a level of 11.7 km^3/year. These details are given to explain how Europe is linked with North Africa, Italy with Libya and Algeria, in addition to the UK with Algeria, and how the pipeline network has been developed so that Italy is at the ends of both the Netherlands and USSR export systems. There are from time to time also revivals in a long-conceived scheme for piping Middle East gas to Europe.

Japan

Japan is the third largest oil consuming country after the USA and USSR, and only in 1976 lost to the USA its position as the largest oil importer. Consumption rose from 143 Mt in 1968 to 269 Mt in 1973, fell to 244 Mt in 1975 but rose to 263 Mt in 1978. Indigenous oil production was roughly 0.5 Mta over that period. Japan has also some hydropower, geothermal power and coal. The coal industry suffers from difficult geological conditons and poor labour relations; production declined steadily from almost 45 Mtce in 1969 to 18 Mtce in 1977. Japan's remarkable economic growth from 1950 to 1973 was based on cheap oil, and is especially vulnerable therefore to adequate supply and reasonable price.

The embargo on oil imports by the Arab producers and the jump in oil prices in 1973 struck Japan at a time when it already had monetary and inflation problems.[239,240,241] The impact was traumatic. The embargo was lifted in December 1973, after a hurried visit to the Arab States by Mr. Takeo Miki, then Deputy Prime Minister. Laws were enacted to reduce imports. The real economic growth in 1974 was minus 1.8%, the first negative growth since 1950, but began to pick up in 1975, though in the second half of the 1970s Japan experienced recession in the traditionally strong industries like steel and shipbuilding, and suffered from the double effect of protectionism abroad and a sharply appreciated currency.

In 1977 there were a number of reports which offered different scenarios for Japan's growth and these are summarized in figure 28.

The Japanese appreciate their vulnerability through dependence on foreign oil imports and that this is lessened in a buyer's market, as in 1976. However, the Institute of Energy Economics have illustrated[336] that it is very difficult for Japan to reduce its dependence. Therefore they have a need to reformulate their energy policy in order to improve the security of supply. A number of possible measures were suggested, all affecting the pattern of demand. Reorganization of the Japanese petroleum industry to strengthen the supply side was one suggestion. Another was for the Government to increase its strategic stockpile and prepare for increased imports of oil products, liquefied petroleum gases (LPG) and chemicals. The industry should tackle the problems of transportation, handling, conversion and pollution arising out of increased coal imports. Similar problems need to be tackled for liquefied natural gas (LNG), including an import centre and integration into industrial systems. Imports of LNG might be about 27 Mt in 1985. Immediate decisions were seen to be necessary on nuclear power construction if even the Institute's total of 27 GW (gigawatts), in 1985, and 50 GW in 2000 were to be achieved and not the higher target of 48 GW in 1985 which was called for in the National Plan. There seemed also to be a need to harmonize Japan's nuclear energy policy with that of the USA. In brief, the Institute brought together suggestions for an energy policy which, whilst accepting a continuing dependence on oil for over half its primary energy needs, would increase the inter-relationship between the

JAPAN — SOME ECONOMIC GROWTH POSSIBILITIES
Average Annual Percentage Growth Rates
Fig. 28

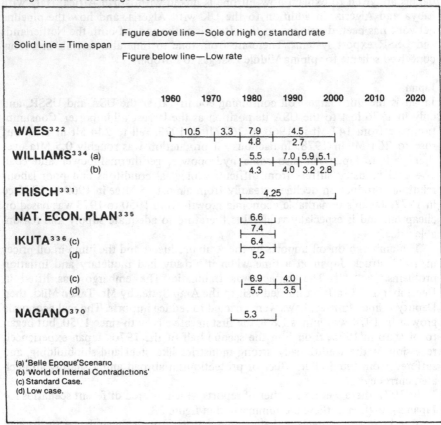

Figure above line—Sole or high or standard rate

Solid Line = Time span

Figure below line—Low rate

	1960	1970	1980	1990	2000	2010	2020
WAES[322]		10.5	3.3	7.9 / 4.8	4.5 / 2.7		
WILLIAMS[334] (a)				5.5 / 4.3	7.0 / 4.0	5.9 / 3.2	5.1 / 2.8
(b)							
FRISCH[331]				4.25			
NAT. ECON. PLAN[335]				6.6			
				7.4			
IKUTA[336] (c)				6.4			
(d)				5.2			
(c)				6.9	4.0		
(d)				5.5	3.5		
NAGANO[370]				5.3			

(a) 'Belle Epoque'Scenario
(b) 'World of Internal Contradictions'
(c) Standard Case.
(d) Low case.

Sources: As referenced

various energy industries and with the whole economy. At the same time, there was an obvious need to ensure, with Government help, that the necessary capital would be available.

One method of increasing security of supply is to diversify the points of supply. The bulk of Japan's oil came from Saudi Arabia and Iran, but high priority, at least in general economic ties, was given to Iraq, and this seemed[341] to have been given an extra boost in January 1977. Oil imports from China began in 1973, but in 1976 were only 4 Mt, with the planned imports for 1977 only 5.18-6.18 Mt. The Japanese hoped for bigger quantities after 1978, when China may have overcome the setbacks, both physical and political, which beset it in 1976. Taching oil, though low in sulphur (0.2%), is waxy and has a high viscosity; this makes it difficult to handle and it has to be blended with other crudes for the current Japanese refineries. The Japanese refiners would therefore prefer not to take Taching oil if the choice was only technical/commercial. Japan's vulnerability rests on the basic lack of indigenous energy resources, and on the structure of its oil industry. The Japanese built up their oil industry, and their manufacturing industry, on the cheap oil of the 1950s and 1960s. Oil supplies some 70% of total primary energy, some 90% of the fuel used in generating electricity, and 80% of that electricity is used in the commercial and industrial sectors — even the aluminium industry depends on oil-generated electricity. Elsewhere cheap hydro-electricity is the basis for siting aluminium plants. This structure[370] magnifies any increase in oil prices, for they affect every sector, and so the whole economy is vulnerable to price increase. Substitution by alternative energy sources still entails imports and both coal, with high costs in meeting environmental standards in use, and liquefied natural gas, having high transport and handling costs, have their disadvantages. Conservation in all aspects seems the only method of lessening dependence on imports, until breeder and/or fusion nuclear reactors are sufficiently established. Therefore it would seem that dependence on oil will be a long-term feature of the economy of Japan.

In early 1979 Japan suffered the temporary loss of all supplies of crude oil from Iran and had only regained some supplies later in that year. Japan had reduced its dependence on Iran from almost 45% of total crude oil imports in 1970 to just over 20% in 1976 and 17% in 1978. Japan's 1978 oil consumption at almost 263 Mt[436] was still less than it had been in 1973, and in the first four months of 1979 imports were at about the 1978 level. Nevertheless efforts were intensified towards energy conservation, diversification of petroleum import sources, research into alternative energy sources, more efficient ultra-high voltage electricity transmission, and a reduction in energy demand by a shift from importing raw materials to importing intermediate manufactured products for final processing. Japan agreed with President Carter, before the OECD 'summit meeting' in Tokyo in June 1979, to limit its oil imports. These had been planned[440,443] to be around 251 Mt in 1979, 258 in 1980, 273 Mt in 1981 and 293 Mt in 1983, with crude oil demand rising by 4.8% p.a. (including stockpiling to 70 Mt by 1982) and petroleum product demand by 4.3% p.a. It was said that 1979 imports would be reduced to 242 Mt, i.e. a reduction of 3.6% below target. At the 'summit meeting' Japan undertook to keep its imports in 1985 within the range of 315 to 345 Mt. Japan has also sought new crude oil sources, (as from Mexico linked, hopefully, with exports of Japanese goods and services), and a stronger position for importing Australian coal by acquiring more mining interests.

Two important new factors in the Far East will affect Japan's future energy demand. First is the increasing competition from the new industrialized and exporting countries of South Korea, Taiwan, Hongkong and Singapore. These countries are successfully competing, even in the Japanese domestic market, in petrochemicals, steel and ships. Horizontal integration between companies in Japan and these developing countries was notable in mid-1979 in electronics and watches. These countries are also beginning to be hit by the problems which plagued Japan in the late 1960s, namely overcrowding, pollution and shortage of skilled labour leading to the escalation of labour costs. Japan will undoubtedly remain the biggest factor, because of its large domestic market and industrial strength, but may be forced by competition into high technology fields and so into greater competition with the USA and Europe. The second factor affecting trading patterns is the speed of industrialization in China. This was an enigma which is discussed later. Japan in 1979 was second to the USA as the biggest oil importer. If however President Carter's July 1979 plan to reduce US oil imports to 225 Mt in 1985 were to succeed, Japan might be very vulnerable as the world's largest oil importer. Many, however, would say in 1979 that the chances of Japan taking appropriate action to avoid that vulnerability

were greater than those of the USA reducing its oil imports to 225 Mt in 1985. A draft 'Vision' of trade in the 1980s, prepared by the Ministry of International Trade and Industry (MITI)[525] and submitted to the Industrial Structure Council in late August, 1979, set out three main targets:
— to increase national spending on research and development from 1.7% of gross national product, to 2.5% in 1985 and 3% in 1990;
— to increase overseas economic cooperation, both in aid to, investment in and imports from developing countries from 1.6% of GNP in 1978 to 3% in 1990;
— to reduce dependence on imported oil from the current 75% of total energy consumption to 50% by the end of the 1980s, whilst also reducing consumption by 15% through conservation.

In October 1979 the Government announced[526] that its aim was for coal imports to be increased, liquefied natural gas imports to increase to 29 Mtoe by 1985, with nuclear plants producing 30 GW of electricity in 1985 and 53 GW in 1990; £8.1 b would be planned to be spent on developing new energy sources by 1990. Thus Japan, despite its major environmental problems in the use of oil and coal, enters the 1980s with clear energy objectives and proved resilience to tackle emergencies. Indeed a journal headline[475] in October 1979 was 'Long term oil strategy succeeding'.

The USSR and Eastern Europe

Many studies on world energy deal only with the world outside the Communist areas, (WOCA). The argument is usually that within the time scales of thier work, usually within the present century, the impact of the USSR, Eastern Europe and China on international energy exchange will be negligible. However, centrally planned economies are able to over-ride, for political purposes, economic factors which dominate other economies. Yet there are resource parameters which cannot be ignored and therefore should be considered if one is to judge the limits of possible action. Furthermore the question as to whether the Communist areas will be in a position to export, or be importing, is very important to the rest of the world when, as seems probable, there will be scarcity and therefore a sellers' market world-wide. All countries will be affected, whether developed or developing.

Published information on energy within the Soviet bloc is scanty, variable in reliability, and badly reported in the West. There is a serious tendency to exaggerate the prospects and the problems, the successes and the failures. The system of quinquennial plans fosters this exaggeration. There is apparent ease in comparison of targets and actuals. Many Western observers put great stress on the over- or under-fulfilment of targets, original or revised. The Soviet oil industry appears to have reached its targets in recent years. The Soviet gas industry appears often to have failed. However, this may merely mean that the oil planners were more realistic and/or knowledgeable and/ or politically astute. There are some statistics which are tightly controlled, like reservoir characteristics, even if not actually classified as secret, like oil reserves. Productions and sometimes targets are given as percentages from a base one year which is changed the next year, or is very imprecise. Hence all available data should be taken as indicative rather than absolute.

The Soviet Union has an abundance of energy resources, coal, lignite, peat, oil shales, crude oil, natural gas, uranium and thorium ores, and hydro-electric power. Coal may have 'geological reserves' totalling possibly as high as 8.7 Tt, including hard and brown coal to 5,500 feet but only 470 Gt may be ultimately recoverable. Some western sources claim [339] only 250 Gt might be termed 'measured'. These reserves would fall within the Soviet reserves classifications A, B, and C1, which are more or less similar to, but not identical with, respectively, the proved, semi-proved and probable categories of others.[351]

The A, B, C1 reserves of oil shales have been estimated[322] as about 6.6 Gtoe, with C2 'possible' reserves at over 10 Gtoe. The bulk of these reserves are in the Baltic States. Most of the output is used as raw boiler fuel for the big Estonian electricpower plants, with a combined capacity of 2.9 GWh. The heat value of Estonian shale is about a fifth that of crude oil and a third that of good quality coal. In 1973 the USSR production was planned to be 37 Mt, made up of 30 Mt in Estonia, 5.8 Mt in the Leningrad Oblast (region), and 1.2 Mt in Kuibyshev Oblast. Their use is local because the low heat value does not allow economic transportation, except in the form of electricity.

The other low quality energy source, also therefore of only local importance, is peat. Total recorded production in 1971 was 54.3 Mt, including 33.7 Mt in the Russian Federated Republics and over 11 Mt in Byelorussia. The quantities used by householders in these and the Ukraine and the Baltic States are not recorded, though use is considerable in rural areas. The 1975 Plan was for a total production of 97 Mt and is therefore considered to be a useful, though local, energy source.

Crude oil resources are covered by the States Secrets Act and information has to be pieced together from many different sources and there can be wide divergencies of opinion. One estimate[339] is that the explored and measured reserves of the A and B categories, proved and semi-proved, are some 5.7 Gtoe, commercial reserves of A, B and C1 categories total 14 Gt and the ultimate recoverable reserves, that is including categories C2, (prospective), D1 and D2, (prognostic or predicted), total 50 Gt. However the Central Intelligence Agency of the USA, considered[350] that the Soviet oil reserves are only 30–35 billion barrels (4.1–4.8 Gt). They estimated that production may peak as early as 1978 at a maximum of 550–600 Mta, and, most significantly, would drop steeply, giving a net import of 175–225 Mt by 1985 for all Comecon or CMEA (Council for Mutual Economic Assistance).* The argument was based mainly on three points:
— declining reserves/production ratio,
— major water encroachment problem, and
— high depletion, low substitution rates, in that the discoveries of new oil were not compensating for the decline of production in the older fields.

The water encroachment problem is not unique to the USSR. In July 1977 concern was being expressed about the Argyll field in the North Sea after only two years' production. In the USSR, however, it has been common practice to assist the inherent production mechanism of most fields by the injection of water, usually at an earlier stage in the production history than is the custom in the USA. This method assists production rates, in combining primary and 'secondary' recovery almost from initial production. Reservoir control has obviously to be more careful than otherwise to avoid damage to the reservoir and loss of possible production by by-passing 'blocks' of oil, as the water, as intended, pushes the oil to the wells faster than otherwise. However, each reservoir reacts differently and though there is a gamble, this is one which is

---

* The members of CMEA are Bulgaria, Czechoslovakia, German Democratic Republic, Poland, Romania, USSR, Mongolia and Cuba. In common usage CMEA or Comecon is often loosely made to cover only the European members.

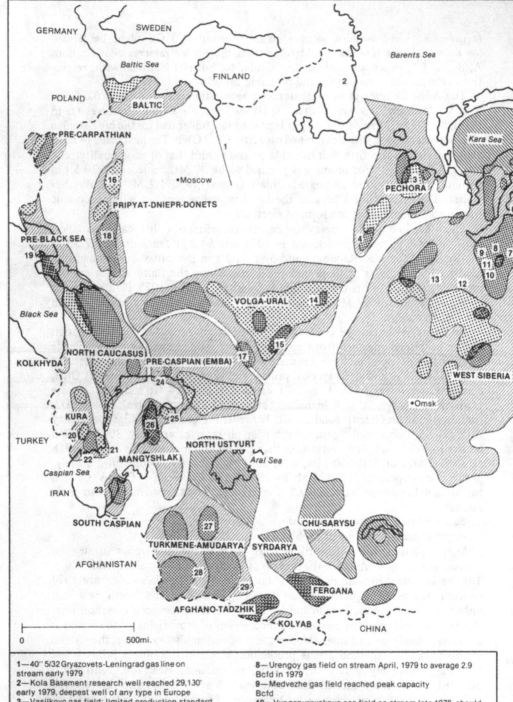

1—40″ 5/32 Gryazovets-Leningrad gas line on stream early 1979

2—Kola Basement research well reached 29,130′ early 1979, deepest well of any type in Europe

3—Vasilkovo gas field: limited production standard

4—Vuktyl has field: new extension of shallow pool found, appraisal of deep pool continuing

5—Antipayutinskoye gas discovery, reserves tentatively estimated at 5.3 Tcf

6—Gydanskoye gas discovery, reserves tentatively estimated at 4.35 Tcf

7—South Samburgskoye gas discovery, reserves tentatively estimated at 10.6 Tcf. New deep oil pools found, Samburg-Urengoy area

8—Urengoy gas field on stream April, 1979 to average 2.9 Bcfd in 1979

9—Medvezhe gas field reached peak capacity Bcfd

10—Vyngapurovskoye gas field on stream late 1978, should yield 1.45 Bcfd in 1979

11—Northern West Siberia: 1978 gas output 8.22 Bcfd, planned 1979 production 11.12 Bcfd

12—Tyumen province, 1978: exploratory & appraisal drilling 2,725,000′ (+50%), oil development drilling 16.4 min ft (+30%), 1,800 oil wells (+23%)

13—Mid Ob: average 1978 output 5,080,000 bpd, 28 fields on stream at year's end, 1979 production planned at 5.7 min bpd

Source: World Oil, 15 August 1979, pp. 166 & 167

Fig. 29

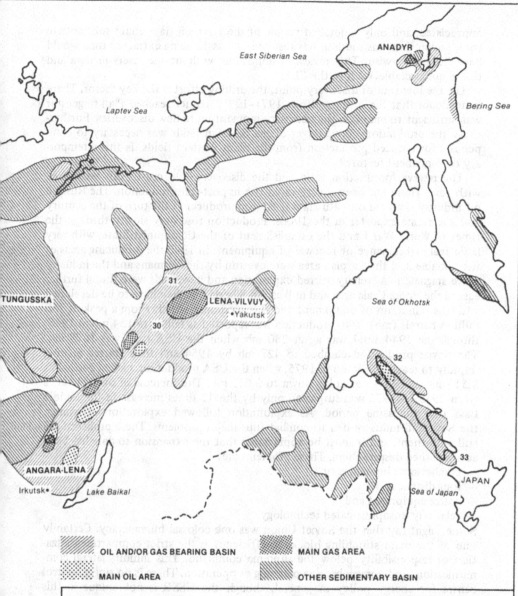

East Siberian Sea

ANADYR

Laptev Sea

Bering Sea

TUNGUSSKA

31

LENA-VILVUY

•Yakutsk

Sea of Okhotsk

30

32

ANGARA-LENA

33

Irkutsk•   Lake Baikal

JAPAN

Sea of Japan

▨ OIL AND/OR GAS BEARING BASIN   ▨ MAIN GAS AREA

▨ MAIN OIL AREA   ▨ OTHER SEDIMENTARY BASIN

14—Perm province: 1978 oil production plan missed by 40,000 bpd
15—Bashkiria: several reefal oil pools found, might allow to maintain production at 800,000 bpd
16—Bielorussia: several oil discoveries
17Orenburg gas field: reached peak capacity of 4.35 Bcfd
18—Dniepr-Donets basin: oil found for first time in Devonian
19—Soviet jack-up for 197' water, 19,680' wells to start drilling in Black Sea in 1979
20—Saatli deep research well: 13 3/8" casing set at 11,645', October, 1978
21—Caspian offshore: 5 jack-ups active (2 explor., 3 appraisal), one semi-sub to start operating end 1979. Gas and condensate production to increase in 1979
22—21,535' well drilled from fixed platform in Bulla Island gas and oil field, deepest Soviet offshore well
23—West Erdekli: offshore oil find, three wells drilling
24—Astrakhan sour gas field: probable reserves 2 Tcf, possible reserves at least 26 Tcf

25—Buzachi peninsula oil fields: 1976-1978 appraisal drilling resulted in evidencing over 1 bln bbls of reserves
26—Mangyshlak: 1978 oil production plan missed by 20,000 bpd
27—Naip field: deeper gas pool found
28—East Turkmenistan & Uzbekistan, 1978: several significant gas discoveries, several new gas fields on stream
29—Gissar foothills gas area: oil found for the first time
30—El Paso/Oxy/Japanese: to go ahead with development of Yakutia gas, construction of 1,500-mile line to Pacific coast, imports of LNG (Japan: 967 MMcfd, USA: 967 MMcfd). Project scheduled to start operating mid-80s
31—Mid Tyungskoye gas field: new pools found
32—Japanese scored second oil strike off Sakhalin, testing 1,600 bpd. Two jack-ups active in 1978
33—North Lugovskoye: gas discovery
34—Chukot: wildcat testing end 1978, TD 10,867'

appreciated, and only a detailed review of the reservoir data could substantiate the opinion that this method has resulted in less oil being extracted than would have been otherwise. This review is impossible without the reservoir data and this is not available, even to the CIA.

On the low rate of discovery point, the drilling effort is the key factor. There is no doubt that the drilling effort, 1971–1975, neither reached Plan target nor was sufficient to maintain the necessary momentum to new discoveries. Furthermore the production above target of the Siberian fields was necessary to compensate for reduced production from the older western fields. Is this a temporary or permanent feature?

The reserves/production ratio and the discovery rate must both be viewed with care. There has been a very rapid rise in post-war production. The Russian oil industry is an old one. Russia was the top producer at the turn of the century and the prime exporter in the 1900s. Production rose only slowly through the times of World War I and the establishment of the Communist State, with very little real maintenance or renewal of equipment. In 1940 the producing areas of the Ukraine and the Caspian area were overrun by the Germans and the industry there stagnated. Adversity stirred exploration and new fields were found further east in the Volga-Urals area and in Western Siberia but they have to be developed with the minimum of equipment and the maximum speed. From a peak of 280 million barrels (mb) 1940 production was reported as falling to 154 mb in 1945, though the 1944 total was again 280 mb when the USA total was 1678 mb. The Soviet production climbed to 427 mb by 1954 and then started moving rapidly to reach 3,609 mb in 1975, when the USA production, having peaked at 3,517 mb in 1969, was going down to 3,052 mb. This increase of over 21 times from 1945 to 1975 was surpassed only by the 41 times increase in the Middle East over the same period. As exploitation followed exploration eastwards, the Soviet oil industry had to combat four major problems. These problems are still with them, and it must be appreciated that the expansion to date has been accomplished despite them. These problems are:
— cumbersome bureaucracy,
— long distances,
— scarce equipment and
— relatively unsophisticated technology.

Some might say that the Soviet Union was one colossal bureaucracy. Certainly one of the main stumbling blocks to efficiency is the strict compartmentalization of responsibility below the supreme command. This inhibits lateral communication, and comprehensive planning or operation. There is no single control centre for energy policy in general, though the USSR is not unique in this respect. There is no central control of fuel policy.[338] Initial exploration for natural resources is in the hands of the Ministry of Geology. Development is in the hands of other ministries, and responsibility for construction of facilities, refining and then distribution rests with yet other ministries. This sub-division of responsibility, with planning under yet other organizations, becomes enhanced by the long distances and different peoples involved and by the differences between the extreme environments of the deserts in the south to the arctic in the north.

The logistic problems of distance were discussed earlier, for coal and natural gas. They have been and are formidable in regions lacking any infrastructure. European Russia has 80% of the energy consumption, and 90% of Soviet energy resources are east of the Urals. Distances also create the dilemma as to whether to move the resource raw material to the consumer, or to up-grade the raw material, e.g. brown coal to electricity, and then transport, or to move industry to the resource. The distance problems have always been with the Russians and, indeed, distance has been the saviour of Russia in three major wars. In the late 1940s and 1950s, massive hydro-electricity generating schemes and transmissions were the most fashionable answer to the question of how to move energy. Coal, however, was the dominant domestic energy source. A lot of research and some development went into attempts to improve mobility, and efficiency in end-use. Work was done on underground gasification, manufacture of gas from both coal and shale, and on the liquefaction of coal. No methods proved commercial and most projects were suspended in the late 1950s as oil production increased. It may well be that these projects, and others like slurry pipelines, will be revived in the 1980s. Long distances add to the transportation cost element, but the central political control does mean that the consuming area does not have to pay an economic rent to the producing area. As effort has not hitherto been cost conscious, and cost-benefit not the ruling factor, Soviet internal prices are not as responsive to scarcity values. In a market economy this would be considered a weakness, and indeed this must have an important bearing on conservation, for the system lacks the normal economic incentives of other countries to conserve scarce commodities.

In contrast to the oil industry, the natural gas industry in the USSR is a very young one, because it is only since 1957 that gas has become a significant energy source. Production had risen from just over 45 km³ to 261 km³ in 1974. Some more recent data are interesting, and gas is an important feature of the energy picture of the whole Soviet bloc. Ultimate recoverable reserves have been variously estimated. One estimate[339] is for 100,000 km³. Reserves in the A, B, and C categories were said[348] by the Minister of the Gas Industry to be 27,000 km³, which the French oil company, Total, divided as to 23,000 km³ in categories A and B and 4,000 km³ in category C. Ruhrgas are said[348] to estimate 23,000 km³ in categories A, B and C1, and 28,000 km³ in category C2. Yet another estimate[339] is for A, B, and C1 as 22,000 km³ in 1974. Total are said to make a regional allocation of their proved and semi-proved 23,000 km³ as:

| Orenburg | 3,000 km³ | Turkmenistan | 2,800 km³ |
| Yakutsk | 1,000 km³ | Tyumen | 17,000 km³ |

The same source[348] reported that Gaz de France quoted hopeful talk, by East Siberian exploration teams, which placed potential reserves in that region at 13,500 km³. Total offered an opinion that by the year 2000 some 23,000 km³ may have been found in offshore areas with 60% being in the Arctic Ocean.

Whether or not these resource and reserve figures are of the right order, there seems little doubt that natural gas has a high potential. The resource suffers, though, from the same mal-distribution between supply and consumption centres as does oil. Thus, despite the eastern wealth, Iranian gas has been imported into 'The Centre', European Russia, since 1970. Gas from coal and shale only adds some 1.5 to 2 km³ to the total natural gas production and it was said[339] that some 19 km³ of natural gas was being flared because of inadequate pipelines – this, one assumes, must be a very speculative figure.

The rapid growth of both the oil and gas industries, the tremendous effort necessary to overcome both distances and hostile environments, the disadvantages of the centralized yet compartmentalized and cumbersome bureaucracy have indeed left the USSR lacking in both modern oil industry technology and equipment. One major strand in Soviet reasoning on the advantages of détente is said[339] to be to improve the chances of importing Western technology and equipment by lessening the list of 'prohibited exports' from the USA. The need for help in the energy field had already become apparent by the time of the Nixon-Brezhnev meeting in May 1972, as the long lead times from plan to completion of distant projects started inhibiting growth. Soviet imports of oil equipment[348] in 1975 can be listed, (with 1974 values in brackets), as from the USA, $176 m, ($43 m), from FR Germany $46 m, ($74,000), from France $22.5 m, ($5.5 m), from the UK $1.9 m ($80,000). The totals from these countries were 1975 $264 m, and 1974 $49 m. The need for equipment and technology will increase as exploration and production for oil and gas moves into more hostile areas, offshore as well as onshore, and this need is an important factor in both energy and political policies.

The USSR has the world's largest continental shelves of up to 2.5 million square miles in extent as well as the large inland sea of the Caspian and the Soviet parts of the almost landlocked seas of the Black Sea and the Baltic. Most of the shelf area is in the Arctic Ocean and it will probably be a long time before the middle half of that area is fully explored or exploited. Offshore production from the shallow water fields off Baku in the Caspian Sea is well established but as the waters are shallow, especially over the Neftkhaniyeh field with the rocks only just submerged, and the reservoirs are at a very shallow depth below the sea bottom, production has been by very simple methods. Some exploration has been done with jack-up rigs in the past five or so years, but the first semi-submersible rig for use in the Caspian was only bought in 1976.[407] Exploration has begun in the Black Sea and in the Baltic and though the latter is not generally considered very prospective there have been some favourable pointers, but less than in the Black Sea, where three gas fields have been indicated. Off Sakhalin, a joint venture with the Japanese has found light, low sulphur oil in what may be a large field; the original test flowed at a rate of 6,000 bd. There are also hopes in the Barents Sea with Norwegian waters to the west of an ill-defined and disputed boundary,[337] and in the Kara Sea to the east. However, with so many onshore prospects still to be exploited, it is probable that offshore development will continue to be slow except where there are specific advantages, like in the Far East, where Japanese technology and equipment could be organized near the expanding Russian Far East industrial centres, and in the Black Sea with Ukraine providing skilled oil workers and a ready, close market for either oil or gas. The Caspian can also be expected to be the training ground for offshore workers because of its established but declining onshore oil industry. In other words these offshore areas mentioned will be keyed into domestic markets.

The domestic energy demand pattern of the USSR has some interesting features.[338] A high proportion of primary energy, some 40%, is used to generate electricity. Household and municipal use at 17% and transport use at 8% are low compared to Western industrialized countries, but use in industry is high at 35%. The 1972 total energy demand pattern in the USA included 32% for transportation, 21% for residential use and 27% for industry. The low shares of the Soviet residential and transport sectors may be expected to increase as income per capita increases — wood still provides 28% of household energy, and mention has already been made of the unrecorded domestic use of peat. Part of the high industrial consumption may be ascribed to waste. There could be a reduction in specific demand in this sector, but, as noted above, there is no price mechanism to help conservation, and exhortation is a weak instrument against the drive for production targets. These targets in the manufacturing sector are still not producing all that is required for increasing the energy supply which they themselves need.

In the 1960s there was a shift to improve efficiency by substituting oil and later gas for coal and the trend has continued, as table 75 shows, but the 1980 Plan falls far short of Khrushchev's plan in 1961 for 1980.

Table 75
USSR — major fuels supply, 1970, 1975, 1976, 1980

	1970	1975		1976		1980	
	Actual	Plan	Actual	Plan	Actual	Plan	Khruschev
Electricity (TWh)	741	1065	1038	1095	1111	1360	2850
Oil incl. condensate (Mt)	353	505	491	520	520	620–640	700
Natural gas (km ^{3})	198	320	289	313	321	400–430	700
Coal (Mt)	524	695	701	715	713	790–810	1190

Sources: Prof. A Nove, *The Times*, London, 5 January 1976
Prof. A Nove & R. Clarke, *The Times*, London, 14 February 1977
Khruschev: Presentation to 1961 Congress, cited in [413]

Roughly half of the 1980 oil production is expected to come from the Tyumen and Tomsk areas of Western Siberia. The productions[346] of these areas in 1970 were only 31 Mt but 148 Mt in 1975, which emphasizes the continuing shift to harnessing Siberia's resources.

The 1976–1980 Plan was outlined[342] in mid-December 1976 and forecast a slower rate of economic growth, a renewed emphasis on heavy industry and a continued pouring of resources into farming. The increase in national income in the Plan was noted as being 26% by 1980. The rise in industrial production of 37%, (40% in heavy and 31% in light industry), compared with the 47% planned and the 43% claimed in the 1971–1975 Plan. The Planning Minister was reported[343] as saying that imports would mainly consist of equipment for the production of fertilizers, and for the development of oil, gas, paper, cellulose and 'certain other industries'. These imports can only be paid for by exports, and oil and gas are the main items. To these domestic needs have to be added the political need to supply the rest of Comecon. Point is added to this argument of the need for exports when notice is taken of the balance of visible trade of the USSR[345] as shown in table 76.

In 1975 total exports of crude oil and products were some 130 Mt, with 54 Mt going to the West, mainly to Western Europe and 76 Mt to Comecon. In 1976 a net export of 140 Mt was estimated[347] to have earned $6.5 billion, but Comecon get favourable terms. In 1977 Comecon members were reported[347] to be paying $9.45 pb (per barrel), which was well below the January Saudi Arabian light crude marker official sales price of $12.09 pb. The increase in world oil prices has undoubtedly helped the USSR. 110 Mt exports in 1970 earned $2 billion, and 130 Mt in 1975 earned $5.9 billion.[345] The USSR hard currency indebtedness at the end of 1975 was estimated to be $8.14 b, with the deficit to developed capitalist countries at under $4 b. Oil export had increased to 148 Mt, including 77 Mt to Comecon and 48 Mt to Western Europe[359] by 1976.

Table 76
USSR — visible trade balances, 1971—1975
(in million roubles)

	1971	1972	1973	1974	1975	1971—1975
Socialist countries	+757	−233	−101	+850	+600	+1873
of which Comecon	+78	−416	−300	+482	+500	+344
Developed West	−117	−1000	−839	+111	−3600	−5445
Developing countries	+554	+658	+1201	+1004	+300	+3717
Total	+1194	−575	+261	+1965	−2700	+145

Sources: *Vleshnaya torgovlya* for 1971—1974
Ekonomicheskaya Gazeta, May 1976 (for 1975)

The oil position is summarized in table 77 for the whole of Comecon.

Table 77
Comecon — oil balances, 1976 & 1980
(megatonnes of oil equivalent)

	1976 (preliminary)	1980 Based on Soviet Plan	Based on CIA prediction
Soviet Union			
Production	520	640	550—600
Consumption	380	500	500
Exportable surplus	140	140	50—100
Available for exports to			
Comecon	70	70	50—70
World Markets	70	70	0—30
Rest of Comecon			
Production	16	18	
Consumption	94	120	
Imports from USSR	70	70	50—70
Imports from world markets	8	32	32—52
Comecon's position with World	+62	+38	−52 or −2

Source: 'Soviet Oil Position', David Lascelles, *Financial Times*, 2 August 1977

The CIA prediction of lower production in the USSR and lower exportable surplus obviously has an important impact. The USSR needs its exports to pay for its own imports and to redress the deficit Comecon balance.

Eastern Europe
The European countries in Comecon were in energy balance in 1960, moved into deficit about 1963, and were dependent on imports to the extent of 11% in

1970, to 20% in 1975 and the deficit by 1980 may well exceed 25% of total energy demand. Table 78 shows an estimate[339] for 1980, which is probably quite close to the Plans of these countries.

Table 78
European Comecon countries — energy balance 1980
(megatonnes of Standard Fuel Equivalent (SFE))

	Production				Consumption				Balance
	oil	gas	coal	total	oil	gas	coal	total	
Poland	0.6	22.3	171	194	36	29	114	179	+ 15
Czechoslovakia	0.7	2.8	72	75	36	9	74	119	− 44
German D. R.	0.3	16.6	86	103	32	20	86	137	− 34
Romania	22.2	28.0	18	68	32	32	18	82	− 14
Hungary	2.8	7.1	10	20	22	11	11	44	− 24
Bulgaria	0.7	1.2	14	16	27	10	17	54	− 38
Totals	27.3	78.0	371	476	185	111	321	615	−139

Source: Jeremy Russell [339].

Table 79
European Comecon countries — an energy consumption forecast, 2000
(megatonnes coal equivalent)

	Oil	Gas	Coal	Hydro	Nuclear	New Energies	Total
Poland	110	45	280	5	45	5	470
Czechoslovakia	115	15	135	5	25	5	290
German D. R.	162	18	168	1	56	5	395
Romania	109	37	23	9	42	5	220
Hungary	63	12	23	2	15	—	115
Bulgaria	53	22	30	3	7	—	115
USSR	1365	1735	1070	210	840	30	5250

Note: Tables 78 & 79 are in the units as reported. Applicable conversion factors are:
1 Mt SFE = 0.718 Mt Coal = 0.0308 EJ
1 Mt SFE = 0.70 Mt Oil
1 tSFE = 840 m³ Natural Gas

Source: 'L'Equilibre Mondial entre Besoins et Resources d'Energie...' J-R, Frisch [331] Paris, France, 1977.

Table 78 shows on the production side that only Romania has significant oil: Poland, the Democratic Republic of Germany, (DRG), and Czechoslovakia have significant coal; Poland, the DRG and Romania have some gas production. On the consumption side, all have big oil and gas deficits; only Poland has an exportable surplus, coal, and that gives the group a coal surplus. It was noted above in table 78 that the USSR exported 70 Mt oil in 1976 to Comecon. This included 12 Mt to the non-European members, (Cuba and Mongolia), and the associates, (North Korea and Vietnam). In addition to the 58 Mt from the Soviet Union, Eastern Europe imported in 1974 some 11 Mt from the Middle East, (7.0 Mt from Iraq, 3.5 Mt from Iran and 0.4 Mt from Kuwait). Table 77 suggests an import in 1980 of 102 Mt of oil. Another estimate[388] suggests 150 Mt. Taking the table 77 figures, the European Comecon countries would need to acquire some 32 Mt of oil from the world markets in 1980, (120 – 12 Mt non European Comecon Consumption – 58 Mt imported from the USSR, – 18 Mt of indigenous production, for the non-European production is and may be negligible). At $13.65/barrel or $100 per tonne the cost would be $3.2 billion. If the suggested additional 30 Mt Comecon consumption were to imply an extra 25 Mt in European Comecon, then the cost would be $5.6 billion. It would seem most unlikely that there could be increased consumption if the USSR available surplus was only 50 Mt, that is the lower CIA estimate, but the lower imports from the Soviet Union would mean a total bill of $5.2 billion. These are considerable sums in relation to the total hard currency earnings which were between $4 and $5 billion in the early 1970s. The alternatives facing the Eastern European countries appear to be:[339]

- reduce personal consumption further and enforce stricter conservation in industry and commerce. Hungary can be taken as an example for such action;[369]
- delay or even reverse the trend to substitute oil for coal, even though some existing plants and some under construction would be under-utilized. Romania plans to reduce the use of hydrocarbons in electricity generation from 50% in 1975 to 35% in 1980;
- step up coal production, despite the heavy investment necessary, again most efficiently accomplished with imported equipment;
- increase the contribution of nuclear energy but this would require Russian help and would not be effective for 10–20 years or more. Romania's first nuclear power station is due for completion in 1980;
- step up investment in Soviet energy sources, particularly in natural gas, on the lines of the contribution of labour and materials to the Urengoy gas export line of 15 km^3/year eventual capacity. The Friendship line reached its planned capacity in 1975, but there seem to be no plans for expansion;
- encourage OPEC to accept technological and industrial aid in return for oil or a low guaranteed oil price.

Increased involvement in the USSR energy field, whether in gas pipelines or oil pipelines or nuclear plants, means greater ties between Eastern Europe and the Soviet economy and therefore closer political ties. The USSR has encouraged the rest of Comecon recently to import Middle East oil, but not only are there the financial limits, but it is doubtful if the USSR would favour too great dependence. Hence one can expect Eastern Europe to have first call on Russia's export surplus oil and gas for some time. Big new oil or gas discoveries in Eastern Europe or offshore in the Baltic or Black Sea would alleviate the situation, as would development in both production and transportation of Siberian oil at a faster rate than even the Soviets expect.

The refining side of the oil industry in the USSR, as in other aspects of that industry, has been built up in a rugged style, to produce fuel oil for industry with minimum gasoline. A change of output pattern to increase the proportion of gasolines, and light ends for the petrochemical industry, would increase the value of exports, but once again, could be effected more quickly by the import of equipment and technology from the West.

The Soviet Union appears to have a circular problem. Supply of oil in five years' time may not be sufficient to satisfy simultaneously:

- increasing domestic demand;
- an export surplus with which to buy equipment and technology to increase its oil supply and raise its value;
- the increasing demands of Eastern Europe and allow them increased industrial production to improve their own hard currency earnings with which to import the equipment and technology which they need to increase their own indigenous energy supplies.

A circular motion can develop into a spiral motion, going up or going down, through quite a minor shift in the containing forces. It is most unlikely that the USSR will spiral down without a fight. What circumstances might allow an upward spiral? Big new discoveries and exploitation of oil and gas resources? If so, how big?

As a guide to an answer let us look at table 79, which gives the view of one analyst[331] as to USSR and European Comecon in the year 2000. The European members of Comecon, excluding the USSR, are expected to consume 416 Mt of oil (612 Mtce @ 1.47 t coal = 1 t crude oil), an increase of almost 6.0% per annum, from 1980 (table 78). Let us assume that their indigenous oil production has declined to 10 Mt from the 18 Mt of 1980, so leaving 406 Mt to be imported. USSR production would have to equal its oil consumption at 929 Mt in order to be self sufficient, that is 18.75 mbd. This implies an average production, 1976–2000, of 725 Mt, and an aggregate production of some 17.4 Gt. If to this total one adds a 15 year life for the 930 Mt production, namely 13.9 Gt one gets a total of, say, 30 Gt. Earlier reference was made to an estimate[339] giving proved, semi-proved and probable reserves at 14 Gt and ultimate recoverable reserves at 50 Gt. The 30 Gt requirement by 2000 would mean that 16 Gt would have to be found by 2000, an average of 666 Mta, or almost 5 bb/year. This would be a formidable, but not impossible task, if the reserves estimates are roughly of the right scale, but probably unlikely. An oil production of 27 mbd (1,336 Mt) rising from the present roughly 11 mbd does not seem an impossible task. The first solution would be to reduce the estimate of consumption which would mean reducing the USSR growth rate of 2.0% p.a. from 1980 to gain any substantial reduction – halving the apparently high 6.0% rate for European Comecon would reduce the import requirement to 223 Mt. Even considering such figures is, of course, a far cry from the CIA thinking.

What other options are open to solve the problem of the gap between oil demand and supply which would seem to be possible before 2000? A scientific breakthrough in the nuclear energy field might be difficult to exploit within 20 years. A technological breakthrough in the coal industry might have a better chance of being effective within the time scale. The development of new energy sources is again difficult to quantify.

There is the possibility of attempting to gain greater access to the hydrocarbon resources of the Middle East. Early in 1978 the import of natural gas from Iran was working well and expansion had been agreed. At that time Soviet policy in the Middle East seemed to be stagnant, as the Middle East seemed to be at the start of rapid development. Iran and Saudi Arabia both seemed strong states, with Iran, in particular building up its military and economic industrial strength on ever-increasing oil production and revenues. There appeared therefore no grounds for any Soviet fears of undue Western influence in its 'soft under-belly', nor any doubt but that oil and gas supplies would be forthcoming when and if required from rising Middle East production. The unrest in Iran in late 1978, the overthrow of the Shah, the chaos throughout 1979 under the rule of Ayatollah Khomeini, and the enhanced 'conservationist' attitude of the other oil producers, created an entirely new situation. Although gas supplies were resumed after a stoppage over three winter months, plans for IGAT II were cancelled and reduction in oil exports had hurt the Comecon importers. The Islamic Republic, despite help in the early stages of the revolution by the Iranian Communists, seemed as anti-Communist as anti-Capitalist. The eventual reaction of the USSR to the Iran of 1980 will depend on the progress of the Islamic Republic, on the status of the Communist Party as an alternative government, and on the real need of the Soviet Union and its Comecon partners for oil imports.

During 1979 Western opinion seemed to harden that the USSR might need to import oil in the 1980s. The loss of IGAT II gas supplies was one setback. Oil production in mid-1979 was said to be running 3–7 Mta below target. Exports to the West in 1978 had fallen by 20%, and in value from Roubles 3.8 b to 3.57 b (£2.6 b), that is by 6% despite a 15% increase in prices. A review[330] in July 1979 noted that the 1979 production was planned to be 593 Mt oil, 404 km^3 natural gas, and over 750 Mt coal, plus hydro- and nuclear-generated electricity, with oil exports of about 150 Mt split roughly 80 Mt to Comecon and 70 Mt to the West. The latter with the proceeds of gas exports to the West provided 50% of the USSR hard currency earnings. These were augmented by hard currency or the equivalent in high quality goods which the USSR sought from its East European partners. Hence though the planned 360 Mt to those partners, 1976–1980, might be raised to 380 Mt, its partners were paying, as, indeed, they had done, towards the finance and construction of the gas pipeline from Orenburg, which could provide them with 15.5 km^3 annually. In July 1979 it was reported that the USSR intended to hold its oil exports under the 1981–1985 Plan, to about 160 Mt, but with Comecon exports rising by 20%; hence exports to the West would fall. The CIA about the same time suggested that the USSR would be a net importer in 1982 to the tune of 35 Mt and production in 1985 would be only 385 Mt, a decrease from the 593 Mt planned for 1979. If this became so, then there would be a very serious shortfall. It is noteworthy that Poland in November 1979, attacked OPEC for its policy of increasing oil prices; Poland would be very badly hurt as it has already had to turn, encouraged by the USSR, to OPEC to make up its oil needs beyond Soviet supplies. Another indication of Comecon concern about future energy supplies was talk about an 'energy bank' between East and West Europe and possible equity investments by Western companies in joint ventures.

Review of the possibilities of coal production being expanded in the USSR and East Europe to fill the notional gap in energy supplies, if oil and gas faltered, seemed in 1979 to offer no great encouragement. Production in the USSR of 1 Gt may be achieved in the 1990s, but the production and transfer of the planned 20 GW from five power stations in the Ekibastuz area of Khazakhstan and the 100 GW from the Kansk-Achinsk area of Central Siberia may only materialize in the late 1980s.

Nuclear energy could contribute a major share, but doubts were expressed[531] as to whether the USSR and its Comecon partners could expand their nuclear generating capacity from the current 13 GW to 140 GW by 1990, with the USSR component being 100–110 GW. The 1971 Plan had set a target for the USSR of 30 GW by 1981, reduced to 18.5 GE in the 1976 Plan and only 12 GW had been installed, by mid-1980 (*Financial Times*, London, 10 July 1980).

Finally it may be noted that in a paper to the Tenth World Energy Conference in September 1977, it was said[372] that the latest energy policy was intended to:

> reduce the share of oil, (with simultaneous absolute increase of its output) in the total increase of energy resources consumption in the country, and sharp increase of the nuclear fuel share, the share of Siberian and Kazakhstan coal, of low pressure natural gas from Toemen gasfields. Later on, from the beginning of XXI century nuclear fuel, (obtained by fission and fusion), will take the leading part in the increase of fuel consumption in the USSR.

The People's Republic of China

It is extremely difficult to use Western economic criteria to judge growth in the centrally planned economy of the People's Republic of China. Lack of firm data and the violent political changes in direction, to which the country seems prone, and the decentralized nature of industry in the regions and even villages make all economic data very hazy. In August 1977, it appeared that 'profit' in the Western sense was being re-instated as the incentive to increase productivity, replacing ideological exhortation. How this will affect industry is not clear. If the Chinese from being busy become productive; if there is less enforced reading of the Thoughts of Chairman Mao and a turn to a profit motive; if a second stage of industrialization begins through competitive entrepreneurism; then a growth might be possible such as the 20 million Chinese in Taiwan, Hongkong and Singapore achieved from 1960 to 1977 and the 100 million Japanese did from 1950 to 1967. The world market in every commodity would then be transformed.[360] The industrious Chinese have cheap human labour, but such a growth would require cheap mechanical energy. Only coal or oil could provide the primary energy, for China does not seem to have an adequate infrastructure to develop and maximise a massive nuclear energy programme. The communes appear to be the dominant social and industrial structure. They are agriculturally based and self-reliant. They decide what they want and then they make it. They import, preferably from nearby, only if the item is cheaper than they can manufacture. Local wages are lower than national scales in the big factories. Some Western observers say that the factories are inefficient because they have not got the drive of seeking cost-effective systems, which was the key to Japanese growth. Moreover the factory workers are said to be restless, because they have no personal incentives for improved productivity within the system. However, in 1979 the Government took a breathing space from an immediate headlong rush into modernization on all fronts which had seemed a possibility in 1978 and a more cautious progression to modernization seemed more probable and more sensible.

Coal The coal resources of China are very large, but, as said earlier, there are few hard facts available even as to the order of magnitude of their recoverable reserves. Current production was taken earlier, as 450 Mta, which was in line with a report[361] of November 1976, with coal providing 63% of total primary energy needs in 1974. The constraints facing the industry are the low quality of much of the coal, limited rail transportation and high exploration and production costs. These are said to be higher than for oil, but this is perhaps a Western-type judgement which does not properly assess the position in China. An Australian Coal Mission to China in 1976, as reported in June 1977, noted[362] that one third of production is in small mines for local consumption, 10% is from open-cast mines, and about half the underground production is by mechanized mining. The overall impression of the Australian mission was that there was every probability that the Chinese coal industry will expand, but that its export capability seems unlikely to be significant before 1985. Even then, they suggested, trade might be confined to steam coal and high-volatile coking coal to nearby countries, Japan, Korea and Vietnam. There seems little to challenge this view. It is difficult to see how any massive export capability could be developed before the early 2000s, and then only if that was Government policy, which would be more obvious in the late 1980s than in the late 1970s. Nevertheless coal must be a very important, if not the most important primary domestic energy source on which industrial expansion must be based.

Oil When the People's Republic was founded in 1949, oil production was only 0.12 Mt (2,400 bd). The Soviet Union then supplied equipment, technology and training in all aspects of the industry as well as oil, which had risen to about 3 Mt by 1960. In 1960 the Sino-Soviet friendship ended. China turned to a selected group of some six Western countries for equipment. Between 1963 and 1972 the worth of this equipment totalled about US $550 million. This was roughly equivalent to the estimated Russian aid in the first ten years of the regime. By 1966 China was virtually self-sufficient in oil for its modest needs of some 13 Mtoe.[330] The strongly nationalistic, even xenophobic character of the regime undoubtedly favoured continued self-sufficiency, and an exportable surplus, if possible. What is then the potential indigenous supply?

No oil reserves figures are published. Most of the producing areas have not been fully delineated. Many of the prospective basins, particularly offshore, have only been very superficially explored. There are some 16 major sedimentary basins, four of them extending offshore, with some forty oilfields and fourteen gasfields indicated. Estimates of recoverable oil vary from 3 Gt to 50 Gt. Japanese sources estimated in 1973 some 20 Gt of recoverable reserves with an increase to 30 Gt expected in the near future. The Chairman of the US Commission for US-Chinese Trade, in 1974 mentioned reserves of 40–50 Gt. Grossling[389] accepted estimates of ultimate recoverable oil, attributed to A. A. Meyerhoff, of 6.8 Gt as a lower limit, but with over one million square miles of prospective territory suggested an upper limit of two to three times that number, say 14–20 Gt. Those who responded to the Desprairies Delphi Poll for his report for the Conservation Commission of the World Energy Conference,[376] with separate figures for China varied in estimate from 5 Gt to 25 Gt. The 1977 CIA report[366] gave a figure of 39 billion barrels (bb), for onshore reserves and 2 to 3 bb for offshore, making a total of 5.75 Gt, (Taching export crude has 7.3 barrels/tonne). *World Oil* journal gave a figure, as of 1 January 1978, of 4.1 Gt for estimated petroleum reserves and 26×10^{12} cubic feet, (736 km^3) of natural gas. *The Oil & Gas Journal* gave a figure, as of 1 January 1979, of 3 Gt of crude oil and 25×10^{12} cubic feet of natural gas, (708 km^3). There is therefore a wide choice from these 'guesstimates' and great difficulty in knowing whether figures quoted are of proved, probable or ultimate reserves. It is, however, noteworthy that the six Chinese oilfields which were listed[368] in 1970 as giants, were from six different basins. It is suggested therefore that whilst proved reserves may be of the order of 6 Gt, the ultimate recoverable reserves may be of the 20–30 Gt bracket, and even more would not be surprising.

A 1977 study by Wolfgang Bartke,[355] based on Chinese published sources, stressed that only relative production data have been given since 1958. There is no central statistical office, so that the Chinese communication media often use conflicting numbers. Hence all Western figures must be viewed with caution, including those published by the United Nations. On the production side, Bartke says:

the forecasts made all over the world since 1974 on China's oil production tend to be exaggerated because they all go back to the figure ascribed to Chou-En-Lai, of 50 million tons of oil produced in 1973 — which has never been confirmed in the Chinese Press. This figure is untenable. China's actual output in 1973 can have been no more than 36 million tons.

The UN production figure[330] for 1973 oil production is 50 Mt. Bartke suggests that China's production could not possibly reach 100 Mta before 1979 and 200 Mta before 1985. The CIA suggest[366] a production of 125 Mta in 1980.

With such a range of reserves estimates, and the dubiety about even recent production, which are normally the hard facts of the industry, there are, naturally, wide variations in forward-looking estimates. In such circumstances it may be helpful to do an exercise of simple extrapolation to see what scale of production could be sustained by what reserves. This is a case where a computer program would have no value, for there are few facts and no analogies. Figure 30 illustrates such an exercise.

CHINA — EXERCISE IN FUTURE CRUDE OIL PRODUCTIONS AS RELATED
TO POSSIBLE RESERVES   Fig. 30

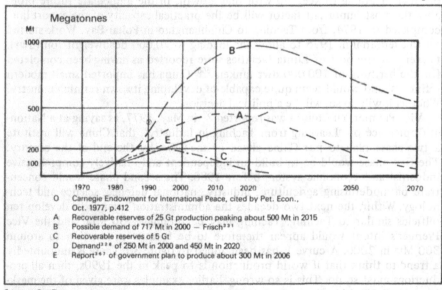

Sources: See Text

The curve to 1985 is based on the Bartke suggestions. Two curves have been drawn. The upper curve, B, used the figure 717 Mt in 2000, suggested by Frisch[331] as a possible demand figure, on the argument that China would try to cover this. The resulting curve has been drawn to give a production of 950 Mt in 2020 and a 'reasonable' decline thereafter. The cumulative production to 2020 would be some 25.5 Gt, and 64.5 Gt by 2070. These figures are well outside any known guesses of reserves. Recoverable reserves of the order of 25 Gt would support a production curve A, peaking about 500 Mta in 2015. Reserves of only 5 Gt would only support a production of the order of that shown in curve C, peaking about 200 Mta (4 mbd) in the mid-1990s. It is suggested that curve A might be used as a working hypothesis.

The implications of the order suggested are that:
- the possibility[373] of production of 400 Mta in 1990 is unlikely;
- demand, as suggested by Frisch.[331] would imply 300 Mt of imports in 2000, which also seems unlikely;
- if demand[328] was about 250 Mta in 2000 and 450 Mt in 2020 (an average annual growth rate of 3.5%) then over this period, curve D of figure 30, there could be export surpluses peaking in 2000 about 150 Mta.

This exercise is most speculative, but it should be looked upon as an illustration of the interdependence of supply and demand and that growth rates *in vacuo* can be most misleading; common sense must rule. Now let us turn again to examine some of the other aspects of the same problem.

In 1978 China exported about 7 Mt[421] and was expected to double that figure in 1982, but many deals seem to be discussed, be mentioned in the Press and then fade from view. The administration appears to have avoided any long term arrangements. It may be that hitherto exports, and the talk of exports, have been governed by the possible surplus between production and refining. In 1974 the refining capacity may have been 39 Mta in twenty-two refineries scattered over the country[355] with those in the areas of recent rapid crude oil production failing to keep up with that growth. In the immediate future probably the most important factor will be the practical capacity of the export line, completed in 1974, from Taching to Chinhuangtao in Pohai Bay. Work started on this terminal in 1975 to allow the loading to 70,000 deadweight ton (dwt). tankers. At the port of Lunta facilities were reported as having been completed for the berthing of 100,000 dwt tankers.[366] China has imported small modern refineries, and would seem quite capable of developing its own refining industry. Whether it will do so, will be a political decision.

Vice Premier Yu Chiu-Li was reported[367] in May, 1977, as saying at a National Conference on 'Learning from Taching in Industry', that China will institute a two-phase plan to put China ahead of the USA by the end of the century. The first phase would be to build an independent and relatively comprehensive industrial and economic system before 1980. The second phase would concentrate on modernising agriculture, industry, national defence, science and technology. Within the next two decades the administration intends to develop ten oilfields similar to Taching. Taching may be producing 30 Mta, and the Vice Premier's intent would appear therefore to be to raise production to around 300 Mta in 2000. A curve E has been added to figure 30. There is undoubtedly a trend to think that if world production is to peak in the 1990s, then all productions must so do. This is so wrong. Earlier examples were given of the masking of individual idiosyncratic production curves. The unique history of China's immediate past and, perhaps, the story of its immediate future could well delay its development as an oil producer/consumer, long past the decline in production elsewhere.

In the immediate past there have been swings between radical and modern economic policies. Growth slowed down in the Cultural Revolution, took off again in the early 1970s, hesitated in 1974 as 'The Gang of Four' launched their anti-Confucius campaign against Premier Chou En-Lai, recovered in 1975, slumped in 1976, and appeared to have stabilized by mid-1977, preparatory to a resurgence of 'progress' in the Western sense. In the latest phase of the societal development of over 600 million people, the People's Republic of China will undoubtedly develop further in a unique manner.

The complete isolation of former times is possibly now impossible because of fast communication and even satellite surveillance. The Chinese need Western technology. The Administration of Chairman Hua said so. The problem is how best to pay for the equipment and the technology. In October, 1977, China was said[386] to have foreign debts due for repayment in 1977 of $1 billion, of which $650 million was due on equipment bought in the early 1970s. Similar annual payments were to be necessary until at least 1980. The 1976 surplus was thought to be about $700 million and imports had been cut in early 1977. The Chinese have had the experience of the difficulties of paying the Russians for the equipment bought before the breakdown of relations in 1964. They remember also the inflation produced by the loans they got in 1949. These memories, added to the ideological fear of dependence on foreigners, make the Chinese chary of borrowing. This is illustrated by the seesaw in their trade with Japan and the easy terms which they were seeking in late October 1977, for the latest deal. Such problems will have to be solved if they are to acquire the equipment and technology which they now seem to admit publicly that they need. Their reluctance to borrow is a factor which has, and may continue to inhibit their industrial growth and that of their oil production.

If the crude oil resources of China are of the order of 25 to 35 Gt, and their financial and technical problems are solved, their impact on the world at the turn of and into the first quarter of the twenty-first century could be very important. On the other hand, if the more unlikely scenario, that China has only about 5 Gt of recoverable oil, then curve C of figure 30, shows a widening gap with a possible demand curve D, and China will be important as a buyer in a sellers market.

The USA and energy self-sufficiency
The national energy-sufficiency policy of the USA is also of vital importance to the rest of the world. The original target date of the Nixon 'Project Independence' moved from 1980 to 1985 and since October 1973 there has been increased talk. Domestic politics between Congress and Administration, have made impossible the evolution of any definition of firm policy, except, and this is important, an obvious strong, general intention to become as energy-independent as possible. The manner in which it strives towards independence is the crucial point. Complete self-sufficiency in energy would be a very inhibiting policy. The USA is not self-sufficient in a number of other resources, and international trade is a most important feature of the diplomatic, political, commercial and social world. The USA is 100% dependent on the import of the platinum group of metals, of sheet-mica, chromium, strontium, cobalt, tantalum, columbium, and almost 100% on import of manganese. The harvesting of metallic nodules from the ocean floor might assist with some of these deficiencies. The USA is 80% dependent on the import of some other resources, with larger quantities often involved, such as asbestos, aluminium, titanium, tin, mercury, bismuth, fluorine and nickel.[244] This list illustrates the USA's interdependence on others. Self-sufficiency might be selective as to fuels, for example the USA might wish to continue to import heavy fuel oils from Venezuela and natural gas from Canada in order to retain the connections. There can be many variations. They may differ with time. They will depend on the results of research and development into alternative energy sources and the degree to which those results are

imposed on the country or allowed to develop in the free market. All new development will need capital and manpower and involve financial and perhaps environmental risks probably needing both support and regulation by the Government. In discussion earlier of the USA energy situation the need for Government support has been noted and, indeed, the USA energy industry is already more dependent on its own government than in many less obviously 'democratic' countries.

It was in the USA that the excessive environmental conservation movement, now world-wide, was initiated by the impending suffocation of Los Angeles. It was in the USA that there was most public fuss and bother, and conflict between industry and the wild-life and clean air conservationists, resulting in some of the most stringent anti-pollution laws which have been enacted. In other countries there has been a more calm and reasonable approach to many aspects of pollution. Even in the USA, examples like the clean-up of central Pittsburgh show that anti-pollution measures can be successful. The concern for possible environmental pollution caused by the production and use of energy has altered the pattern of fuel use and substitution.[204] However, the quantification of the values of environmental circumstances is still very vague indeed. So much talent is available in the USA to be applied to attempting greater meaning in such matters that a great amount of literature is being and will be produced. Much will be of great interest. However, care must be taken to scrutinize any results, whether of fact or opinion, as to the groundwork. No attempt should be made to apply them beyond their basic assumptions. There are many basic differences between the forms and cures for environmental pollution even in the industrial or urban areas, of the USA, Japan and Europe and greater ones with other countires. Environmental consideration will affect demand and supply in all countries but in different ways according to local circumstances.

Similar argument applies to energy conservation, as had been noted earlier. Claims may be well founded that energy conservation estimates for the USA may be derived from an economic model in conjunction with basic heating information.[245] The basic data and model must be changed for other places.

In summary, on the USA self-sufficiency policy and its impact on the world energy supply and demand pattern, a few important points can be made. In every chapter of this book there has been mention of the situation in the USA because the USA is the best-known, most publicly discussed and documented energy economy. Many lessons can be learned but these are not always as universally applicable as some might think. The degree of self-sufficiency towards which the USA will strive and the methods adopted will depend on the reaction of the USA Government to internal and external pressures and its resultant actions. The shifts in supply/demand pattern in the two major, world-wide aspects of environmental and energy conservation will be peculiar to each individual country and not as the USA develops its reactions. Admitting though that complete energy self-sufficiency is not an acceptable goal, there seems to be considerable confusion at this time owing to the wide variation in forecasts of possibilities. With such variations in possibilities, we agree with the opinion of the 1975 Report of the US Academy of Sciences[25] that a spot supply/demand energy forecast for the USA is extremely difficult, yet, unless there is some further crisis, the pattern of the US energy mix in 1980 will be basically similar to that of 1974.

The most important single factor in the disposal of the crude oil which might be available on the world market may well be the import policy of the United

States. This is a comparatively new factor. A 1968 study[397] of US interests in the Middle East stated that 'the Middle East is still the principal supplier of petroleum to Europe and Japan and is likely to hold this position for the forseeable future'. There was no mention of the possible needs of the USA. Schurr and Homan[400] in 1971 were aware of the possibility, but said 'the degree of dependence (of the USA on oil imports from the Middle East and North Africa) can be kept at a very low level relative to consumption, if the United States chooses to continue to pursue a policy designed to achieve such a result'. The big question is have they had such a policy and will they so choose now to pursue such a policy? It has been mentioned that the Nuclear Energy Policy Study Group (NEPSG)[320] on whose work much of the Carter Administration's Plan appeared to have been based, ignored the oil import implications of their policy recommendations. Yet in the earlier consideration of USA oil imports, it was noted that these might be of the order of 700 Mt (31 EJ), in 2000, and there was little sign of moving away from an increase in imports from the Middle East.[412] A July 1977 estimate, quoted earlier, suggested 1985 imports at 600–650 Mt. In the first nine months of 1977, crude oil and produce imports were running[401] at a rate of 440 Mta, (up 65 Mt from the same period of 1976). Indeed there seemed a change in pattern, as power companies turned to fuel oil to generate electricity, because of the fears of natural gas shortages, which seemed at variance with the Carter Plan to switch to coal.

Table 80 gives the crude oil imports into the USA, in megatonnes and as percentages, from the main supplying countries from 1971. The pattern was disrupted in October 1973 to mid-1974 by the Arab oil embargo. Despite this, there are some obvious trends. Imports from Canada will cease from 1980 according to the declared policy of the Canadian Government. The increasing weight of African imports is noticeable; those from Algeria are included in the total OECD data, being 9 Mt in 1974, 13 Mt in 1975 and 21 Mt in 1976. If the optimists, and particularly the Mexican optimists, are correct then Mexico should be seen appearing as a crude source. The OECD figures do not seem to coincide with the UN data for the Caribbean. The increase in share and quantity of the Middle East and of Saudi Arabia in particular in 1976 is noticeable, as are the increases in 1976 over 1975 of Iran and the United Arab Emirates. If as suggested earlier, OPEC, outside the Middle East, and the minor Middle East producers maintain their productions, then increased USA supplies would be expected to come from Saudi Arabia. Unless there is a marked change in political thinking in Iraq, it is probable that Iraq's production for export will go to Eastern Europe preferentially.

US imports of petroleum products are shown by origin in table 81. The OECD figures for 1975 and 1976 do not conform strictly with UN useage. The bulk of the product imports from the Caribbean are from the Netherlands Antilles and the US Virgin Islands. There may be some future increase here, as more oil from South America becomes available outside Venezuela. Export products from Venezuela cannot be expected to increase until, perhaps, their very heavy oils are exploited. Whatever method of extraction is developed it may involve semi-refining, at least, on site. Mexico can be expected to try to gain the added value of exporting products rather than crude, but cannot be expected to have either the funds or industrial effort available for extensive refining for several years.

Returning, for the moment, to the shorter term and more general energy aspects in the USA; in September 1977, there were three interesting developments.

Table 80

USA imports of crude oil by country of origin, 1971–1976

(megatonnes)		1971	1972	1973	1974	1975	1976
Total world		84	111	161	174	207	304
AFRICA	Total	9	24	42	48	60	101
	Libya	3	5	6	0.3	11	25.7
	Nigeria	5	13	26	35	36.8	53.8
CARIBBEAN	Total	16.6	14.7	23.3	20	19.9	*
	Venezuela	16	13	18	17		18.2
MIDDLE EAST	Total	17.6	21.5	32.3	49.3	55.5	110.6
	Iran	5	7	11	23	13.7	28.2
	Iraq	0.5	0.3	0.3	–	–	1.5
	Kuwait	2.8	1.8	2.0	0.2	0.2	0.1
	Oman			0.3	0.03		
	Qatar		0.17	0.35	0.82	0.9	3.8
	Saudi Arabia	5	9	15	22	34.8	60.7
	UAE	4	3.7	3.4	3.3	5.7	16.2
FAR EAST	Total	5.6	8.2	10	14	18.7	26.5
	Indonesia	5.5	8.0	10	14	18.7	26.5
CANADA		35	42	51	40	33	21
(percentages of total crude oil imports)							
AFRICA	Total	10.7	22	26	27.6	29	32.2
	Libya					5.3	8.5
	Nigeria	6	12	16	20	17.8	17.7
CARIBBEAN	Total	20	13	14.5	11.5	9.6	
	Venezuela	19	12	11	9.7		6
MIDDLE EAST	Total	21	19	20	28	26.8	36.4
	Iran	6	6	6.8	13.2	6.6	9.3
	Qatar						1.2
	Saudi Arabia	6	8	9	12.6	16.8	20
	UAE					2.7	5.3
FAR EAST	Total	6.7	7	6.2	8	9	8.7
CANADA		42	38	32	23	16	7

* other South & Central America 12 Mt

Sources: 1971–1974, UN Series J, 1950–1974
1974–1976, OECD, *Quarterly Oil Statistics*, 2nd Qr., No. 3. 1977 includes NGL and Feedstocks

Table 81

USA — Imports of petroleum energy products

	1971	1972	1973	1974	1975[1]	1976[1]
Imports of petroleum energy products (including bunkers) by main countries 1971—1976 (in megatonnes)						
Caribbean	60	65.5	73	70	—	32[2]
Venezuela	38	37	41	35	15.5	24.6
Canada	7	13	17	14.5	7.1	6.5
Europe	7	9.5	16	10.4	0.5	4.4
GRAND TOTAL	118	133	158	137	91.7	75
TOTAL IMPORTS CRUDE & PRODUCTS	202	244	319	311	—	—
TOTAL CONSUMPTION	628	679	714	683	—	—
Imports of petroleum energy products (including bunkers) by main countries 1971—1974 (in US barrels)						
Caribbean	416	455	508	488		
Venezuela	262	258	288	241		
Canada	49	93	118	101		
Europe	49	66	109	73		
GRAND TOTAL	819	924	1099	953		

Sources: 1971—1974, US Bureau of Mines, 'Energy Perspective 2', Dept. of Interior,[3] 23 June 1976
1975 & 1976 (1) OECD
(2) This figure is given as Other South and Central America. OECD gave no figure for Caribbean.

The Carter Administration circulated a Bill which would speed-up the bureaucratic procedures and enable nuclear plants to start generating power six years after they were proposed. A Federal Appeals Court overruled a lower court decision which had inhibited drilling off the US Atlantic coast. Forty oil companies paid in total in 1976 some $1.1 billion for the right to drill and they now have a better chance to start operations. In contrast, the restraints on the coal industry were increased rather than lessened in the mid-months of 1977. Hence the US policy on energy was still very much under indecision. Meantime also, as might be expected, new figures were offered as practical targets. The US General Accounting Office (GAO), in July 1977, suggested that the US would have to increase oil imports to 10.3 mbd by mid-1980s, but in October revised this figure to between 12 and 13 mbd, (600–650 Mt). This is twice the target of the Carter Plan for 1985. Table 82 summarises some of the forecasts of US oil imports.

Table 82
USA — Some estimates of future oil imports
(megatonnes and exajoules)

	1980		1985		1990		2000	
	Mt	EJ	Mt	EJ	Mt	EJ	Mt	EJ
OECD[329] Base Case	455	(20)	477	(21)				
Accelerated Policy			205	(9)				
Dept. Interior, USA[323]			500	(22)	523	(23)		
Carter Plan, April, 1977			295	(13)				
US Oil Company[364]							705	(31)
US General Accounting Office			600–650	(26.4.-27.6)				
Exxon[488]			570	(25)	650	(28.6)		
Carter 1979[516]			425	(19)				

Sources: as referenced

These are the numbers which are so important to the rest of the world. There is a range of 445 Mt (9 mbd) of oil in 1985 which could otherwise be available to the rest of the world, and the figures for 1990 and 2000 indicate possible later figures.

NUCLEAR POWER

Nowhere in the energy industries is the power of the governments more obvious than in the nuclear energy field. Nowhere are the figures for the near future, but particularly in the far future, more variable with the author. The nuclear energy industry men see their role as providing, as quickly as possible, the means to fill the increasing gap between high demand and the rapidly diminishing supplies from fossil fuels. The coal men see their own 'enormous' resources as the world's energy sheet anchor and nuclear energy as required only to supplement their efforts. The petroleum industry men normally deny the speed of decline of their available resources but welcome nuclear energy and, indeed, coal. These they consider as rightful base load resources for primary energy needs in order to release petroleum for exploitation of its chemical characteristics. Table 83 sets out the forecasts by the International Atomic Energy Agency made early in 1973 and in 1974 of the needs for nuclear energy as they saw them at those times, so illustrating the impact of the late 1973 events.[253]

It is interesting that in both exercises the demand is kept the same, without any acknowledgement that energy conservation might or, more importantly, should reduce demand.

The European Economic Community in mid-1974 saw a 10% reduction in energy demand in 1985 from that forecast in early 1973. The increasing need for nuclear power, however, is undoubted as illustrated by the EEC Commission's energy sub-committee forecasts of mid-1974,[254] table 84.

These two tables can serve as a record of opinion on the potential of nuclear power in 1973 and 1974. Table 84 gives the 1974 estimates for 1990 of world total primary energy demand as 155,000 TWh, of which the nuclear share at 24% would be 37,000 TWh. Converting these numbers to oil equivalents, they become roughly 12.44 Gtoe and 3.0 Gtoe, (1 GWe = 6 TWh, 1 TWh = 3,600 TJ, 1 GToe = 44.86 EJ).

In 1979 the Uranium Institute Report[515] suggested low and high estimates for 1990 of 410-530 GWe nuclear capacity. These estimates were made before the Kemeny Commission Report on the Three Mile Island incident which resulted in the suspending of current projects and cast doubt on any new reactors being ordered in the USA before 1983. The revolution in Iran resulted in probably three if not the four power plants under construction or planned

Table 83
Forecasts of the role of nuclear energy in the world
(excluding China)

	1980	1985	1990	1995	2000
Installed nuclear capacity (10^9 kW)					
Early 1973	0.3	0.7	1.4	2.4	2.6
1974	0.3	0.9	1.9	3.4	5.3
Nuclear electricity generation (10^{12} kWh)					
Early 1973	1.8	4.2	8.5	14.4	22.0
1974	2.0	5.5	11.7	20.7	32.7
Nuclear heat generation (10^{12} kWh)					
Early 1973	5.6	12.7	25.4	42.5	62.7
1974 - for electricity generation	5.9	16.5	34.8	56.4	80.0
- for other uses		1.0	2.3	6.0	14.1
Total primary energy demand (10^{12} kWh)	96	122	155	191	236
Nuclear share - early 1973(%)	6	10	16	22	27
1974(%)	6	14	24	33	40

Source: 'The Role of Nuclear Power in the future energy supply of the world', J. A. Lane *et al*; IAEA, *IX WEC Preprint* 4.1-22, 1974, Tables 5.1 or 5.2

Table 84
EEC forecasts of energy source supply 1985
(million tonnes oil equivalent)

Source	1973 Forecast	%	1974 Forecast	%
Coal	175	10	250	16
Natural gas	265	15	375	24
Oil	1120	64	655	41
Nuclear	160	9	260	17
Hydro	40	2	35	2
TOTAL	1750	100	1575	100

Source: EEC as reported in *Petroleum Times*, 78–88, p. 12
28 June/12 July 1974

being halted. The repercussions of the Three Mile Island incident elsewhere may also have an inhibiting effect on new construction. Hence in late 1979 the lower of the Institute's figures for 1990 seemed more likely. Adding an estimate[531] for the USSR and Eastern Europe of 140 GWe would make a total for the world (excluding China) of 550 GWe of world nuclear capacity, which is roughly 0.26 Gtoe.

In brief, estimates of the possible contribution of nuclear power to world energy supply some 16-11 years ahead, were reduced over ten times within the five years, 1974-1979. This is a most important factor for the future. In the personal forecast of world energy supply and demand, as seen on 7 December 1979, and illustrated in figures 30 and 31 and table 87, the world total energy demand of 12.44 Gtoe would not be reached until about 2005 and the nuclear contribution of 3 Gtoe not until 2025, some 35 years after the 1974 forecast (table 84).

One of the major influences on the contribution of nuclear energy will, as in the past be the attitude of the USA, despite the intention of others to continue their programmes for both thermal reactors and the development of fast, breeder reactors. The US nuclear industry reacted quickly to the strictures of the Kemeny Commission, indeed they moved before its publication. The electricity industry started moving towards policing itself, which was the practice in the UK since about 1960. Three important new organizations were set up after the Three Mile Island incident; an Institute of Nuclear Power Operators for training operators was set up in Atlanta, Georgia, on the initiative of Chauncey Starr of the Electric Power Research Institute, according to David Fishlock (*Financial Times*, 23 November 1979); a Nuclear Safety Analysis Centre was set up at Palo Alto at the US Government Nuclear Research Establishment; a new mutual insurance scheme was being organized.

THE DEVELOPING COUNTRIES

The energy patterns and future trends within the hundred and forty or so developing countries are very difficult to assess. A review[532] of 90 of these countries in the period 1970–1976 illustrated the wide diversity of these countries, even when grouped as in table 85.

Table 85
Ninety less developed countries in 1976

		(% of total LDC totals)				
	No of countries	Popu-lation	Consumption		Net Oil Imports %	GNP/ capita US $
			Energy	Oil		
Oil importing developing countries						
Upper income	7	10	24	30	48	1210
Middle income	32	16	16	20	36	540
Lower income	25	49	20	9	16	140
Total oil importing developing countries	64	75	60	59	100	370
OPEC oil producers	13	16	24	23	–	960
Non-OPEC oil producers	13	9	16	18	–	860
Total developing countries reviewed	90	100	100	100	100	490

Source: ref. 532.

The source gives details of the analysis by countries, but this summary clearly shows the disparities between even these five groups in all the items listed. These and the variations in social systems, political power, religions, and so on, had militated against any real progress in such matters as the provision of technical aid from the industrialized countries, as illustrated in the lack of success, or apparent tangible results arising from the UN Conference on Science and Technology for Development, when 4,000 delegates from 135 countries met in August 1979. The task of the analyst to discern trends and look ahead to the possible needs of this very large group of countries in the energy field is a daunting one, but is a challenge which must be taken up in the 1980s.

10
Conclusion

There are large energy resources of many kinds in many parts of the world. The measure of these resources is inaccurate and lacks sufficiently widespread knowledge to reconcile the different concepts and technologies to gain a comparability which is warranted by the growing interdependence of energy resource utilization. Much work lies ahead in evaluating even true proved reserves of even the common energy resources of coal, oil, natural gas and uranium in every-day use. Indeed this might well be considered the first priority in energy resource evaluation at this time. So much theory is being based on so little fact that decisions dangerous to many people may be made unless there is greater knowledge and greater understanding more widely spread. Moving back into the more speculative areas of probable reserves and so to resources, there is a great need for greater rigour, discipline and scientific integrity to be deployed in the exercises which must be done as an unending continuance. Meantime, in the field, the explorer must be given every support in his quest to find new resources and, with the engineer, convert them to reserves at the highest possible recovery factor, no accountant or politician ever found any new energy resource deposit. I repeat, the first desk priority is to improve our knowledge of the working inventories, that is the proved reserves.

The second desk priority is to obtain greater understanding of energy needs and costs, just as the top field priority must be to eliminate waste, and so transport, store and use energy with the maximum efficiency and minimum cost.

Finding, extracting, transporting, storing and converting of the proved reserves to end-use forms one half of the economic cost. The economics of energy use are the other half of the total cost and are equally difficult to determine precisely. Particularly is this so in the area of difference between demand and need, where conservation of energy is possible, and in the area of acceptable environmental conservation costs, where the behavioural sciences are too immature to give adequate guidance. Yet the total cost of the energy mix and its components should be analysed from the individual through country and region to the world, with due consideration of the physical, economic and political environments, before the energy supply and demand position can be assessed and so the availability of the world's energy resources evaluated, which is the third area for study, and the final objective.

The methods of analysis and assessment are under constant improvement. The use of computers and mathematical models has improved enormously the opportunities for application of probability methods. However, it is the range of the tides and not the length of coastline which is the critical factor in estimating tidal power potential. The volume of sedimentary rock in an explored basin is a figure of little significance in total without some estimate of characteristics and these can be very difficult to determine without detailed exploration. The input into any analysis must be relevant and reliable, or consciously and knowingly accepted as unreliable. This book has attempted to indicate the degrees of unreliability of many of the apparently precise basic figures, yet one can only use those figures which are available and apply best judgement.

Judgement is, however, least reliable in times of transition. Although the energy scene is always dynamic perhaps the present is so hyper-dynamic that any

spot assessment of world availability would be dangerous. A number of factors are undergoing major change simultaneously, affecting all aspects of both supply and demand. Amongst these are:
- dominance of a single resource, petroleum
- shifting control and systems of that resource
- shifting centres of wealth
- shifting centres of political power
- changing environmental considerations
- speed of technological change
- challenge to the ethic of economic growth.

Dominance of petroleum

Coal was the dominant single source of energy in international trade from 1800 to about 1930. Outside Europe, the USA and a few scattered places, energy until about 1900 was supplied predominantly by manpower and the burning of wood, peat and dung, as it still is in some places. Coal had a limited use to make heat directly or through manufactured gas. The position of petroleum as the dominant energy source is very different. Petroleum created the 'mobility revolution' and air, road and sea transportation are entirely dependent on oil. Its products are pervasive in every field, through the young petrochemical industry, to plastics and fibres. Liquid petroleum being far more easily transportable has also a wider distribution than coal ever had. Yet by its use in transportation it has made the world effectively smaller through faster communication. The increasing volumes involved in achieving and maintaining that dominance have a number of important implications. Concentration of reserves in the Middle East means that whilst the loss of supply from one country in 1951 could be made up from others, the reduction of supplies by the Arab producers in 1973 could not be filled from any other source. A production of 150 Mt of crude oil from the North Sea in 1955 would have equalled that of the Middle East, but in 1974 it would have been only 13.9%. The effort required to find alternate sources for these volumes has been illustrated earlier. Furthermore the size and importance of these volumes of oil in the life of nations has brought governments into the energy game. As yet many governments have an immaturity and lack of knowledge which leads to uncertainty and instability at this crucial time of transition from the end of the first era of cheap energy in October 1973, to an era of costly energy. The earlier discussions on energy resources emphasized the imperfections of our knowledge. Even so it would seem that petroleum must remain the dominant energy source for a decade or so. By the end of the century petroleum will, indeed must, become merely one non-dominant element in a mix of many different energy sources. Perhaps the second era of cheap energy may dawn early in the twenty-first century.

Shifting control and mechanism

The system of virtual control by a small number of major integrated oil companies which developed in the first half of this century, though occasionally challenged and less frequently beaten, as by Mexico, was a stabilizing factor in the energy field as supply grew. Its control of the world scene has now broken though its power was still visible in 1973 when industry was able to avoid the chaos which would have resulted from the complete implementation of the political use of oil supply by the OAPEC countries. The system of large concessions of exploration and exploitation rights which had been of enormous, mutual benefit to under-developed countries and, of course, also to the consumers up to the 1930s, was eroded in the 1940s and virtually disappeared in the 1960s. Different systems, more or less attuned to differing local circumstances have and are being developed in many parts of the world. But all are in a state of flux, even the rules in the North Sea are under constant change. The international oil industry still virtually controls the majority of the tanker transportation of the world and, to a lesser degree, pipelines. The refining phase of petroleum was never, internationally, as closely held by the majors as was supply. The growth in the 1950s of scattered refineries located for national reasons, often losing economies of scale and distribution efficiency, and the increase in products' volumes, diminishing the advantages of consumer based refineries, has again resulted in a position of greater freedom of choice of location and control. In the products marketing field there is growing interference by government, even in the non-centrally controlled economies. Thus in all phases of the industry the mechanisms are in a particularly changeable situation, aggravated by the so obvious shift of actual control of supply to the producing countries. Yet even here there is not the uniformity which the statistician or engineer would welcome. The trend towards 100% take-over of the oil companies, even in the Persian Gulf, is moving at different speeds in different countries as they appreciate in different ways, the degrees to which, or the methods by which they may retain the optimum expert services of the companies. Once again, there are many aspects in the shift in control and mechanism of the industry and 'the water is very muddy'.

Shifting centres of wealth
The possible impacts of the newly-acquired wealth of the petroleum producing countries, the availability of investment funds and some aspects of the international monetary system have been discussed in the previous chapter on supply and demand. Some aspects of inflation were mentioned when discussing Japan. It must not be forgotten that, in October 1973, the world in general was suffering from economic and industrial recession in varying degrees and the oil shortage and price rise had a far greater effect than if a world-wide boom had been the circumstance. Usually in recession there is a general unease at least, if not fear, as to security, employment and progress. The world is still in this state and the obvious shift of wealth and financial power is so new that instability is enchanced. Furthermore, the success of the OPEC countries in increasing the monetary value of their energy resources has triggered other countries with mineral and agricultural resources as main assets to form producer associations and seek greater returns than those which they have had and consider to have been unjustly low. This development has created further uncertainty as to economic growth patterns.

Shifting centres of political power
Since 1946 the world political picture has been a shifting kaleidoscope with the birth of many new nations from the wombs of the old empires. These nations, many only 'in their teens' are still fighting for their own individual national identity but are increasingly conscious of their political power as the 'third world' or as African countries or as Afro-Asian countries or whatever grouping may give them most political strength on any specific question. Such political identities as individual countries or groups can have a profound affect on energy supply and demand. During the protracted discussions on the law of the sea and

the seaward extension of territorial rights, many of the new nations have discovered the complexities of potential energy resources. The anti-apartheid
attitudes directed at South Africa have had an influence on its energy policy but
also are used politically in a way that confuses the African scene as well as other
relationships, like the development of Rhodesia.

The shift in economic power to the oil producers is also a shift in political
power. Another centre of significant power has been established, though how
long the common interests will act for cohesion of such disparate components
is debatable.

It might be said that the unsettled conditions in mainland south-east Asia
are no worse than usual. Indonesia seems to have regained political stability and
is developing as both producer and consumer of energy, as are many other
countries of south-east Asia outside the immediate area affected by the Vietnam
wars. It might be said that these areas have little real impact on the world energy
position. Statistically this may be so, but as with the eastern Mediterranean
area, any disorder in a sensitive area might invoke other greater problems.

This is no place for a deep discussion on the balance of power between the
USA, the USSR and China. However there can be little doubt that the relationships between these three powers is by no means as smooth as appearances may
seem. The growing military and naval power of the Soviet Union requires a great
deal of energy for itself whilst the implication of its use is impossible to assess
here.

Despite the hopeful or, perhaps, wishful thinking that dispersion of political
power in many centres, the multi-polarity theme, and 'détente' between the
USA and USSR and later the involvement of China with the rest of the world
would lead to stability, this has not been translated into fact. Although the
world, one hopes, is not in a pre-war era, the political situation does not give
confidence that any energy forecast at this time would be very useful.

Changing considerations of environmental conservation
The explosion of public interest in the environment in the 1960s led to many
displays of excessive emotion and muddle-headed thinking. The energy industries were slow to react with knowledge, scientific integrity and 'sweet reason'.
Standards change with time. The mountains of coal waste at pitheads in remote
valleys in Wales or Pennsylvania were not considered pollution a hundred or
even fifty years ago, nor the sulphurous fumes from plants even in Japan ten
years ago, for wealth was being created. Except perhaps in the nuclear field,
much of the excessive emotion had been cooled by the early 1970s but this new
factor in the energy field will remain. The many attempts by many bodies to
quantify an acceptable environmental conservation cost and give it a true value
in total cost illustrates what is now a generally acknowledged social phenomenon. Again, however, there must be a wide range in the costs applicable in
any one place and grave problems in reliable estimations because of the major
differences in environments.

The speed of technological change
Future technological change is one of the most difficult to assess of the many
dynamic factors affecting availability of energy resources. Whilst, as was said at
the beginning of the last chapter, the classic economic models under-estimated
the impact and speed of technological advances and no clear quantification is
yet possible, perhaps it never will be. Age-old adages like 'necessity is the mother

of invention' still hold true, whether plainly said or wrapped in current jargon. Most of the attempts to apply lessons from the past too often use an arithmetical approach rather than the historical or research approach which requires consideration of the whole environment which produced each figure. In this broad context much more is implied in the word 'environment' than the impact of the Clean Air Act on the consumption of coal in central London or central Pittsburg. Examples of the importance of this approach, when attempting to discern trends in technological advance, have been given earlier. However, perhaps the present day speed of the acceptance of innovation in most countries and across the world should be stressed here. It may be that the textile industries have energy cost figures showing variations in input into articles of clothing as fashions change but the fast speed of communication, when within less than a year young women's fashions encircle the globe, must have a major but not necessarily simple effect.

Changing ethics

The 1960s seemed to be years of social unrest throughout the world, with many challenges thrown down to the established concepts and ideas of living. Much has been spoken and written on this phenomenon. Some unrest spilled over into and/or from concern with the environment. Many ideas were polarized in discussion sparked by *The Limits to Growth* report by the Meadows team to the Club of Rome. It was argued that mankind could only survive by curbing the growth in population and abrogation of the ethos that progress could be measured only by increasing economic growth and material comfort. Public debate was stimulated, some aspects have been discussed earlier and there are some signs of change. Japan led the world in economic growth in the 1960s, but is now planning on a much slower future growth. Its former fast economic growth had aided a decline in population growth rate, and perhaps this will hold. It may, however, be difficult for Japanese to change their habits of intense activity and dedication which created the resurgence since 1950. If they do, then they again may lead but to a less frenetic scramble for material riches. There are stirrings in almost every country against the more obvious anomalies in the distribution of wealth and comfort. However it is difficult to imagine that the bulk of the people in the developing countries will not take as their goal the highest 'standards' set by the affluent developed countries, even if the more developed countries come to accept different standards. The second report to the Club of Rome, *Mankind at the Turning-Point* by Mesarovic and Pestel discusses such problems. There are few obvious signs of real austerity campaigns in any country based on philosophical grounds, yet the reaction against waste is a sign of change.

There are many ideas as to the ideal and possible types of society which should or will evolve over the next thirty to fifty years. Are the nations which have developed into industrial societies nearing the end of an evolutionary path, with the costs of maintenance increasing faster than productivity? Have basic standards to be revised? Will energy scarcity drift nations into increasingly institutional macro-planning? If multi-national research into nuclear fusion is successful, will this lead to multi-national construction and planning of supply allocation? In contrast, if small solar energy units were developed early, would they allow micro-planning and greater individual freedom? In the petroleum field, the increasing difficulties and therefore costs of exploration and distribution from hostile environments, and, in the natural gas sector, the high cost of ocean transport, and the rigidity and cost of collection and distribution systems, tend to macro-planning and control. There must be an economic limit to the declining 'productivity' of capital investment. Has that limit been already reached? The past few years have seen the oil and gas companies increasingly arguing for premium use rather than volume growth. Will that tendency spread to a closer definition of need, rather than desire? Will efficiency be redefined and include societal costs? The income difference between employed and unemployed is marginal in some welfare states. Diseconomies of scale have succeeded economies of scale in some industries. Will antagonism to very large organizations and the power they can wield, whether political, industrial, commercial or financial, lead to a resurgence of individual self-sufficiency? Can 'self-employed' and 'small business' regain a social value in the democracies? Have the Chinese, with the commune system of self-sufficiency noted earlier, leap-frogged into a life pattern to which the centralized industrialized developed countries may have to adapt? Will the trend to substitute capital for labour be reverted? Can that reversal be effected with the minimum loss of comfort and mobility, which cheap energy has given in excess to many in the 'Western World', and to which the developing countries seem to aspire?

Answers to these questions will have a profound effect on the future world energy supply and demand pattern. The equilibrium in the year 2050, or even 2000, need not necessarily be similar to the equilibrium achieved in 1977. The current balance is based on current economic and political actualities, rationalized by the economists into current economic hypotheses. The wide divergencies in estimates of economic and energy growth rates, and the paucity of firm data behind many assumptions, emphasize the shaky foundations on which forecasts are built.

In the concluding chapter of the second supplement to the first edition of this book the stand was taken that the 1977 energy studies with their predilection for the 'scenario device' left the decision-maker with a minimum of guidance. It was assumed that the reader appreciated the dynamism of the subject, the shifting bases for calculating growth and the increasing complexities of the social and political factors involved. On that assumption a preferred diagram of the world energy picture as seen on 5 December 1977, was constructed and presented then as figure ten, based on table 19. These are repeated within figure 30 and table 87.

The bases have shifted. Continued slow world economic growth has lowered the base for future projections. Technological innovations, such as micro-electronic development, have however emphasized that technology is a man-made resource and there is every indication that technology is only in an early phase of development which should exhibit exponential growth.[534] It is impossible to be precise in this area, but whilst long lead times and slow substitution patterns may still hold, there are, nevertheless, some areas where demand for energy may be amenable to reasonably fast change.

A major shift in 1978/1979 was the growing feeling amongst oil producers, but particularly amongst the OPEC members in the Middle East, that they should only produce the amounts of oil which would provide for their own energy needs and funds for their own development, as they saw the need. The corollary was that if higher prices on lower volumes could provide adequate funds then that route was preferable, because it not only preserved their natural resources for the future but reduced investment and overcame any lack of indigenous labour and management skills. At the end of 1979 it seemed that some oil producers had forgotten the lesson that Dr. Mossadegh of Iran had learned in 1953/1954 – a resource only has value when it becomes a supply. A rate of production of the order of 1.5 Gta (30 mbd) from the Middle East, which would be higher than many forecasts since mid–1979 and more commonly taken as total OPEC production, would, as discussed earlier in the Production chapter, only consume 30 Gt of Middle East proved reserves of 51 Gt or the 68 Gt if the proved and prospective reserves of Halbouty and Moody were taken.[482 A] This would leave at least 21 Gt untouched proved reserves. Would these or the remaining 48 Gt of proved and prospective reserves, let alone the possible reserves and the additional resources, even all be produced? The world's dependence on Middle East oil as in 1980 must have been diminished during the twenty years to 2000 by the exploitation of other energy resources, if the industrialized countries were to survive and the non-industrialized countries were to develop.[535]

Although practical conservation of energy was very slow to take hold, particularly in the USA, there were during 1979 signs that consumption was being curtailed. In September 1979, even after the removal of some restrictions imposed by some states, there was not a significant jump in gasoline sales in the USA. In November the European members of OECD confirmed their intention of holding oil imports down; the EEC states had agreed to keep total oil imports below the 1978 total of 471 Mt at least until 1985, and in early December the members agreed on individual allocations for 1980. Indeed the forecasts of EEC energy prospects as seen at that time by the EEC Commission can serve as an illustration of the changes in thinking which have taken place amongst the industrialized nations during 1979, and can be compared with table 84.

In 1973 the EEC forecast for energy supply required in 1985 was 1750 Mtoe but revised in 1974 to 1575 Mtoe with an allowance of 260 Mtoe for nuclear power, whereas the 1979 forecast is for only 204 Mtoe in 1990.

Coal supplies and coal-based gaseous and liquid fuels as alternatives to petroleum had not shown by 1979 signs that there was as yet the political will to overcome the many constraints inhibiting the contribution which most analysts consider that coal can and must make. Even in Britain with large coal reserves some were saying that significant coal imports would be necessary in 2000. Such attitudes become self-fulfilling. In a world climate of increasing shortages of energy supplies for the 1980s and 1990s only vigorous positive action to harness all possible sources should be tolerated by any government with the interests of its people as a priority consideration.

The growth of nuclear power in the late 1970s was inhibited specifically both by the Carter Administration's fear of the proliferation of the capability of nuclear weapons manufacture and by the strength of the anti-nuclear lobby. In the USA the incident to a reactor at Three Mile Island, Pennsylvania, despite its positive safety lessons, caused the US regulatory body to withhold permission for expansion, and was taken by the anti-nuclear people as vindication for their point of view. In Europe where the need for additional energy sources is more pressing, there was less concern, yet, as noted above, there has been a marked scaling down of what is considered to be the role of nuclear power. In table 84 the nuclear contribution in 1990 is under 15% of the total energy consumption.

Political factors became increasingly obvious, indeed dominant, during 1978/1979. The overthrow of the Shah of Iran, in particular, and the ensuing chaos, started the 1980s with a major query as to how far militant Islam might affect the whole of the Middle East and North African petroleum supplies.

On the financial side, the dangers to the world economy of the shift of money to the producers of petroleum which were so vividly clear in the mid-1970s did not materialize. However, the Iranian chaos and particularly the confrontation in November 1979, with the USA, led to the freezing of funds and threats of cancellations of debts, which shook confidence in the safety of investment of petrodollars. Threatened punitive action against the 'windfall' profits of oil companies and redistribution by the USA Government in particular, not only re-inforced the obvious growing importance of governments in energy affairs but also indicated the fragility of the world financial establishments.

Table 86
Energy prospects of the European Economic Community
(million tons oil equivalent)

	1973	1978	1990
COAL			
Production	200	174	194
Net imports	19	26	57
Consumption	222	204	251
OIL			
Production	12	63	87–147
Net imports	589	472	497–572
Consumption	593	535	644–659
GAS			
Production	114	135	115–130
Net imports	3	31	121
NUCLEAR			
Production	14	29	204
Net imports	—	—	—
Consumption	14	29	204
OTHERS			
Production	25	32	39
Net imports	2	3	4
Consumption	27	35	43
TOTAL CONSUMPTION	973	967	1,393

Source: EEC Commission, 1978 provisional, as reported in the *Financial Times*, London, 7 December, 1979.

The future energy positions of the centrally planned economies, and especially the USSR and the People's Republic of China, enter the 1980s as objects of greater rather than less controversy and of greater importance, because of the greater sensitivity of the positions of the rest of the world.

Despite all these doubts and unknowns, it is considered necessary to construct another world energy picture from 1950 to 2050 to embody the personal opinion of the author as at 7 December 1979. This revised view is illustrated in figure 32 based on the figures of table 87.

Five major factors developed during 1978 and 1979 which affected the world energy forward pattern, namely:
— the slow growth in supply and political will towards growth in coal;
— the trend of OPEC members to reduce production but increase prices to satisfy restricted national needs;
— the inhibition of nuclear power expansion;
— the resulting need for demonstration of a stronger and earlier development of inexhaustible energy sources;
— a growing acceptance in the industrialized countries of much lower economic growth rates than in the past.

Consideration of the composite diagram and table of the two world energy supply and demand pictures as seen on 5 December 1977 and 7 December 1979, should include the same nine areas of thought as were postulated for the former picture.

1. The OPTIMISM of the continuing increase of energy supplies, despite the lowered targets, must not be confused with complacency.
2. The steady rise in COAL production requires much more effort and political will than has been apparent since the first reaction to curtailment of supplies and the higher price shock of 1973/1974.
3. The CRUDE OIL lens has been narrowed to 2000 but then continued longer as a significant contribution into the next century. It presupposes:
   — availability of 272 Gt over the 1950–2050 period,
   — a recovery of some order from the pricing chaos of 1979,
   — a return to normal financial and commercial practice by Iran,
   — the regaining of trading stability which allows the industrialized, the non-oil-producing developing countries, the oil-producing developing countries and the centrally planned economies to grow in harmony, perhaps through multinational discussion and agreement,
   — vigorous exploration, world-wide, with technological advances overcoming the problems of more difficult environments and crude oil types, and timely development of production capacity within OPEC.
4. The development of massive transportation schemes for NATURAL GAS so that waste can be minimized and also full advantage be taken of the unconventional resources in tight formations, geopressured zones etc. The very long lead times and high costs of such schemes, and the necessary stable political climate, have inhibited development e.g. the 1979 cancellation of a number of major projects in Iran, and so postponed the earlier expected peak of exploitation.
5. The warning from Iran of the dangers of too rapid modernization/industrialization was appreciated by other oil-producing developing nations, including Mexico and Venezuela, but, perhaps, the advantages which accrued to the world economy from the pre-1950 oil industry being self-financing had been forgotten in 1979. The substitution of government-to-government trading for trading in petroleum through integrated international oil companies has also removed a valuable buffer. Security of supplies is not necessarily increased. Governments are not necessarily easy or willing trading bodies, as the difficulties on a number of goods between members of the European Economic Community illustrated vividly in 1979. Governments can ignore trading obligations more easily than companies. Governments have overt political as well as commercial aims. Governments can have protocol restrictions

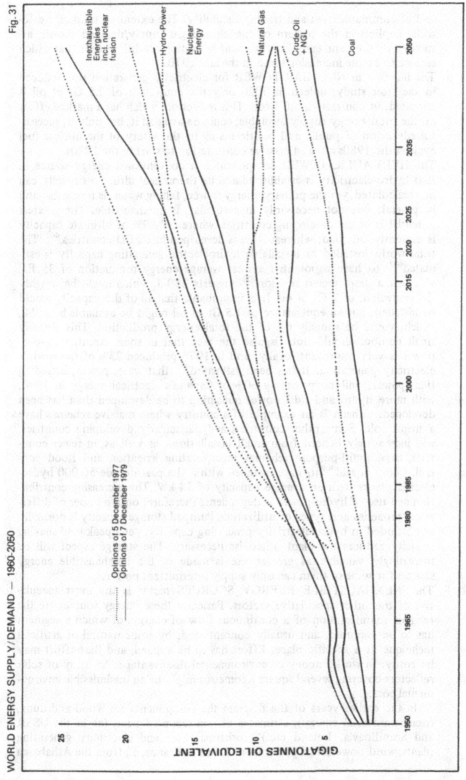

WORLD ENERGY SUPPLY/DEMAND — 1960–2050

GIGATONNES OIL EQUIVALENT

Fig. 31

Inexhaustible Energies incl. nuclear fusion

Hydro-Power

Nuclear Energy

Natural Gas

Crude Oil + NGL

Coal

Opinions of 5 December 1977
Opinions of 7 December 1979

to free communication and trading flexibility. The extent of trading, world-wide, implicit in the pattern of the oil and gas supply/demand shown, assumes a marked and early improvement in trading conditions over that which seemed to be the increasing trend in the late 1970s.

6. The increase in NUCLEAR POWER for electricity generation was reduced in the later study, indeed in 2020 only the equivalent of 2.5 Gt of oil is allocated, in contrast to 6.0 Gtoe. This reduction, which has a marked effect on the total energy supply available, could be avoided if, but only if, successful education of public and politicians as to the safety of the nuclear fuel cycle in the 1980s allowed massive construction to start in the 1990s.

7. The HYDRAULIC POWER component is a conventional energy source in that hydro-electricity is established and its theoretical ultimate capacity can be recalculated, yet the primary energy source, falling water, is renewable and in general, but not necessarily in particular, is inexhaustible. The greatest potential is in the developing countries, where only 7% of ultimate capacity is currently installed, whereas 46% is developed in OECD countries.[408] The total world installed or installable hydroelectric generating capacity is estimated[409] to have a potential annual average energy production of 35 EJ, although a more recent revision[408] suggests 25 EJ, which might be roughly the equivalent of 2 Gt of oil. It is improbable that all of this capacity would be installed, but an equivalence of 1.5 Gt of oil might be available by 2050, which would be roughly 6% of the total energy production. This globally small number should not disguise the fact that in some countries hydro-power is very important locally, and in 1976 produced 23% of the world's electricity generation. It has been estimated[410] that water power, including tidal power, will be providing 60% of Canada's electrical energy in 1990, with more hydro and tidal power remaining to be developed than has been developed. Canada is an example of a country where massive schemes have a major role. Some other countries, and particularly developing countries, will increasingly benefit from small installations, as well as, in fewer countries, large multi-purpose schemes incorporating irrigation and flood control. China is said[409] to have installed within the past decade 50,000 hydro-electric plants with an average capacity of 34 kW. The increasing contribution and use of hydro-power is dependent, therefore, on a number of different approaches and types of utilization. Pumped storage capacity is normally not included in hydro-electricity generating capacity, yet a peak load shaving capacity reduces the plant otherwise necessary. The storage aspect will be increasingly valuable as greater use is made of the inexhaustible energy sources for some of them can only supply intermittent power.

8. The INEXHAUSTIBLE ENERGY SOURCES sector is the most speculative of the other speculative sectors. Basically these energy sources are the tangible manifestation of a continuous flow of energy, of which a segment has to be captured, and usually concentrated, by some natural or artificial technique at a specific place. Effort has to be applied, and that effort may be costly, in work, money or environmental disadvantage. An array of solar reflectors covering several square kilometres might be an inadmissible environmental cost.

In the earlier years of the diagram the components are wood and dung, (non-commercial energy), estimates of non-recorded peat, (as in the USSR and Scandinavia, Ireland etc.) geothermal heat and electricity generating plants, wind power, tidal power from Rance in France, oil from the Athabasca oil sands of Canada, oil from coal at the SASOL plant in South Africa, and solar energy. These sources will be augmented over the years with the commercial utilization of biomass, increased geothermal exploitation, more tidal stations, (though only 25 have been designated[403] as suitable for big plants), more windmills, new wave-power installations, more and new solar energy conversion devices, and the increased exploitation of heavy oils, oil sands and oil shales. These form a very mixed bag of sources with very big differences in the timing and depth of market penetration in each country. Some will need changes in life style or habit; some will move down from being a luxury to being a 'necessity'; some will be introduced at the 'poor end' and will be gradually accepted by the richer social strata or nations.

The greatest expansion is expected to be in the use of solar energy. No scientific breakthroughs, only technological breakthroughs are involved. If, for example, the cost of the photovoltaic equipment, now used on space craft for the direct conversion into electricity of sunshine and sunlight, can be reduced some 200 times, then solar energy becomes almost universally viable. Such a transformation has occurred in the computer/calculator field over the past twenty years.

The principles behind nuclear fusion are more or less understood, but the necessary experimental circumstances are extremely costly and require multinational cooperation. Progress is therefore slow and the solution would seem, at this stage, to necessitate very large central power stations. Both nuclear fusion and fast breeder fission reactors have the potential of providing almost unlimited amounts of energy. The fast fission breeders are within sight of commercial use. They have therefore been assumed to contribute to the 'conventional' nuclear energy sector. Fusion techniques have still to be proved and therefore have been included in the inexhaustible energy sector. Fusion has a number of theoretical advantages over breeders in that it has lower risk potential, both physical and political, lower radioactivity in its operation and in its wastes, and, with sound initial engineering design, should have fewer environmental problems.[411] Hence it is here assumed that research and development will continue and be successful for major expansion in the second half of the twenty first century.

Some would argue that the contributions of these inexhaustible energy sources as shown to 2025 are over-conservative estimates, yet by 2025 they provide 19% of world energy. Although the unconventional oil resources are very large, so large as to be taken here in the inexhaustible category, and the exploitation techniques are known, and, indeed, oil sands are being commercially produced, yet progress is slow. Oil shales will not be contributing before 1985, except in the USSR and China, and the development in the USA, even if it starts within the next few years, will have to be on a massive scale, with environmental problems, if it is to be significant. The Tar Belt of the Orinoco basin in Venezuela will similarly have to compete for capital with many other schemes in Venezuela as the shales will have to in the USA. Even the Oil Sands of Athabasca are not expected to be contributing more than some 50 Mta by 2000, though Canada will be short of conventional oil long before that,[414] unless the Arctic springs a surprise. In the later years a growth rate of 3.35% pa is shown from 2025 to 2050 for the inexhaustible sources in total.

9. The AVERAGE ANNUAL GROWTH RATES IN ENERGY which can be calculated from the world totals used in the table 87 are:

Table 87
World energy supply & demand 1950 – 2050
Estimates of 7 December 1979 and 5 December 1977
(gigatonnes oil equivalent, rounded)

	World Totals (1)		Coal & Lignite		Crude Oil & NGL		Natural Gas		Nuclear		Hydro & Nuclear		Hydropower		Inexhaustibles		World Totals (2)	
	1979	1977	1979	1977	1979	1977	1979	1977	1979	1977	1979	1977	1979	1977	1979	1977	1979	1977
1950	1.8		1.1		0.5		0.2				0.03				0.25		2.1	
1960	3.0		1.5		1.1		0.4				0.06				0.25		3.3	
1965	3.8		1.54		1.57		0.6				0.08				0.25		4.0	
1970	5.0		1.65		2.4		0.9				0.11				0.3		5.3	
1974	5.9		1.7		2.9		1.1				0.14				0.38		6.2	
1978	6.7		1.8		3.1		1.2		0.15				0.4		0.45		7.1	
1985			2.0	2.6	3.2	3.85	1.5	1.75	0.25	0.4			0.45	0.3	0.6	0.6	8.0	9.5
2000			3.25	3.4	3.7	4.25	2.1	3.25	1.1	2.0			0.7	0.7	1.0	0.8	11.8	14.5
2010			3.8		4.0		2.4	3.0	1.8				0.9		1.7		14.4	
2015			4.0		3.9		2.4		2.1				1.1		1.9		15.4	
2020			4.2	4.8	3.7	2.9	2.5	2.7	2.5	6.0			1.1	1.0	2.2	1.6	16.4	19.0
2025			4.4	5.5	3.5	2.3	2.6	2.3	3.0	6.7			1.2	1.05	2.9	2.1	17.6	20.0
2040			6.0		2.0		2.0		4.9				1.4		5.4		22.0	
2050			7.0	8.4	1.2	1.0	1.5	1.0	6.5	6.6			1.5	1.5	7.5	7.5	25.2	26.0

(1) Commercial energy only
(2) including the inexhaustible energy sources, wood, dung etc. in early years, solar etc. later.

Sources: 1960—1974 UN Series J, 1978 BP Statistical Review, 1985—2000 various amended/calculated D. C. Ion.

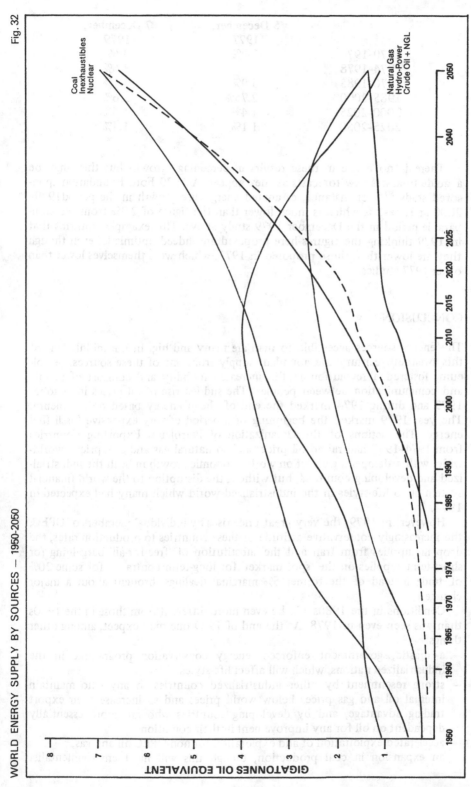

Fig. 32

WORLD ENERGY SUPPLY BY SOURCES – 1950-2050

GIGATONNES OIL EQUIVALENT

Coal
Inexhaustibles
Nuclear

Natural Gas
Hydro-Power
Crude Oil + NGL

	5 December, 1977	7 December, 1979
1970–1974	3.9%	3.6%
1974–1978		3.6%
1970–1985	5.9%	4.1%
1985–2000	2.75%	2.6%
2000–2025	1.4%	1.6%
2025–2050	1.1%	1.4%

There is no magic in these results in percentage growth but they may be a guide to assess new forecasts as they appear. A 1979 Ford Foundation sponsored study,[508] for instance, estimated energy use growth in the period 1979–2000 at 1.1%–2%, which is much lower than the figure of 2.3% from the comparable period in the December 1979 study above. This example confirms that in 1979 thinking the figures here proposed are indeed optimistic, even though they are lower than those proposed in 1977, which were themselves lower than other 1977 studies.

CONCLUSION

The energy sources accessible to man are many and big. In the middle half of this twentieth century the abundant supply from one of these sources, petroleum, fostered a revolution in the increased mobility and comfort of people and communication between peoples. The sudden rise of oil prices in October 1973 and during 1974 marked the end of cheap energy based on petroleum. The year 1979 marked the beginning of a period of very expensive fossil fuel energy. The actions of the Organization of Petroleum Exporting Countries from 1974–1978 had raised oil prices and so natural gas and coal prices, worldwide, with a dampening effect on world economic growth in both the industrialized and developing countries, but without the disruption to the world financial system or to life-styles in the industrialized world which many had expected in 1974.

However, in 1979, the very great price rises by individual members of OPEC, the increasingly conservative attitude of these countries to production rates, the drop in supplies from Iran and the substitution of 'free-for-all' bargaining for short-term supplies on the 'spot market' for long-term contracts, for some 20% of trade instead of the former 5% marginal dealings, brought about a major change.

Conditions in the 1980s will be even more different from those in the 1970s than was seen even in 1978. At the end of 1979 one may expect, among other things:

– a drastic, government enforced, energy conservation programme in the industrialized nations, which will affect life-styles,
– strong resentment by other industrialized countries to any who maintain internal oil and gas prices below world prices and so increase their export trading advantage, and by developing countries who are more essentially dependent on oil for any improvement in their conditions,
– accelerated exploitation of and exploration for non-OPEC oil and gas,
– an expansion in coal production, though this will meet environmentalist

opposition in many countries and cannot be rapid because of long lead times,
— a drive by nuclear power advocates to overcome the opposition of the anti-nuclear lobby with their arguments which are more often of attitude than scientific or even technological,
— a drive for the development of the inexhaustible energy sources increasing as comfort and mobility are threatened.

Unfortunately the measures to substitute alternative energy sources for petroleum may take at least all the 1980 decade to make a significant impact and certainly will lead to changes in life-style for many people. Nevertheless, ingenuity particularly in the industrialized world, should be able, given the effort, to adapt the exploitation and use of the world's energy sources to meet the needs if, but only if stable conditions are regained. The energy sources which could be available are indeed many and big, but their real, effective size and character must be appreciated. One major teaching of this book is that resources have no value except when transposed into supplies. The great oil reserves of the Middle East, for example, can only have economic and political leverage if they are likely to be produced in the future. The oilfields of Iran could be as valuable to Iranians in the 1990s as the peat bogs of Ireland were to the Irish in the 1960s.

At times of crisis and transition, miracles are sought, forecasts are demanded, and wild figures are thrown around which make wise decisions difficult and irresponsible actions most likely.

The grave danger is that such actions, even in 1980, could precipitate economic, financial, political and even military reactions which could lead to chaos. 1980 will undoubtedly be a most critical year in world history and only wise and cooperative management of energy supplies may avoid great trouble.

APPENDIX I

References

1974 WEC Survey is the *Survey of World Energy Resources* by the World Energy Conference, 1974

IX WEC Preprint is a preprint of the Ninth World Energy Conference, Detroit, 1974.

IX WPC Preprint is a preprint of the Ninth World Petroleum Congress, Tokyo, 1975, published in the *Proceedings* in October 1975.

[1] 'The Plain Man's Guide to Plate Tectonics', E. R. Oxburgh, *Proc. Geol. Assoc.* Vol. 85, Part 3, 1974, pp. 299–357.

[2] *Marine Science Affairs*, p. 16 and Table 11-1, p. 19, January 1969, Washington D.C.

[3] McKelvey & Wang, Notes Accompanying Oceanic Maps, *U.S. Geol. Surv. Circ. 694*, 1974.

[4] Reported in *Geotimes*, V. 19, n. 9, pp. 18, 19 Amr. Geol. Inst. September 1974.

[5] 'The Role of Gas and the International Gas Union in World Energy', L. J. Clark, *IX WEC Preprint* 1.2-6 1974.

[6] 'Energy Development and its Social Impact', O. Groza et al, *IX WEC Preprint* 1.2-15, 1974.

[7] *Energy: From Surplus to Scarcity*, ed. K. A. D. Inglis, chap. 5, p. 65: D. G. Leslie, Applied Science Pub. for Inst. Petr. London 1973.

[8] 'Power Plants and Environmental Interference in Congested Areas', B. v. Gersdorff & W. Sommer, *IX WEC Preprint* 4.2-5, 1975.

[9] 'Utilization of Energy', C. T. Chave, & W. L. Kennedy, p. 9, *IX WEC Preprint,* Positon Paper 6, 1974.

[10] Statement of M. King Hubbert to the Sub-Committee on the Environment of the Committee on Interior and Insular Affairs, House of Representatives, US 93rd Congress, 4 June 1974.

[11] *Energy: From Surplus to Scarcity*, ed. K. A. D. Inglis, Chap. 10, p. 155, A. R. Ubbelohde, Applied Science Pub. for Inst. Petr. London 1973.

[12] 'Solar Energy', B. J. Brinkworth, *Nature*, vol. 249, no. 5459, 21 June 1974.

[13] (not given)

[14] *The Rotation of the Earth, a geophysical discussion*, W. H. Munk and G. J. F. Macdonald, Camb. Univ. Press, 1960; quoted in IX WEC Preprint 3.1-10, 1974.

[15] *WEC Survey*, 1974, pp. 251–252.

[16] 'Wave Power', S. H. Salter, *Nature*, vol. 249, No. 5459, 21 June 1974.

[17] 'Geothermal Power', T. Leardini, *Phil. Trans. R. Soc. Lond.* A. 276, 507–526, 2974.

[18] *WEC Survey*, 1974, p. 165.

[19] G. C. Hardin, Fuels Symposium, *Amer. Assoc. Petr. Geol.* Annual meeting, Oklahoma, April 1968.

[20] 'Thermonuclear Energy', C. M. Braams, *IX WEC Preprint*, 4.1-11.

[21] W. C. Gough & B. J. Eastland - *Sci. Amer.* Vol. 224, No. 2, p. 50, 1971.

[22] Conversion, Chapter IX.

23 *WEC Survey, 1974*, p. 16.
24 'Resources of Oil, Gas & NGL in the United States and the World', T. A. Hendricks, *U. S. Geol. Surv. Circ.* 522 Washington, D. C., 1965.
25 *Mineral Resources and the Environment* - Table 3, p. 98, National Academy of Sciences, Washington, D. C. 1975.
26 'Organic-rich Shale of the USA and World Land Areas', *US Geol Surv. Circ.* 523, Washington DC, 1965.
27 *Phil. Trans. R. Soc. London.* A. 276, p. 608, 1974.
28 'Population, Energy & Development', H. Barnet, *IX WEC Preprint*, 1.1-4, 1974.
29 *Energy from Geothermal Resources*, Report prepared for the sub-committee on energy, Committee of Science & Astronautics, US House of Representatives, 93rd Congress, Second Session, May 1974, p. 38.
30 *Nuclear Energy and Fossil Fuels*, M. King Hubbert, Pub. no. 45, Table 6, p.. 22, Shell Dev. Co. Houston, Texas, reprinted from *Drilling & Production Practice*, 1956.
31 'Natural Sources of Nuclear Fuel,' S. H. U. Bowie, *Phil. Trans. R. Soc. Lond.*, A. 276, 495-505, 1974.
32 *Journ. Geol. Soc. Lond.* Vol. 130, pp. 387-391, K. C. Dunham & E. G. Poole, 1974.
33 *Methods of Estimating Reserves etc.*, W. L. Lovejoy & P. T. Homan, p. 13, Resources for the Future Inc., Washington D. C. 1965.
34 'New aspects in theory and practice of water flooding as applied in USSR oilfields', V. D. Shashin, *IX World Petr. Cong. Preprint*, Special Paper 4, 1975.
35 'Communication problems in reserve concepts & environmental control', D. C. Ion, *Proc. Inst. Exploration & Economics of the Petroleum Industry*, S. W. Legal Foundation, Dallas, vol. 8, pp. 31-57, Matthew Bender & Co. Inc. New York, 1970.
36 'The Significance of the World's Petroleum Reserves', D. C. Ion, *Proc. VII, World Petr. Congress*, Vol. 2, 1967.
37 *Future Petroleum Provinces of the USA* ed. Ira H. Cram, Amer. Assoc. Petr. Geol. Memoir 15, 1971.
38 idem. p. 5.
39 US Energy Resources etc. *A National Fuels & Energy Study* for 93rd Congress, Serial No. 93-40, (92-75), part 1, 1974.
40 *Future Petroleum Provinces of USA*, Nat. Petr. Council, pp. 133-138, 1970, and earlier papers.
41 A. R. Martinez, *Proc. VI World Petr. Cong.* vol. 1, New York, 1959.
42 'An estimate of the world's recoverable crude oil resources', J. H. Moody, *IX WPC Preprint*, P. D. 6(2), 1975.
43 *USGS Bulletin* 1142-H, Washington D. C., 1962.
44 'The sun may yet help to solve Britain's energy problems', K. Owen, *The Times*, London, 1 November 1974.
45 'America's Energy Potential', M. K. Udall, *Report of Sub-Committee on Environment*, Committee on Interior etc. US House of Representatives, Oct. 1973.
46 *Solar Energy for the Terrestrial Generation of Electricity*, Hearing before sub-committee on Energy, Committtee on Space & Aeronautics, US House of Representatives, June 5, 1974.
47 *Newsletter, July 1974*, (CGLO (74)NL7), p. 10, Commonwealth Geological Liaison Office, London.

[48] Solar Energy, 14, 21–28, 1972, F. Bassler, reported by B. J. Brinkwater, *Nature*, vol. 249, no. 5459, p. 72, 21 June 1974.

[49] *Science* 185, 940, Aug, 2, 1974, reported in *The Times*, London 12 Aug. 1974.

[50] *Boll. Di Geogisica Teorica ed Applicata*, vol. IV, No. 14, G. Facca & F. Tonani, 1952.

[51] 'Hydrology of Neogene Deposits in the Northern Gulf of Mexico Basin', Paul H. Jones, *Bull. GT2*, April 1969, Louisiana Water Resources Inst. Louisiana State Univ.

[52] idem, p. 88.

[53] *Report* prepared for the sub-committee on Energy, Committee on Space and Aeronautics, US House of Representatices, May 1974.

[54] *US Energy Prospects: An Engineering Viewpoint.* Report prepared by a Task Force, Chairman, W. Kenneth Davis, Nat. Acad. Eng. Washington DC, USA, 1974.

[55] *Soviet Geothermal Electric Power Engineering*, p. 11, Advanced Projects Agency, Dept. of Defense, Report 2, Washington DC, Dec. 1972.

[56] 'Powerful hydro-electric stations and the role they play in a comprehensive utilization of hydraulic resources', D. M. Yourimov et al. *IX WEC Preprint*, 1.2-26, 1974.

[57] 'Hydro (incl. Tidal) Energy', K. B. Vernon, *Phil. Trans. R. Soc. Lond.* A. 276, p. 486, 1974.

[58] Professor Scorer, Imperial Coll. Lond. reported by K. Owen, p. 19, *The Times*, Lond. 5 Nov. 1974.

[59] *1974 WEC Survey* and 'Survey of World Energy Resources', E. L. Nelson, et al. *IX WEC Preprint*, 1.2-33, 1974.

[60] 'Coal Resources of the USA', *US Geol Surv. Bull.* 1275, 1 Jan., 1967.

[61] *'US Energy Outlook, Coal Availability'*, Report of Coal Task Force, Ch. E. H. Reichl, Nat. Petr. Council, Washington D. C., 1973.

[62] *1974 WEC Survey*, App. 2, p. 315.

[63] 'Energy Policy in the People's Republic of China', G. C. Dean, pp. 33–55, *Energy Policy*, vol. 2, no. 1, March 1974.

[64] 'World Energy and the Nuclear Electric Economy', J. W. Simpson & P. N. Ross, *IX WEC Preprint*, 6.1-18, 1974.

[65] 'World Oil Reserves', Paul D. Torrey, *Proc. VI World Petr. Congress*, 1963.

[66] Reported by M. K. Hubbert in *Resources and Man*, chap. 8, p. 194, Nat. Acad. Sci., 1969.

[67] V. V. Semenovich, *Proc. VIII World Petr. Congress*, vol. 2, pp. 293–315, 1971.

[68] 'Technical Progress and its role in the development of the oil industry in the USSR,' A. A. Karayev, *IX WEC preprint* 3.1-19, 1975.

[69] 'The World-wide Search for Petroleum Offshore, A Status Report for the Quarter Century, 1947-1972', N. L. Berryhill, *US Geol. Surv. Circ.* 694, Washington DC, 1974.

[70] idem. fig. 6.

[71] 'Baffin Bay to the Bahamas', *Amer. Assoc. Petr. Geol.* Special Bull. vol. 58/6, Part II of II, June 1974.

[72] 'Geology of Giant Petroleum Fields', ed. Michel T. Halbouty, *Amer. Assoc. Petr. Geol. Memoir* 14, p. 5333, 1970.

[73] 'Eometamorphism and oil and gas in time and space', K. K. Landes, *Amer.*

Assoc. Petr. Geol. Bull. vol. 51, no. 6, pp. 828–841, 1967.

74 'The North Sea, – a new major oil province in changing world', R. Bexon, Preprint 26 Annual Tech. meeting, Petr. Soc. Can. Inst. Mine, 10–13 June 1975.

75 'Estimate of World Gas Reserves', T. D. Adams & M. A. Kirkby, *IX WPC Preprint*, PD 6 (1), 1975.

76 See (72), Table 2, p. 505.

77 *Development of the oil and gas resources of the United Kingdom*, Dept. of Energy, London HMSO, 1975.

78 'Recovery of oil from Athabasca oil sands and from heavy oil deposits of Northern Alberta by in-situ methods', R. Mungan & H. J. Nicholls, *IX WPC Preprint*, PD 22(2), 1975.

79 'The Oil sands of Alberta', H. J. Webber, *Journ. Can. Petr. Technology*, Oct–Dec. 1967.

80 *An Energy Policy for Canada, Phase I,* vol. II, Table 1, p. 32, Ministry of Energy, Mines and Resources, Ottawa, 1973.

81 'The Major Tar Sand Deposits of the World', P. H. Phizackerly & L. O. Scott, *Proc. VII World Petr. Congress,* Vol. II, PD 13(1), 1967.

82 'US Energy Resources, a review as at 1972', *M. K. Hubbert, Committee on* Interior and Insular Affairs, US Senate, Serial No. 93-40, Part I, p. 189, Washington D. C., 1974.

83 *Oil & Gas Journal,* pp. 44–45, 13 Aug. 1973.

84 *UN Publication 67.II.B.20,* ST/ECA/101, 1967.

85 *US Energy Outlook – Shale availability,* U. S. Nat. Petr. Council, Washington, D. C. 1973.

86 'Satellite Surveys for Energy Resources and Environmental Assessments', V. E. McKelvey, *IX WEC Preprint*, 1.2-27, 1974.

87 'The Energy Crisis and the World Economy', Lincoln Gordon, *IX WEC Preprint* 1.3-13, p. 2, 1974.

88 'Environmentally induced changes in the production, distribution and consumption structure of US bituminous coal', J. P. Brennan et al, *IX WEC Preprint* 2.6-12, 1974.

89 'The Impact of Resources Opencast Recovery on the Environment shown for the Brown Coal District of the Rhineland', E. Gaertner, *IX WEC Preprint*, 2.3-2, 1974.

90 'The Consequences on the Environment of Building Dams', ICOLD, *IX WEC Preprint* 2.6-14, 1974.

91 *Oil and Gas Journal,* vol. 72, no. 26, Newsletter, 1 July 1974.

92 *Interim Report,* Dome Petroleum Ltd., 30 Sept. 1974.

93 Reported in *The Times,* London, 6 Aug. 1974.

94 'How much oil – how much investment?', Energy Economics Div., Chase Manhattan, New York, March 1975.

95 Editorial, *Fortune,* vol. xc, no. 6, p. 110, Dec. 1974.

96 '60 billion barrel tertiary recovery potential claimed', R. E. Snyder, *World Oil,* vol. 179, no. 7, pp. 70–73, June 1974.

97 See (25) p. 92.

98 *Wairakei Power Station,* New Zealand, Electricity, 1971.

99 'Geothermal Energy and its Uses', Energy Section, Resources and Transport Div., U. N., *IX WEC Preprint* 2.1-17, 1974.

[100]   *1974 WEC Survey*, Appendix 2, Table 2.

[101]   *'World coal resources and future potential'*, G. Armstrong, *Phil. Trans. R. Soc. Lond.* A276, pp. 439-452, 1974.

[102]   'Fuel and Energy Resources etc.' N. V. Mel'Nikov et al. *IX WEC Preprint* 1.2-23, 1974.

[103]   '1972: The Year of the Arab', ed. F. G. Gardner, *Oil and Gas Journal*, vol. 70, No. 52, pp. 82-121, 25 Dec. 1972.

[104]   'Summary Petroleum and selected Mineral Statistics for 120 countries, including Offshore Areas', J. P. Albers, et al *U. S. Geol. Surv. Prof. Paper 817*, 1973.

[105]   *1974 WEC Survey*, Appendix 5, Table 5.

[106]   *1974 WEC Survey*, Table ix-2, p. 104.

[107]   *'Technical Progress in the gas producing industry of the USSR'*, A. D. Sedyk, *IX WEC Preprint, 1.2-4, 1974.*

[108]   'Fuel and power economy of Soviet Union', P. S. Neporozhny et al. *IX WEC Preprint*, 1.2-4, 1974.

[109]   'Innovations in the High Arctic', C. H. Hetherington and H. J. Strain, *IX WPC Preprint*, SP2, 1975.

[110]   'The availability of indigenous energy in Western Europe, 1973-1998, etc. *Proc. 1st World Symp. Energy & Raw Materials*, June 1974.

[111]   'Oil Sands – Canada's First Answer to the Energy Shortage', R. D. Humphreys et al, *IX WPC Preprint* PD 22(1), 1975.

[112]   'Alberta's Oil Sands in the Energy Supply Picture', G. W. Govier, *Proc. Can. Soc. Petr. Geols., Symp.* September 1963.

[113]   *'An initial appraisal by the Oil Shale Task Force, 1971-1985'*, Nat. Petr. Council, Washington D. C., 1972.

[114]   *1974 WEC Survey*, Appendix 6, Table 6.

[115]   'Population, Power and Pollution', K. K. Murthey, *IX WEC Preprint* 1.2-1, 1974.

[116]   *Energy in the World Economy: A Statistical Review of Trends in Output, Trade and Consumption since 1925.* J. Darmstadter et al, pub. for Resources for the Future Inc., Johns Hopkins Press, Baltimore, USA, 1971.

[117]   idem, p. 26.

[118]   *1974 WEC Survey*, Appendix 2, Table 2.

[119]   'Energy in the UK', Energy Tech. Div., Dept. of Energy, UK, *IX WEC Preprint*, 1.2-32, 1974.

[120]   *1974 WEC Survey*, Table III-10, p. 65, quoting *Minerals Year Book 1971*, USBM.

[121]   *UN Statistical Papers*, Series J. No. 17, Table 2.

[122]   See (108) Table V, p. 5.

[123]   'A forecast of energy supply and demand in South Africa', D. J. Kotze, *IX WEC Preprint* 1.2-14, 1974.

[124]   Reference given in (123), as Van Rensburg et al. Coal Advisory Board, 1969.

[125]   'Coal Mining and the Environment – an overview from a developing nation', *IX WEC Preprint*, 3.2-4, 1974.

[126]   'Arctic Oil and the World – one perspective', D. C. Ion, Amer. Assoc. Petr, Geol. Memoir 19, *Arctic Geology*, p. 619, 1973.

[127]   *Petroleum Economist*, vol. XLI, No. 11, Nov. 1974.

[128]   *Petroleum Times*, vol. 78, No. 1996, 15 Nov. 1974.

[129]   *Twentieth Century Petroleum Statistics*, 1974, pp. 19-23, (Authority,

US Bureau of Mines), de Golyer and MacNaughton, Dallas, Texas.

130   *Amer. Assoc. Petr. Geol. Bull.* Vol. 55/7, p. 986, July 1971.

131   *Ahead of his time, Michel Halbouty speaks to the people,* ed. J. A. Clarke, Gulf Pub. Co., Houston, Texas, 1971.

132   *Chronology of Venezuelan Oil,* A. R. Martinez, George Allen & Unwin, 1969.

133   See (129) p. 8, authority, US Bureau of Mines.

134   *Oil and the Romanian State,* M. Pearton, Clarendon Press, Oxford, 1971.

135   'Natural Gas', C. P. Coppack, *Phil. Trans. R. Soc. Lond. A. 276,* pp. 463–484, 1974.

136   *World Oil,* 15 Aug. 1974.

137   'Natural Gas in Italian Energy Economics', G. M. Sfligiotti and G. Marruzo, *IX WEC Preprint,* 1.2-11, 1974.

138   *World Oil,* 15 Aug. 1974, p. 122.

139   Reported in *The Times,* Lond. 16 June 1975.

140   *World Oil,* Int. Ed. p. 148, July 1974.

141   'The Impact of Natural Gas in Victoria, Australia', N. A. Smith & B. B. Bennett, *IX WEC Preprint,* 6.2-6, 1974.

142   'The production of liquid fuels from Coal in Europe and Africa', H. Pichler et al, *IX WPC Preprint,* PD 22(4), 1974.

143   'Energy Conversion', A. A. Baker, *IX WEC Preprint,* Position Paper No. 4, 1974.

144   'The New Energy Sources', W. Kenneth Davis and Simcha Golan, *Industrial Research,* 15 Nov. 1974, New York.

145   'Fusion Power, an assessment of its potential impact in the USA', G. L. Kulcinski, *Energy Policy,* vol. 2, no. 2, pp. 104–125, June 1974.

146   'Converter Reactor Alternatives', W. Kenneth Davis, presentation at the Atomic Industrial Forum Conference on Energy Alternatives, Washington, DC, 19 Feb. 1975.

147   'Economics of Nuclear Power', W. Kenneth Davis, presentation at *International Symposium on Nuclear Power Technology and Economy,* Taipei, Taiwan, 13 Jan, 1975.

148   'Liquids from coals in the USA', G. H. Hill, et al, *IX WPC Preprint* PD 22(5), 1975.

149   'Fluidized Bed Gasification and combustion for Power Stations', D. H. Archer et al. *IX WEC Preprint* 4.1-18, 1974.

150   'Conversion of Solid Fuels into other energy forms', L. Grainger, *IX WEC Preprint* 3.1-9, 1974.

151   Reported in the *Commonwealth Geological Liaison Office, London, Newsletter* (75, NL. 2), p. 12, Feb. 1975.

152   '*Oil in the UK'*, Pamphlet, Information Services, Inst. Petr. Lond. 1974.

153   'Economy of scale in Refining, Storage and Distribution', W. F. Brown, *IX WPC Preprint,* RP 16, 1975.

154   *Energy Policy,* vol. 2, no. 3, pp. 244-248, Sept. 1974.

155   'The Hydrogen Economy – A State of the Art', The Dean, School of Eng. & Experimental Design, Univ. Miami, *IX WEC Preprint* 5.1-18, 1974.

156   'World Petroleum Energy Model', R. J. Deam, J. Leather and J. G. Hale, *IX WPC Preprint* SP11, 1975.

157   *IX WEC Preprints,* Division 4, 1974.

158   *IX WEC Preprints,* 3.2-5, 4.1-7, 5.1-3, 1974.

159   'Coal as a fuel for MHD Generators', C. Brobrowski et al *IX WEC Preprint* 4.1-9, 1974.

[160] 'MHD Electrical Power Generation – An international Status Report', A. E. Sheindlin and W. D. Jackson, *IX WEC Preprint*, 4.1–13, 1974.

[161] 'Gases for Clean Energy' PD 17, 'Removal of sulphur from petroleum oils and gases', PD18, 'Environmental Protection in the petroleum and petrochemical industries', PD24, 'Air Conservation and Automotive Transport' PD23, (Total 20 papers) *IX WPC Preprints* 1974.

[162] 'Transportation of Energy', John E. Robb, *IX WEC Preprint* Position Paper 5, 1974.

[163] *BP Statistical Review of the Oil Industry – 1974*, British Petroleum, London 1975.

[164] 'VLCCs in Japan', J. S. Kagami, *IX WEC Preprint* 5.1–10, 1974.

[165] 'Developments in Petroleum Transport by Sea', R. P. Lescohier, *IX WEC Preprint*, 5.1–14, 1974.

[166] 'Marine Transportation of Crude Oil and Products', S. Yamagi, *IX WPC Preprint*, Review Paper 10, 1975.

[167] *BP Statistical Review of the Oil Industry – 1961*, British Petroleum, London 1962.

[168] see (163) p. 15, quoting John I. Jacobs & Co. Ltd.

[169] C. H. Tung of Island Navigation, Hong Kong, as reported in *The Times*, London, p. 9, 19 March 1974.

[170] *'Energy in Crisis'*, Peter Hill and Roger Vielvoye, Robert Yeatman Ltd, 1974.

[171] *Petroleum Economist*, vol. XLI, No. 8, p. 305, Aug. 1974.

[172] 'The Role of LNG in Energy Logistics', W. L. Culbertson and J. Horn, *IX WEC Preprint* 5.1–15, 1974.

[173] *Outlook for Natural Gas, a Quality Fuel*, ed. P. Hepple, Inst. Petr. Lond., Applied Science Pub., Ltd 1972.

[174] 'Liquefied Natural Gas', R. Boudet, *IX WPC Preprint*, Review Paper 11, 1975.

[175] 'Coal Transportation Economics', J. G. Montford and E. J. Wasp, *IX WEC Preprint* 5.1–16, 1974.

[176] 'Pipeline Transportation of Fluid and Solid Energy Sources', K. Schiffauer, *IX WEC Preprint* 5.1–6, 1974.

[177] 'New Developments in Pipeline Design, Construction and Operation', R. E. Watkins, *IX WPC Preprint*, Review Paper 9, 1975.

[178] 'Transport of energy on road and rail', K. Bauermeister, *IX WEC Preprint*, 5.1–7, 1974.

[179] 'Submarine Power Cable between Denmark and Norway', E. L. Jacobsen, *IX WEC Preprint* 5.1–1, 1974.

[180] Nordel – Brochure of the Nordic Agency for Co-operation in Electric Power, Nordel Secretariat, Helsinki, Finland.

[181] Panel 21, Papers 1, 2.3 and 4, *IX WPC Preprints*, 1975.

[182] See (181) paper 5.

[183] 'Etude de stockage souterrain de gaz dans les mines de charbon', J. Josse, et al, *IX WEC Preprint* 6.1–7, 1974.

[184] 'Le stockage souterrain de l'Energie', E. Schlumberger and A. Luxo, *IX WEC Preprint* 6.1–12, 1974.

[185] *Petroleum Economist*, vol. XLI, No. 8, August 1974.

[186] *Energy Conservation*, Central Policy Review Staff. H. M. Stationery Office, Lond. July 1974.

[187] 'Integration of pumped storage schemes etc.' S. Nieri Barillari, *IX WEC Preprint* 4.1–20, 1974.

188 'Assessment of advanced concepts in energy storage and their application on electric utility systems', R. Fernandes et al, *IX WEC Preprint*, 6.1-17.

189 'Energy storage and its role in Electric Power Systems', J. L. Haydock, *IX WEC Preprint*, 6.1-21.

190 'The nuclear steam storage plant, an economic method of peak power generation', P. V. Gilli and G. Beckman, *IX WEC Preprint*, 4.1-10, 1974.

191 'Fuel Cells, past, present and future', R. S. Tantram, *Energy Policy*, vol. 2, no. 1, pp. 55-67, March 1974.

192 *Science* 185, 440, Aug. 2, 1974, reported in *The Times*, London, 12 Aug. 1974.

193 'Consideration of possible improvement in the conversion and use of energy', Energy Section, Dept. of Econ. & Soc. Aff. UN Sec. *IX WEC Preprint* 6.1-19, 1974.

194 *UN Statistical Yearbook 1972.*

195 'Population increase and Distributional Change in Japan', *IX WEC Preprint* 1.1-2, 1974.

196 'Forecast of population and activities in the Paris area in the year 2000', R. Courbey, *IX WEC Preprint*, 1.2-5, 1974.

197 'Reduction of Labour, and Economy in Energy', H. Rolshoven, *IX WEC Preprint* 1.1-1, 1974.

198 *'Limits to Growth'*, D. L. Meadows et al for the Club of Rome, Project on the Predicament of Mankind, Universe Books, N. Y. 1972.

199 *UN Publication*, UN. E/C.7/40/Add.1, 5 Dec. 1972.

200 'The effect of prices and economic growth on consumer's energy requirements, Britain', T. A. Boley and D. L. Walker, *IX WEC Preprint* 1.3-5, 1974.

201 *'California's Electricity Quandary'*, Chap. 1, Rand Corporation, (R-1084, NSF/GSRA) Sept. 1972.

202 *Middle Eastern Oil and the Western World: Prospects and Problems'*, Sam H. Schurr et al. Statistical Appendix, American Elsevier, 1971.

203 *Oil and Gas Journ.* p. 146, 11 Nov. 1974.

204 'Future Patterns of Interfuel substitutions etc.' K. C. Huffman et al. *IX WEC Preprint* 1.3-14, 1974.

205a 'Energy Budgets', R. F. Chapman, *Energy Policy*, vol. 2, no. 2, June 1974.

205b 'Energy Budgets 2', R. F. Chapman, *Energy Policy*, vol. 2, no. 3, Sept, 1974.

206 'Energy Budgets, 3', D. J. Wright, p. 315, *Energy Policy*, vol. 2, No. 4, Dec. 1974.

207 'Methodology of analysis of the energy economy', ECE, St/ECE/Energy 1, *UN Publication*, 1963.

208 'Nuclear Power's Contribution to Energy Growth', W. Kenneth Davis, Presentation to *Atomic Industrial Forum Conference*, New Orleans, 3 March 1975.

209 Personal communication from Bruce C. Netschert, and also 'Energy Utilization and Pollution Aspects of two space-heating Alternatives', M. C. Cordaro, Bruce C. Netschert, and J. R. Mahoney, *IX WEC Preprint* 6.2-5, 1974.

210 Reported in *The Times* London, p. 17, 24 March 1975.

211 'Development of Electric-on-the road Vehicles in F. R. Germany', H. G. Moller, *IX WEC Preprint* 6.1-4, 1974.

212 'Very High speed steel wheel train, TGV001, M. Gaudichou, *IX WEC Preprint*, 6.1-1, 1974.

213 'The emerging role of future transport systems', R. A. Rice, *IX WEC Preprint* 6.1-14, 1974.

214 'Improvements in guided land passenger transport systems', A. L. Fairbrother and S. F. Smith, *IX WEC Preprint* 6.1-2, 1974.

215 *Petroleum Economist*, vol. XLII, No. 3, p. 102, March 1975.

216 'L'Isolation Thermique des Habitations etc.', J. J. Dubois et al *IX WEC Preprint*, 6.1-13, 1974.

217 'Cost structure of electricity supplies for space heating', Od Todnem, *X WEC Preprint*, 6.1-19, 1974.

218 'Perspectives techniques et economiques du chauffage a distance en Suisse', C. Zanger et al, *IX WEC Preprint* 6.1-15, 1974.

219 'Role of District Heating in increasing the efficiency of fuel combustion and decreasing air pollution in large populated areas', E. J. Sokolov et al. *IX WEC Preprint* 6.2-1, 1974.

220 'A study concerning different energy supply alternatives, etc.,' J. Mikola et al. *IX WEC Preprint* 6.1-10, 1974.

221 'Making energy value for money', W. Short, NIFES, *Nature*, vol. 249, no. 5459, p. 715, 21 June 1974.

222 'Seawater desalination by low temperature waste energy', R. Saari, et al, *IX WEC Preprint*, 1974.

223 'Ecological Aspects of the development of the energy economy in the Polish Iron and Steel Industry', Z. Falecki, et al, *IX WEC Preprint* 6.2-4, 1974.

224 'Possible Energy Patterns for the British Steel Corporation etc.' R. S. Barnes & D. M. Cowie, *IX WEC Preprint*, 6.2-3, 1974.

225 'Nuclear fission as a general source of energy', L. R. Shepherd, *Nature*, vol. 249, June 21, 1974.

226 *'Scarcity and Growth'*, H. J. Barnet and C. Morse, Resources for the Future Inc., Johns Hopkins Press, 1965.

227 *Petroleum Economist*, vol. XLII, No. 3, p. 85, March 1975.

228 'The role of petrodollars in international finance', Hussein Najdi, *Europe & Oil*, vol. 15, pp. 26-30, Feb. 1975.

229 *Oil and Gas Journ.* vol. 71, no. 53, 31 Dec. 1973.

230 'Study predicts strong rise in OPEC's surplus', Guy de Jonquieres, New York, 17 June, *Fin. Times*, Lond. p. 5, 18 June 1975.

231 'Energy and development policies in Iran: A Western View', Guy de Carmoy, *Energy Policy*, vol. 2, no. 4, pp. 293-306, Dec. 1974.

232 Reported in *The Times*, Lond. 26 Feb. 1975.

233 *Time Magazine*, 24 March 1975.

234 London Business News, *Sunday Times*, Lond. p. 45, 30 March 1975.

235 'How energy relates to world monetary problems', H. A. Merklein, *World Oil*, vol. 180, no. 1, pp. 92-98, Jan. 1975.

236 Merklein, *World Oil*, Vol. 180, no. 2, Feb. 1975.

237 *The Times*, Lond. 6 June 1975, p. 19.

238 *'Capital Requirements of Energy Supply'*, Edward Symonds, Vice-Pres. First National City Bank, New York, presented at Bergen Conference 1974.

239 'Japan's uncertain energy prospects', J. Surrey, *Energy Policy*, vol. 2, no. 3, pp. 204-230, Sept. 1974.

240 'The Japanese economy after the Oil Crisis', Dr. Toshihiko Yoshino, Yamaichi Research Institute, a presentation to the *Roy. Soc. for Asian Affairs*, 2 April 1975.

241 'The Petroleum Industry in Japan', K. Miyamori, *IX WPC Preprint*, RP1a, 1975.
242 *Petroleum Times*, vol. 78, no. 1997, p. 16, 29 Nov. 1974.
243 *The Times*, Lond. p. 18, 17 June 1975.
244 'The current status of the US mining industry', J. D. Morgan Jr., *Resources Policy*, vol. 1, no. 2, Dec. 1974.
245 'Energy Conservation Policies of the Federal Energy Office: Economic Demand Analyses', T. A. Ferrar & J. P. Nelson, *Science*, vol. 187, no. 4177, p. 644-646, 21 Feb. 1975.
246 'US dependence on OPEC growing despite Oil Tax', *Fin. Times*, Lond., p. 5, 29 May 1975.
247 *Energy Policy*, vol. 2, no. 4, pp. 316-329, Dec, 1974.
248 *Development of the Oil and Gas Resources of the United Kingdom*, Dept. of Energy, HMSO, Lond. 1975.
249 'Britain's Offshore Challenge', *Fin. Times*, Lond. p. 16, 11 Dec. 1974.
250 'Britain's oil and gas prospects', D. C. Ion, article commissioned by the Central Office of Information, Lond.
251 Statement by the Sec. of State for Energy on Depletion Policy, *Hansard*, 6 Dec. 1974.
252 *Fin. Times*, Lond. p. 8, 27 Dec. 1974.
253 'The role of nuclear power in the future energy supply of the world', J. A. Lane et al, *IX WEC Preprint*, 4.1-22, 1974.
254 *Petroleum Times*, vol. 78, no. 1988, p. 12, 28 June/12 July 1974.
255 R. Bexon, as reported in *Petroleum Times*, vol. 78, no. 1994, pp. 6-7, 18 Oct. 1974.
256 'Energy Resources are not all fuel reserves', D. C. Ion, *New Scientist*, vol. 70, No. 998, p. 222, 29 April 1976.
257 'Contributions to the assessment of world coal resources, or coal is not so abundant', G. B. Fettweis, Paper for Energy Resources Conf. of Int. Inst. for Applied Systems Analysis, (IIASA), Laxenburg, Austria, May 1975.
258 'Whither Uranium', S. H. U. Bowie, Presidential Address, Inst. Mining & Metallurgy, London, 20 May, 1976.
259 Proc. Int. Coal Exploration Symposium. Miller Freeman, 18-21 May 1976, London.
260 'Seismic Surveying and mine planning: their relationship & application', A. M. Clarke, NCB, idem.
261 'Consumer coal criteria as a guide to exploration', R. A. Schmidt et al., idem.
262 *UN Statistical Papers, Series J,* No. 18, 1970-1973, UN, New York, 1976.
263 *Petroleum Economist*, vol. xliii, No. 5, May 1976.
264 'World Coal Supply and Demand', H. E. Collins, see[259].
265 *World Coal*, vol. 2, Nos. 1-5, Jan-May 1976, London.
266 'The Changing role of coal', L. Grainger, NCB, Ann. Conf., Combustion Engineering Assoc., Slough, Berkshire, Eng., November 1975.
267 See[258].
268 *Nucleonics Week,* 12 February 1976, p. 10.
269 *Economist,* 29 November 1975, p. 97.
270 'Uranium resources and the scope for nuclear power', R. D. Vaughan, see[266].
271 Energy Report, Chase Manhattan Bank, April 1976, New York.
272 *Science*, vol. 189, No. 4203, p. 621; vol. 190, No. 4218, pp. 961, 964; vol. 191, No. 4222, p. 47; vol. 192, No. 4235, p. 120. Amer. Assoc. for Advancement of Science, Washington.

273 *Financial Times,* London, 1 April 1976 and *The Times,* London, 5 June 1976, p. 4.

274 *Science,* vol. 190, No. 4227 & vol. 191, No. 4227, p. 551.

275 'Uranium: resources, production and demand', Nuclear Energy Agency, OECD, and Int. Atomic Energy Agency, December 1975, OECD, Paris, 1976.

276 'USA Uranium Resources, an analysis of historical data', M. A. Lieberman, *Science,* vol. 192, No. 4238, 18 April 1976.

277 'West Germany encourages nuclear debate', Michael Kenward, *New Scientist,* vol. 170, No. 1003, pp. 521–523, 3 June 1976, London.

278 Letters, *Scientific American,* April 1976.

279 *Science,* vol. 191, No. 4230, p. 931, 5 March 1976.

280 *Science,* vol. 191, No. 4232, p. 1162, 19 March 1976.

281 *Economist,* 6 December 1975, p. 74.

281 'The Slagging Gasifier; maybe an answer to coal's prayer', David Fishlock, *Financial Times,* London, 8 June 1976.

282 'The European Community and the Energy Problem', Europ. Documentation, trade union series, periodical, 1975/2.

284 *The Economist,* vol. 259, No. 6927, Special Survey, p. 7, 29 May 1976, London.

285 *Petroleum Economist,* vol. xliii, No. 4, p. 145, April 1976, London.

286 *Petroleum Economist,* vol. xliii, No. 6, p. 208, June 1976, London.

287 Int. Symposium on Uranium Supply & Demand, Inst. of Uranium, 15–17 June 1976, London.
   a. 'North American Uranium Resources, Policies, Prospects and Pricing', G. M. MacNabb, Uranium Canada Ltd.,
   b. 'Australian Uranium – will it ever become available?', A. J. Grey, Pancontinental Mining Ltd.
   c. 'Uranium demand and security of supply – a consumer's point of view', Prof. H. Mandel, Rheinische-Westfalishes Elektrizitatswerke A. G., Essen.
   d. 'The pattern of uranium production in South Africa', R. E. Worrell & S. A. Young, Nuclear Fuels Corp., South Africa.
   e. 'Key issues affecting the future development of the Uranium Industry', J. Kostuik, President, The Uranium Institute.
   f. 'Economic and political environment of the Uranium Mining Industry', A. Petit, Commissariat a l'Energie Atomique, Paris.
   g. 'Alternative Sources of Uranium', A. von Kienlin, Urangesellschaft, Frankfurt.

288 *Nuclear Power,* W. Patterson, Penguin Books, 1976.

289 *Financial Times,* London, 21 June 1976, (a) Special Iran Survey, p. 21, (b) p. 23.

290 'Nuclear Power; the achievements, the problems and the myths', Sir J. Hill, Melchett Lecture to the Inst. of Fuel, London, 20 November 1975.

291 Sir Derek Ezra, Chairman NCB, as reported, *Financial Times,* London, 24 June 1976.

292 'Venezuela', Anibal Martinez, Energy Policies of the World, Center for the Study of Marine Policy, University of Delaware, USA, 1975.

293 'Nuclear Energy will dominate power supply in Spain', David Fishlock, *Financial Times,* London, 29 June 1976, p. 23.

294 'The hazards of rushing to build a nuclear fast reactor', Cheshire, Surrey

& Dombey, The University of Sussex, *The Times*, London, 28 June 1976.

295 'Energy R & D in the United Kingdom, A Discussion Document', W. Marshall, for Advisory Council on Research & Development for Fuel and Power, (ACORD), National Energy Conference, London, 22 June 1976.

296 'New high output underground mines in Great Britain, with specific reference to the Selby project', P. B. Rees, *Mining Magazine*, pp. 503–515, June 1976.

297 *World Coal*, July 1976, p. 15.

298 'USSR labours under a huge deficit to the West', Prof. A. Nove & D. Malka, *The Times*, London, p. 21, 22 July 1976.

299 'Westfield – the development of processes for the production of SNG from coal', D. Hebden & C. T. Brookes, Inst. Gas Eng., Ann. Meet. Edinburgh, April 1976.

300 *Scientific American*, May 1976.

301 'National Coal Board Mining Research Achievements', Sir D. Ezra, 24 Feb. 1976. Speech given at Stanhope, Bretton.

302 'Progress on five coal research projects – International cooperation on new Technologies', L. Grainger, British Embassy, Washington D. C., USA, 24 February 1976.

303 *Financial Times*, London, 3 June 1976.

304 *Report of the Commission of enquiry into the Coal Resources of the Republic of South Africa*. Govt. Printer, Pretoria, 0001, S. A.

305 Proc. Conference on the Resources of Southern Africa, Johannesburg, September 1975, Associated Scientific & Technical Societies of South Africa.

306 *World Oil*, 5 August 1975, p. 56.

307 *BP Statistical review of the world oil industry*, 1975, British Petroleum Co. Ltd., London.

308 *Petroleum Economist*, vol. xliii, No. 5, May 1976, p. 177.

309 *Financial Times*, London, 6 July 1976, p. 8.

310 'United Kingdom Energy Prospects and the Plan for British Gas', A paper presented to the National Energy Conference, 22 June 1976, London.

311 *Petroleum Economist*, vol. xliii, No. 7, July, 1976.

312 13th. World Gas Conference of the International Gas Union, June 1976, London.

313 *Oil & Gas Journal*, 29 December 1975.

314 *Financial Times*, 4 June 1976, Special Survey, World Gas Industry.

315 'The Development of the USSR Gas Industry', S. A. Orudzhev, USSR, Paper IGU/N.4-76, 13th World Gas Conference, June 1976, London.

316 'The prospective role of Iranian natural gas in the international energy market', M. Shirazi and H. Tahmassebi, Iran. Paper IGU/N.s-76, 13th World Gas Conference, June 1976, London.

317 *Energy International*, July 1976, p. 25, vol. 13, No. 7, Miller Freeman Pub. Inc.

318 *Time*, European Edition, 2 August 1976, p. 46.

319 *The Future of the World Economy, a Study of the Impact of Prospective Economic Issues and Policies on the International Development Strategy*, U. N. Sponsored study, Oxford Univ. Press., 1977.

320 *Nuclear Power Issues and Choices*, Report of the Nuclear Energy Policy

Study Group, sponsored by the Ford Foundation, administered by the Mitre Corp., Ballinger Pub. Co., Cambridge, Mass. USA, 1977.

321 *Options, A IIASA Mews Report*, Spring, 1977, pp. 2, 3.

322 *Energy: Global Prospects, 1985-2000.* Report of the Workshop on Alternative Energy Strategies, McGraw Hill Book Co., New York, USA, 1977.

323 *Energy Perspectives 2,* U.S. Dept of Interior, Washington D. C., USA, 1976.

324 *Concise Report on the World Population Situation in 1970-1975 and its Long Range Implications,* U. N., New York, 1974.

325 *Perspectives energetiques pour le tiers monde, 2000-2025,* P. Daures and J-R. Frisch, Projet Developpement et Cooperation Dag Hammerskjold published by le Groupement Prospective de l'Energie de la Maison des Sciences de Homme, Paris, Jan. 1975.

326 *World Oil,* June 1977, Gulf Pub. Co., Houston, Texas, USA.

327 *Science,* 10 June 1977, vol. 196, No. 4295, Editorial, Amer. Assoc. Advan. Science, Washington, DC, USA.

328 Executive Summary of Draft Report on Demand & Conservation, prepared for the Conservation Commission of the World Energy Conference, for presentation at the Tenth World Energy Conference, Istanbul, Sept., 1977, and circulated to selected delegates at that conference, by the Energy Research Group, Cavendish Laboratory, Cambridge Univ., England.

329 *World Energy Outlook,* OECD, 1977.

330 *World Energy Supplies, 1950-1974,* Dept. of Econ. & Social Affairs, UN Statistical Papers, Series J, No. 19, New York, 1976.

331 'L'Equilibre Mondial entre Besoins ·et Ressources d'Energie a l'Horizon 2000: evolution et regionalisation du probleme', J-R. Frisch, *Revue de l'Energie,* Paris, March 1977.

332 'Energy Plan offers prizes to all', E. Symonds, *Pet. Econ.* p. 173, vol. xliv, No. 5, May, 1977.

333 'Widespread Recovery in Demand', D. O. Croll, *Pet. Econ.* p. 257, vol. xliv, No. 7, July, 1977.

334 'Projected energy requirements up to the year 2000', K. R. Williams, Int. Symp. on Uranium Supply and Demand, Uranium Institute, London, June 1977.

335 'Japan – Can official targets be met?', A. Lumsden, *Pet Econ.*, p. 88, vol. xliv, No. 3, Mar. 1977.

336 'Policy Choices in the Age of Diversified Energy Sources', Toyaki Ikuta, Inst. of Energy Econ., – Supp. to *Energy in Japan,* No. 36-1, Dec. 1976, Tokyo.

337 'The North Sea Countries', D. C. Ion, pp. 97-217, *Energy Policies of the World,* vol. II, ed. Gerard J. Mangone, Elsevier, New York, USA, 1977

338 'The Soviet Union', Robert W. Campbell, pp. 218-310, *Energy Policies of the World* vol. II, ed. Gerard J. Mangone, Elsevier, New York, USA, 1977.

339 *Energy as a factor in Soviet foreign policy*, J. Russell, pub. for Roy Inst. Int. Affairs by Saxon Howell/Lexington Books, 1976.

340 *Energy Demand Studies: Major Consuming Countries,* First Tech. Report Report of Workshop on Alternative Energy Strategies, M. I. T. Press, Cambridge, Mass. USA, 1976.

341 'High Priority goes to Iraq ties', C. Smith, *Fin. Times,* p. 7, 26 Jan. 1977.

342 *The Economist,* p. 89, 20 Dec. 1975, London.

343 'The Soviet Five-Year Plan', D. Lascelles, *Fin. Times,* p. 6, 7 Jan. 1976, London.

344 'Modest Targets in Russia's Five-Year Plan', Prof. A. Nove, *The Times*, p. 14, 5 Jan. 1976, London.

345 'A Soviet $10 billion debt to the West', Jan Zoubek, *Fin. Times*, p. 24, 9 June 1976.

346 'Export potential in doubt', *Pet. Econ.* p. 204, Vol. xliii, No. 6, June 1976.

347 'Russian Oil Exports', *The Economist*, p. 107, 5 Feb. 1977, London.

348 'Soviet Union faces future gas and oil dilemma', M. Swiss, *Energy Int.*, June, 1977.

349 'Soviet Oil Production', D. Lascelles, *Fin. Times*, p. 4, 2 Aug. 1977, London.

350 'CIA too gloomy over Russian Oil Prospects', J. Russell, *The Times*, p. 23, 27 Jul. 1977.

351 'Classification of Oil & Gas Resources', Report of an Expert Group, Natural Resources Forum, pp. 397-403, Vol. 1., No. 4, Jul, 1977, Reidel Pub. Co, USA.

352 *The Soviet Energy Balance*, I. F. Elliot, New York, USA, 1974.

353 *Pet. Econ.* vol. xliii, No. 6, Jul. 1976.

354 *Fin. Times*, p. 4, 30 Aug. 1977, London.

355 *Oil in the People's Republic of China – Industry Structure, Production, Exports*, W. Bartke, a publication of the Inst. of Asian Aff., Hamburg, 1977.

356 *Science*, vol. 197, No. 4300, p. 241, 15 July 1977, & No. 4301, p. 353, 22 July 1977.

357 *An Alternative Energy Strategy for UK*, Nat. Centre for Appropriate Tech., Jun. 1977, Machynlleth, Powys, Wales.

358 *Science*, vol. 197, No. 4298, p. 9, 1 Jul. 1977.

359 'Russia still burdened by a huge trade deficit with the West', Prof. A. Nove, *The Times*, p. 17, 31 Aug. 1977, London.

360 Asia Survey, *The Economist*, 7 May, 1977, London.

361 'China; a reassessment of the economy', The Brookings Inst., as reported in *World Coal*, Nov. 1976, p. 45, Miller Freeman Pubs. San Francisco, USA.

362 *World Coal*, pp. 31-34, Jun. 1977, (see 361).

363 'Carter's Final Plan gets mixed backing', S. Fleming, *Fin. Times*, p. 1, 22 Apr. 1977, London.

364 'Energy to 2000; Caltex Corp. takes a long look ahead', James M. Voss, *World Oil*, p. 68, May 1977, Houston, USA.

365 'Oil: America gets to grips with its greatest domestic challenge', Louis Heren, *The Times*, 18 Aug. 1977, London.

366 *China's Oil Production Prospects*, Central Intelligence Agency, USA, (ER77-10030U) Jun. 1977.

367 *Fin. Times*, p. 4, 9 May 1977, London.

368 *Geology of Giant Petroleum Fields*, ed. Michael T. Halbouty, Memoir 14, Amer. Assoc. Pet. Geologists, Tulsa, Oklahoma, USA, 1970.

369 'Hungary's energy policy and the world economy trends', Geza Szili et al, Tenth World Energy Conf. Preprint 1.4-8, 1977.

370 'Japan's future energy structure and the effect of selection of fuels by the power industry', Y. Nagano, Tenth World Energy Conf. Preprint 2.5-6, 1977.

371 Private communication, H. E. Collins, London, 1977.

372 'Structural Changes in the Pattern of Fuel and Energy Balance of the USSR and the Role of Nuclear Power', P. S. Nephorozhny et al., Tenth World Energy Conf. Preprint 3.1-11, 1977.

373 Study by Carnegie Endowment for International Peace cited by *Pet. Econ.* p. 412, vol. xliv, No. 10, Oct. 1977.

374 Reported in *The Times*, 17 Oct. 1977, London.

375 'Population of the World and its Regions', N. Keyfitz, Research Memo, IIASA, Laxenburg, Austria, cited by Dr. W. Hafele, Dir. Gen., IIASA, in paper to Tenth World Energy Conf., Istanbul, Sept. 1977.

376 'Oil Resources – 1985 to 2020', Executive Summary, prepared by Dr. P. Desprairies, Institut Francais de Petrole, France, Executive Summary of Draft Report on Demand & Conservation, prepared for the Conservation Commission of the World Energy Conference, for presentation at the Tenth World Energy Conference, Istanbul, Sept., 1977, and circulated to selected delegates at that conference.

377 'Natural Gas Resources, 1985 to 2020', Executive Summary, prepared by W. T. McCormick et al., Amer. Gas. Assoc., USA, Executive Summary of Draft Report on Demand & Conservation, prepared for the Conservation Commission of the World Energy Conference, for presentation at the Tenth World Energy Conference, Istanbul. Sept., 1977, and circulated to selected delegates at that conference.

378 'Larger Role for Natural Gas', E. S. Tucker, *Pet. Econ.* pp. 344-347, vol. xliv, No. 9, 1977, London.

379 'A tangle of gas lines', Ray Dafter, *Fin. Times*, p. 31, 28 Oct. 1977, London.

380 *Development of the Oil & Gas Resources of the UK, 1977*, Dept. of Energy, Apr. 1977, London.

381 *Pet. Econ.*, vol. xliv, No. 9, Sept. 1977, London.

382 *Statistical Handbook*, Canadian Pet. Assoc., Calgary, Alberta, Canada.

383 *Fin. Times*, p. 6, 25 Oct. 1977, London.

384 'Italian Gas Plan succeeds at last', R. Betts, *Fin. Times*, 19 & 21 Oct. 1977, London.

385 *World Oil*, 15 Aug. 1977, Houston, USA.

386 'China and foreign credit', Colina MacDougal, *Fin. Times*, 27 Oct. 1977, London.

387 Reviews by D. C. Ion of 'Optimal Development of the North Sea's Oil-fields', by Prof. P. Odell & Dr. K. Rosing, Kogan Page, 1977, in *New Scientist*, Dec. 1976 and *Marine Policy*, Apr. 1977.

388 'Alternative Methods of oil supply forecastings' M. A. Adelman & H. D. Jacoby, Supply Analysis Group of the M. I. T. World Oil Project. Paper prepared for the Int. Energy Agency, Workshop on Energy Supply, Nov. 1976, Paris, France.

389 *Window on Oil – a survey of world petroleum resources*, B. Grossling, Financial Times Ltd., 1976, London.

390 'World Energy Outlook', Exxon Corpn., New York, 1977.

391 'Oil Resources in the next half century', D. C. Ion, pp. 70-86. Proceedings of Institute of Petroleum Summer Meeting, 1956, London.

392 'LNG 5 call for gas investment to plug energy gap', M. Swiss, *Energy Int.*, pp. 36-38, Nov. 1977.

393 *Pet. Econ.* p. 313, vol. xliv, No. 7, Aug, 1977, London.

394 *Offshore Engineer,* Aug. 1977, Inst. Civil Engineers, London.

395 *Pet. Econ.* p. 459, vol. xliv, No. 11, 1977, London.

396 'Energy Sources Availability in Mexico to Satisfy its Demands', J. Eibenschutz et al., Tenth World Energy Conf. Preprint 1.4-7, 1977.

397 *United States Interests in the Middle East,* ed. G. Lenowski, A Special Analysis for the Amer. Enterprise Inst. for Public Policy Research, Oct. 1968, Washington, DC, USA.

398 *Oil & Gas Journal,* 14 Feb. 1977.

399 *Oil & Gas Journal,* 27 Dec. 1976.

400 *Middle East Oil and the Western World; prospects and problems.* Sam. H. Schurr & Paul T. Homan, et al., Elsevier, New York, 1971.

401 'US oil demand is still rising', John Wyles, quoting API figures and AP-DJ reports on Secretary Brown's remarks, *Fin. Times,* 27 Oct. 1977, London.

402 'The prospect of biogas as one of the sources of energy in Nepal', A. B. Karbi & B. A. Coburn, *X WEC Preprint* 4.6-3, Istanbul, 1977.

403 'Unconventional energy resources', Prof. P. Auer et al., for EPRI draft report for the Conservation Commission of WEC, Executive Summary of Draft Report on Demand & Conservation, prepared for the Conservation Commission of the World Energy Conference, for presentation at the Tenth World Energy Conference, Istanbul, 1977, and circulated to selected delegates at that conference.

404 *Energy or Extinction — the case for nuclear energy,* Sir Fred Hoyle, Heinemann, London, 1977.

405 'Magnetohydrodynamic Power Generation', IAEA, *X WEC Preprint* 3.6-1, Istanbul, 1977.

406 'OPEC: Oil Report', *Pet. Econ.* London, 1977.

407 'Soviet Union — Offshore prospects and problems', B. A. Rahmer, *Pet. Econ.,* vol. xliv, No. 5, pp. 191, 192, May 1977.

408 'Hydraulic Resources', Ellis Armstrong, Outline of Preliminary Report for the Conservation Commission, WEC, Executive Summary of Draft Report on Demand & Conservation, prepared for the Conservation Commission of the World Energy Conference, for presentation at the Tenth World Energy Conference, Istanbul, Sept., 1977, and circulated to selected delegates at that conference.

409 'Survey of World Energy Resources', WEC, London, 1976.

410 'The Canadian energy situation in 1990 and beyond', 'Undeveloped hydro-electric resources', J. E. Warnock; Paper prepared for the Canadian National Energy Forum, Halifax, N. S., Canada, April, 1977.

411 *Fusion & Fast Breeder Reactors,* W. Hafele et al., IIASA, RR-77-08, November, 1976, (revised July, 1977).

412 'Public Opinion and Energy Use', Philip H. Abelson, *Science,* vol. 197, No. 4311, Editorial, 30 Sept., 1977.

413 'Soviet Resources Policy and the Tenth Five-Year Plan', Daniel L. Papp, *Resources Policy,* vol. 3, No. 3, Sept., 1977, IPC Sci. & Tech. Press.

414 'Status and challenges in the recovery of hydrocarbons from the oil-sands of Alberta, Canada', C. W. Bowman & G. W. Govier, *X. WEC Preprint* 1.2-6, Istanbul, 1977.

415 *Financing the International Petroleum Industry,* Norman A. White et al., Graham & Trotman, 1978, London.

416 *International Classification of Mineral Resources,* UNESCO, E/c.7/104, 1979, (79-07655).

417 *Plan for 2000,* National Coal Board, London.
418 'Potential World Petroleum Resources', P. W. J. Wood, Conference on World Energy Economics IV, Council for Energy Studies, London, 26–28 Feb. 1979.
419 'Frontiers of world exploration', Keith F. Huff, *O & GJ,* v. 76, N. 9, pp. 214–220, 1978.
420 *Giant oilfields and world oil resources,* R. Nehring, Rand Corp. Report, R-2284 – CIA, p. 162.
421 UN Statistical Papers Series J., No. 21, *World Energy Supplies, 1972–1976,* UN, 1978.
422 Survey of Energy Resources 1978, World Energy Conference, London, 1978.
423 *World Coal,* vol. 4, No. 4, April, 1978.
424 'World Energy Resources, 1985–2020', World Energy Conf., IPC. Sci. & Tech. Press, 1978.
424a *World Energy – Looking Ahead to 2020,* Report by Conservation Commission, World Energy Conf., IPC Sci & Tech. Press, 1978.
425 'World Energy Supplies', D. C. Ion, *Proc. Geol. Ass. 90* (4), 193–202, London, 1979.
426 *Fin. Times,* p. 2, London, 2 Mar., 1979.
427 *Fin. Times,* p. 3, London, 2 Jul., 1979.
428 *Fin. Times,* London, 15 Nov., 1977.
429 *Fin. Times,* 16 May 1979, quoting Chmn. Consolidation Coal Corp. to US Senate Committee, London.
430 *World Coal,* vol. 4, No. 11, Nov. 1978.
431 *World Coal,* vol. 5, No. 4, Apr. 1979, quoting from 'Energy Futures for Canadians', (EP 78-1), Fed. Dept. Energy, Mines and Resources, Ottawa.
432 *Fin. Times,* London, 7, 21 & 22, Feb. 1979.
433 *World Coal,* vol. 5, No. 3, Mar. 1979.
434 *Fin. Times,* London, 19 Jun. 1979.
435 *Fin. Times,* London, 10 Jan. 1979.
436 *BP Statistical Review, 1978,* London, 1979.
437 *Fin. Times,* London, 22 Nov. 1978.
438 *New Scientist,* p. 557, London, 17 May 1979.
439 *Fin. Times,* London, 1 Dec. 1978.
440 *Fin. Times,* p. 36, London, 26 Jun. 1979.
441 *O & GJ* – World-wide Oil Issue – 25 Dec. 1978.
442 'Development of the Oil & Gas Resources of the UK', Dept. of Energy, HMSO, London, July, 1979.
443 *Pet. Econ.,* vol. xlvi, No. 6, p. 255, June, London, 1979.
444 'World Uranium Resources – an international evaluation', OECD, NEA/ IAEA, OECD Paris, 1979, reported in 443, p. 255.
445 *Pet. Econ.,* vol. xlvi, No. 9, p. 362, Sept. London, 1979.
446 'Batteries, Fuel Cells, A Hydrogen Economy', M. Barak, Future Energy Concepts Conference, Pub. No. 171, Inst. Elec. Eng., London, Feb. 1979.
447 'Hydrogen Energy System Concept and Engineering Application', T. Veziroglu, idem.
448 Future Energy Concepts Conference, idem.
449 'Satellite Solar Power Stations', P. J. Collins, idem.
450 *Technology for a Changing World,* John E. Davis, Intermediate Techn. Pub. Ltd., London, 1978.

451 Intervention at Solar Energy Symposium, R. Soc. London, By Sir B. Lovell.
452 'Hydrogen Fuel stands by for take-off', W. Sweetman, *New Scientist*, London, pp. 818-820, 7 Jun. 1979, which discusses a Lockheed Corp., USA, proposal that US, UK, FR Germany and Saudi Arabia should build and operate a fleet of liquid hydrogen powered freighters as an experiment.
453 *Fin. Times*, London, 1 Dec. 1978 and 1 Aug. 1979.
454 *Int. Atomic Energy Agency Bulletin*, vol. 21, No. 2/3, June, Vienna, 1979.
455 *World Coal*, vol. 5, No. 7, July 1979.
456 'USA Petroleum and the World Energy Perspective', D. C. Ion, Presentation to Discovery Forecasting Symposium of Amer. Assoc. Petr. Geols., Houston, 1979.
457 *Losing Ground*, Erik P. Eckholm, W. W. Norton & Co. Inc., New York, 1976, quoted by Sumitro Djojohadikusumo, in Epilogue of WEC Istanbul, 23 Sep. 1977.
458 'Energy from the Grass Roots', David Fishlock, *Fin. Times*, London, 22 Aug., 1978.
459 *Energy International*, p. 25, vol. 14, No. 6, June, 1977.
460 'Green Petrol', David Fishlock, *Fin. Times*, London, 4 Jun., 1979.
461 *Fin. Times*, p. 34, London, 27 July, 1979.
462 *Pet. Econ.*, vol. xlv, No. 2, London, Feb. 1978, quoting from 'Clean Fuels from Biomass and Waste', US Inst. Gas Techn., Chicago, 1978.
463 *Pet. Econ.* vol. xlv, No. 8, pp. 340-341, London, Aug. 1978.
464 *Pet. Econ.* vol. xlvi, No. 2, p. 74, London, Feb. 1979.
465 *The World Energy Book*, (a) Map 19, (b) p. 194, Kogan Page, London, 1979.
466 *The Economist Measurement Guide and Reckoner*, p. 89, London, 1975.
467 *The World in Figures*, p. 270, The Economist, London, 1976.
468 *Time Magazine*, vol. 114, No. 6, p. 51, Europ. Ed., 6 Aug. 1979.
469 *Energy International*, vol. 16, No. 8, pp. 25-27, June 1979.
470 'Nuclear is a safe power', P. Beckman, Prof. Elec. Eng., Colorado, USA, in *Daily Telegraph*, London, 7 August 1979.
471 'The years that the locust hath eaten: oil policy and OPEC development prospects', Walter J. Levy, *Foreign Affairs*, Fall, 1978.
472 'The years that the locust hath eaten: energy and the fate of the nation', Walt. W. Rostow, Address to Amer. Assoc. Petr. Geols., Houston, 2 April, 1979.
473 'Probable future progress in seismic techniques', S. E. Elliot & P. G. Mathieu, *X WPC Preprint* PD3, paper 4, Heyden & Son, London, 1979.
474 'World Tanker Survey', *Pet. Econom.*, vol. xlvi, No. 9, London, Sep. 1979.
475 *Pet. Econ.*, vol. xlvi, No. 9, London, Sep. 1979.
476 'The principles of classification and oil resource estimation', A. M. Khalimov & M. V. Feigin, *X WPC Preprint* PD12, paper 1, Heyden & Son, London, 1979.
477 World reserves of oil and gas as discussed in PD12, Tenth WPC, Bucharest, 1979.
478 'Environmental Impact of Renewable Energy Sources', A review paper prepared for Commission on Energy and the Environment by the Energy Techn. Unit, Dept. of Energy, Nov. 1978, revised March, 1979, HMSO, London.

[479] *Fin. Times,* p. 11, London, 13 Jan. 1977.
[480] *Fin. Times,* p. 14, London, 6 Jan. 1977.
[481] *Fin. Times,* p. 14, London, 14 Sep. 1977.
[482] *Twentieth Century Statistics, 1978,* DeGolyer & MacNaughton, Dallas, USA, 1978.
[482a] 'World Ultimate Reserves of Crude Oil', M. H. Halbouty & J. D. Moody, *X WPC Preprint* PD12, paper 4, Heyden & Son, London, 1979.
[483] 'Petroleum Prospects of Deep Offshore', H. D. Hedberg, J. D. Moody & R. M. Hedberg, *AAPG Bulletin,* 63/3 March, 1979, Tulsa, Okla., USA.
[484] 'Proved and Ultimate Reserves of Natural Gas and Natural Gas Liquids', A. A. Meyerhoff, *X WPC Preprint,* PD12, paper 5, Heyden & Son, London, 1979.
[485] Report in *Fin. Times,* p. 29, London, 28 Jun. 1979.
[486] 'World Producing Capacity of Hydrocarbons', F. R. Parra, *X WPC Preprint,* RTD1, Paper 3, Heyden & Son, London, 1979.
[487] *World Oil,* p. 21, vol. 189, No. 1, July, 1979.
[488] *Energy Outlook, 1979-1990,* Exxon, USA, New York, USA, Dec. 1978.
[489] *Potential Supply of Natural Gas in the US,* Potential Gas Committee, Potential Gas Agency, Colorado School of Mines, Golden, Col. 80401, Apr., 1979.
[490] *Statistical Handbook,* Canadian Petr. Assoc., 1978 Ed., Calgary, Alta., Jul. 1979.
[491] 'Technology and Economics of oil sands operations', W. L. Oliver, *X WPC Preprint,* RTD 2, paper 3, Heyden & Son, London, 1979.
[492] *Fin. Times,* p. 3, London, 12 Oct. 1979.
[493] 'The Romanian Petroleum Industry', G. H. Pacoste et al., RP 1, *X WPC Preprint,* Heyden & Son, London, 1979.
[494] Proc. Tenth WPC, Discussion Summary, RTD1, vol. 2, Heyden & Son, London, 1979.
[495] *Fin. Times,* p. 1, London, 18 Oct. 1979.
[496] 'Ways to improve oilfield development schemes based on operation experience analysis', M. M. Ivanova et al., *X WPC Preprint,* PD11, paper 2, Heyden & Son, London, 1979.
[497] 'Technical and economic evaluation of operations for enhanced recovery of crude oil', C. W. Perry et al., PD11, paper 1, idem.
[498] 'Western Monopoly on solar energy', A. Agarwal, *New Scientist,* vol. 84, No. 1177, 18 Oct., London, 1979.
[499] M. Ryle, *Nature* 267, 1977.
[500] *Energy and Human Needs,* S. E. & J. S. Curran, Scot. Acad. Press, 1979.
[501] 'Wave Power Developments', A. Waugh & M. Swiss, reporting on Wave Energy Conf., 22-23 Nov., 1978, London, *Energy International,* vol. 16, No. 2., 1979.
[502] 'Fluid Mechanical Aspects of Wave Energy Projects', J. Lighthill, *Journ. Soc. Underwater Techn.,* June 1978, London.
[503] 'Wave Energy and the Environment', K. Probert & R. Mitchell, *New Scientist,* pp. 371-373, 2 Aug., London, 1979.
[504] 'Energy Review: US Coal', P. Cheesewright, *Fin. Times,* London, 12 Oct. 1979.
[505] *Energy; The next twenty years.* A report spons. Ford Found., admin. Resources for the Future, Study Group Chmn. Hans H. Lansberg, Ballinger Pub. Cambridge, Mass. USA, 1979.

506 *Uranium: Resources, Production and Demand,* Joint Report, NEA/IAEA, OECD, Paris, Dec. 1977.

507 'Dutch gasfield 10% bigger than supposed', *Fin. Times,* p. 3, London, 17 Oct. 1979.

508 'Disillusionment with US nuclear power policies', *Fin. Times,* London, 23 Oct. 1979.

509 'Oil from Coal', F. W. Richardson, *Chem. Techn. Rev.* No. 53, Noyes Data Corp, London & New York, 1975.

510 'Coal Conversion Technology', L. Howard-Smith & G. J. Weisener, *Chem. Techn. Rev.* No. 66, Noyes Data Corp. London & New York, 1976.

511 *World Coal Letter,* Vol. 1, No. 20, 28 Sept. 1979.

512 *Fin. Times,* London, 24 Jul. 1979.

513 'Europe's Electricity Pool Keeps the Power Flowing', David Fishlock, *Fin. Times,* London, 29 Jun. 1979.

514 'Digging an Electricity Link across the Channel', David Fishlock, *Fin. Times,* London, 20 Aug. 1979.

515 *The Balance of Supply and Demand, 1978-1990,* Uranium Institute, Feb. 1979, Mining Journal Books Ltd.

516 'Carter's Energy Package', *Fin. Times.* London, 17 Jul. 1979.

517 'Mini hydro plants boost China's power supply', *Energy International,* vol. 16, No. 11, San Francisco, 1979.

518 'OPEC Oil Revenues', J. Buxton, *Fin. Times,* 20 Sep. 1979, reporting on study by Dr. S. Ghalib, Chase Economic Group, New York.

519 'Sharp drop in aid from OPEC', J. Buxton, reporting on OECD Estimates, *Fin. Times,* London, 9 Aug. 1979.

520 Latin America and Caribbean Oil Report, *Pet. Econ.,* London, 1979.

521 Gen. Alfonso Ravard, Chmn, PETROVEN, reported in *Fin. Times,* London, 5 Sep. 1979.

522 *Fin. Times,* London, 3 Aug. 1979.

523 'The scope for energy conservation in EEC', F. Roberts, *Energy Policy,* vol. 7, No. 2, Jun. 1979.

524 Japan Survey, *Fin. Times,* London, 2 Jul. 1979.

525 'Plan to cut oil import needs', *Fin. Times,* London, 25 Aug. 1979.

526 'Japan boosts energy funds', C. Smith, *Fin. Times,* London, 2 Oct. 1979.

527 *Fin. Times,* London, 1 Nov. 1979.

528 Estimates of Prof. C. Robinson at Brit. Assoc. Advan. Sci., Edinburgh, reported in *Fin. Times,* London, 5 Sep. 1979.

529 UK Dept. of Energy submission to a Public Enquiry, reported in *Fin. Times,* London, 18 Sep. 1979.

530 'The looming Soviet factor in the world oil equation', A. Robinson, *Fin. Times,* London, 3 July, 1979.

531 West Berlin Economic Research Institute study as reported by L. Collitt, in *Fin. Times,* London, 30 Aug. 1979.

532 'Energy and LDCs in 1970s', A. R. Parra, *Pet. Econ.* vol. xlvi, No. 10, London, 1979.

533 'Developments in the Economics of Petroleum Refining', P. H. Frankel & W. L. Newton, *X WPC Preprint,* SP6, Heyden & Son, London, 1979.

534 *Current Issues in Energy – a selection of papers,* Chauncey Starr, Pergamon Press, 1979.

535 Classification of World Oil Resources, D. C. Ion, Part 1, Proved Reserves, *Pet. Econ.,* vol. xlvi, No. 12, 1979, Part II, Additional Resources, *Pet. Econ.,* vol. xlvii, No. 1, 1980.

[536] OPEC Oil Report, Second Edition, 1979, *Pet. Econ.*, London, 1979.
[537] *Natural Resources Economics*, C. W. Howe, John Wiley & Sons, New York, 1979, quoting Donnell in *Future Supply of Nature-made Petroleum & Gas*, ed. R. F. Meyer, Pergamon Press, New York, 1977.
[538] Oil Companies and Governments, J. E. Hartshorn, p. 305, Faber, 1962.
[539] *Nature* vol. 249, No. 5459, p. 275, 21 June 1974.
[540] *Coal-Bridge to the Future*, Report of the World Coal Study, ed. Carol L. Wilson, Ballinger, Cambridge, Mass, 1980.

APPENDIX II

Abbreviations

b	billion = thousand million = 10^9 = billion (US) = milliard or billion (Fr), = milliarde (German).
bb	billion US barrels, usually of oil
bl	US barrel, usually of oil
bd	US barrels per day
Btu	British thermal unit = 1.055 kJ
G_{av}	Average annual generation of electricity based on productivity potential of the power sites involved.
G_{95}	Annual generation of electricity obtainable from 95% of time being available
g/t	grammes per tonne
W	Watt
We	Watt electrical
Wh	watt hour
hp	Horsepower, being unit of rate of doing work = 500 foot pounds per second
J	Joule = 0.9478×10^{-3} Btu
t	tonne = metric tonne
oe	oil equivalent
ce	coal equivalent
mill	one thousand part of a dollar US, as in 10 mills/W
mbd	million US barrels of oil per day
mta	million tonnes per annum
ppm	parts per million
psi	pressure in pounds per square inch
Q	quad or 10^{18} Btu = approx 25 Gt crude oil
tcf	trillion cubic feet = 10^{12} cubic feet

The international System of Units is used where appropriate, with the Units and symbols most common being:

k	kilo = 10^3, as in kg = kilogramme, kt = thousand metric tonnes
M	mega = 10^6, as in megatonne, Mt
G	giga = 10^9, as in gigawatt, GW
T	tera = 10^{12}, as in teratonnes, Tt
E	exa = 10^{18}, as in exajoules, EJ.

APPENDIX III

Conversion factors

Many discrepancies between data on energy resources result from use of different factors. In this book, most world and regional conversions are based on resource-based averages. These have been calculated, as for world crude oil, from figures commonly quoted in US barrels, by giving due weight to the proportions of the total estimated to belong to the different types of crude oil. Some of the factors used by other authorities quoted in the book are also given.

ENERGY CONTENT

Unit = Joule = 0.9478×10^{-3} Btu.   1 Btu = 1.055×10^3 J.

Coal:

Anthracite	33.5 GJ/t	Average high rank coal	= 29.3 GJ/t
Bituminous	29.3	Average low rank coal	= 14.7
Sub-bituminous	25.1		
Brown coal & lignite	14.7	Eastern US coal	= 26.9
Peat	8.3	Western US coal	= 18.74

Resource-weighted world average coal = 25 GJ/t

Crude oil:
US average crude   45.4 GJ/t
Resource-weighted world average crude oil = 42.33 GJ/t

WORLD CRUDE OIL / COAL CONVERSION FACTOR ON RESOURCE-WEIGHTED AVERAGE

ENERGY CONTENT   1 tonne oil = 1.6932 tonnes coal equivalent.

UN Conversion Factor, Statistical Papers, Series J, No. 18, 1 tonne oil = 1.47 tce
BP Statistical Review of the world oil industry, 1 tonne crude oil = 1.5 tce
Conservation Commission, World Energy Conference, 1 tonne crude oil = 1.67 tce

Natural Gas:
Resource weighted world average, 1 m^3 = 34 MJ

Uranium:
In conventional thermal nuclear reactor   0.86 TJ/kg U
In fast breeder reactor   51.75 TJ/kg U

OTHER COMMON FACTORS.

Crude oil:
Resource weighted world average, 1 tonne crude oil = 7.33 US barrels
Common usage is 1 t = 7 bl, and thence 1 mbd = 50 Mta

Natural Gas:
10^9 m^3 = 1 km^3 natural gas = 0.86 Mt crude oil
10^3 m^3 = 1.286 t coal

Hydro/nuclear electricity:
1000 kWh = 95 m^3 natural gas = 0.082 toe = 0.123 tce = 3.6 GJ

Natural Gas Liquids:
1 US barrel/day = 42 tonnes per annum

General:
1 kg = 2.205 lb
1 m^3 = 35.315 cubic feet
1 kilocallorie = 3.968 Btu
Thousand (10^3) often written as M in USA
Million (10^6) often written as MM in USA.

INDEX

Entries in **bold** type give detailed breakdowns in page order of the contents of the chapters indicated.

pipelines, 199
reserve definitions, 7
refining, 280
waterflooding, 17, 273

Venezuela
coal, 104
conservation, 50, 114
policy, 265
oil sands, 38

water power, see hydraulic energy
  or wave power or tidal power
wave power, 180–181

wind power,
  power, 179–180
  resource base, 9
  resources, 29
wood, 182
World Energy Conference
  energy resources surveys
    coal (1968), 30, (1974), 30
    hydraulic energy (1978), 88
    uranium (1974), 39
World Petroleum Congress
  oil definitions, 7
  reserve estimates, 22

Yugoslavia, 160